I0885040

Morphology of the Auditory and Vestibular Organs in Mammals, with Emphasis on Marine Species

Galina N. Solntseva

RUSSIAN ACADEMY OF SCIENCES
A.N. Severtsov Institute of Ecology and Evolution

Morphology of the Auditory and Vestibular Organs in Mammals, with Emphasis on Marine Species

Galina N. Solntseva

Co-published by
Pensoft Publishers & Brill Academic Publishers
Sofia–Moscow–Leiden–Boston
2007

MORPHOLOGY OF THE AUDITORY AND VESTIBULAR ORGANS IN MAMMALS, WITH EMPHASIS ON MARINE SPECIES

Galina N. Solntseva

Editor-in-Chief
Professor V.M. Belkovitch

Translated by Dr N. N. Dergunova
Linguistic Editor: Teresa Ott

First published 2007
ISBN: 978-954-642-280-4 (Pensoft Publishers)
ISBN: 978-90-04-16202-0 (Brill Academic Publishers)

Russian Academic Monographs No 4

© PENSOFT *Publishers*

All rights reserved. No part of this publication may be reproduced, stored in a retrieval system or transmitted in any form by any means, electronic, mechanical, photocopying, recording or otherwise, without the prior written permission of the copyright owner.

Pensoft Publishers
Geo Milev Str. 13a, Sofia 1111, Bulgaria
e-mail: pensoft@mbox.infotel.bg
www.pensoft.net

Brill Academic Publishers
P.O. Box 9000, 2300 PA, Leiden, The Netherlands
e-mail: cs@brill.nl
www.brill.nl

CONTENTS

To the blessed memory of an unforgettable teacher and splendid person, Academician Vladimir Evgenievich Sokolov; to Professor Elena Alexandrovna Baburina, who was always a kind tutor to me; and to my beloved friend, my mother, Julia Vasilyevna Sporysheva. Thanks to the unselfish help and support of these remarkable people, I have been able to perform my creative research.

PREFACE

The study of the sensory systems of mammals has a very long history, during which the views of researchers and their approaches to the material under investigation have changed. Substantial progress in the study of the auditory system in recent decades has led to a complete revision of the traditional ideas concerning the methods of and significance of acoustic signalling in the animal world. Works in the field of mammalian bioacoustics have demonstrated the exceptional importance of hearing for the spatial orientation and communication in all representatives of this class.

Recent studies on the acoustic system in mammals of air and water habitats have shown that hearing is dominant among distant analyzers, and a range of forms (bats, dolphins) show special means of acoustic orientation (i.e., echolocation), thanks to which both aquatic and aerial species make the best use of the specific nature of their habitat.

The study of the peripheral part of the auditory system has been of special interest for many researchers for a long time. However, most of the evidence concerning the morphology and auditory structural physiology was obtained from a limited number of species. This prevented resolution of many important questions about the structural organization of the peripheral part of the auditory system in different ecological groups. In addition, the relation between the structural organization of the auditory organ and the acoustic features of the habitat was not documented.

As a result of long-term research on the auditory organs of different evolutionary lineages of terrestrial mammals, a great amount of morphological, physiological, and biomechanical data has been obtained. The study of the auditory organs of marine mammals (pinnipeds, cetaceans), which appear to represent completely different lineages of placental animals, started more than three centuries ago. However, these works were anatomical and fragmentary in nature. Such an approach to the study of the auditory organs created difficulties in understanding the location of the auditory apparatus as a whole. This inevitably led to incorrect interpretations of the mechanism of acoustic signal reception under water. In order to examine in detail the structure of auditory reception in marine mammals (cetaceans), a comparative morphological study of the outer, middle and inner ears was necessary in mammals belonging to various ecological groups.

In these studies, particular attention was paid to marine mammals (pinnipeds, cetaceans), which have historically been among the least studied groups, in order to examine echolocation abilities (in dolphins) and the impact the aquatic habitat renders on the development of morphological adaptations in the auditory organ of semi-aquatic and aquatic species. These studies revealed the structural adaptations in the outer, middle, and inner ears that were characteristic of every ecological group. They also showed the significance of these adaptations in relation to the acoustic features of the habitat.

The diverse morphological material used confirmed a general biological pattern of divergent development dependent on habitat. As a result, along with adaptive specializations in representatives of various ecological groups of mammals, the structure of the outer ear, which is in direct contact with the environment, undergoes the greatest morphological changes in semi-aquatic and aquatic species. The middle ear maintains a common basic structural pattern in most mammals; in species that are remote phylogenetically but similar in ecological specialization, parallelisms in the structure of particular elements of the auditory ossicles are evident, as well as in the way these elements are joined together and attached to the middle ear cavity. The inner ear acquires some species-specific morphological features, yet the structure of the receptive cells in the auditory and vestibular analyzers is similar in most species, with the exception of echolocating mammals.

As a result of these studies, the understanding of the process of functional division for frequency selectivity between the peripheral and central auditory systems made marked progress. It appears that some functions previously ascribed to the brain can be carried out by sound-transmission and sound-receptive apparatuses. Thus, the differentiation of biologically important frequencies in echolocating species occurs even at the level of the outer, middle and inner ears. These discoveries have led to a recognition of the roles played by the peripheral auditory system in acoustic information processing in a wide range of species possessing low-, middle- and high-frequency hearing [Bogoslovskaya, Solntseva, 1979].

These studies show that the widest range of adaptive potential exists in all parts of the peripheral auditory system, based on its structure and function variations, including even the formation of new, additional structures in pinnipeds and cetaceans not inherent in terrestrial species. As the data demonstrate, only a comparative analysis can reveal the wide range of adaptive transformations of the peripheral auditory system, which could have undergone a complicated process of creation and separation from the other structures in the course of evolution.

In my study of the structural and functional organization of the peripheral part of the auditory system in postnatal ontogeny, I felt it necessary to undertake a prenatal study of the system at the early stages of ontogeny. However, as there was no information available on the prenatal development of the peripheral part of the auditory system in pinnipeds and cetaceans, questions concerning the developmental mechanism of the auditory organ in mammals as a whole could not be resolved.

As a result, I have been the first to carry out detailed comparative and embryological studies on the peripheral auditory system with the use of unique embryological collections representing the orders Pinnipedia and Cetacea.

This monograph is a description of my comparative studies of the peripheral auditory system in pre- and postnatal ontogeny undertaken for the first time in representatives of various ecological groups of mammals that demonstrate low-, middle- and high-frequency hearing. This study has enabled me to describe in detail the hearing and vestibular organs; to determine stages in the formation of individual structures, including the morphological adaptations discussed earlier; to locate the sources of the origin of structural adaptations in semi-aquatic, aquatic and echolocating species; and also to disclose the features of similarity and distinction in the formation of structures at the early developmental stages in different species. Based on the results obtained, a common mechanism of the development of the peripheral auditory system in mammals has been established.

This book is a result of my long-term studies, started in 1966 at the N. K. Koltzov Institute of Developmental Biology of the USSR Academy of Sciences. Since 1993, this study has been continued at the A. N. Severtsov Institute of Ecology and Evolution of the Russian Academy of Sciences (RAS). A Russian version has recently been published [Solntseva, 2006], and now the time has come for its English translation.

For their dedicated assistance over many years, I wish to extend my thanks to Academician D. S. Pavlov, Director of the A. N. Severtsov Institute of Ecology and Evolution; to Professor A. V. Yablokov, Corresponding Member of the RAS; to Professors V. M. Belkovitch, L. S. Bogoslovskaya, O. F. Chernova and V. A. Zemsky; and to Dr. L. M. Mukhametov, Dr. I. V. Smelova, as well as to her colleague, a renowned specialist in marine mammals, Professor J. A. Thomas (Moline, Illinois, USA), for their valuable advice and help in this work. Special thanks go also to T. N. Sidorova, A. N. Nikonova, Y. A. Lishnevsky and my son V. A. Solntsev.

The study of the development of the peripheral part of the auditory system was made possible through collections of embryonic material of pinnipeds and cetaceans. A unique embryonic collection of *Stenella attenuata* was presented by Dr. William F. Perrin (La Jolla, California, USA). Other embryonic material from members of the orders Pinnipedia and Cetacea was kindly offered for study by colleagues from several institutes of the Russian Ministry of Fishing Industry (Moscow, Vladivostok, Kaliningrad, Archangelsk, Astrakhan). I am grateful to all my colleagues for the provision of embryological material.

The study of the prenatal development of the vestibular apparatus was supported by the International Science Foundation (Project NF 3000).

Part I

PERIPHERAL PART OF THE AUDITORY SYSTEM OF MAMMALS IN POSTNATAL ONTOGENY

The auditory system of mammals possesses a number of original patterns of structural and functional organization that distinguish it from other reception systems. This distinct auditory system is the most noticeable within the study of evolutionary and adaptive changes. These unique traits are explained by the peculiarities of vertebrate hearing origin and development and are caused by the specifics of receiving and processing acoustic signals.

A mammal's auditory receiver is more sensitive than that of organisms belonging to other classes due to a specialized sound-detecting apparatus — the auricle — inherent only in mammals. Because of their highly sensitive auditory systems and acoustic signal generation organs, mammals are able to communicate with each other even across large distances.

During the process of evolution, marked changes in auditory system development occurred in a number of vertebrates. Morphological reorganization affected all the parts of the peripheral auditory system, including the outer and middle ears as the earliest phylogenetic formations. For example, a mammal's middle ear contains three ossicles (the malleus, the incus and the stapes), as opposed to the one auditory ossicle (the *columella*) present in lower vertebrates. A mammal's auditory ossicles provide a special mechanical system capable of enhancing sound pressure on an oval-shaped membrane of the inner ear.

The main feature of a mammal's inner ear is the cochlea, which is larger than that of other vertebrates, due to the marked lengthening of its basal, or basilar, membrane, on which receptor cells are located. The basilar membrane in an amphibian is rounded, and the basilar membrane in a reptile is oblong; in a mammal, this membrane looks like a long, narrow fascia.

From the inner ear, acoustic information travels through an acoustic nerve toward the brain centers of the auditory system — the cochlear nuclei of the medulla, the superior olive and the posterior tubera of the mesencephalon quadrigeminal plate — into the inner body and the auditory cortex of the cerebral hemispheres. The auditory centers of the mesencephalon, the more highly organized sections of the auditory system and the cerebral cortex are notable for having the most complicated structural organization.

Every species has its own world of acoustic signals, which are related both to habitat peculiarities and the features of the auditory system. In addition, several mammalian species are noted to be especially sensitive to the sounds of a specific frequency. For example, the greatest sensitivity to high frequency or ultrasonic signals is found in many species of insectivores, bats, rodents, pinnipeds, cetaceans and primates. Mammals with echolocation abilities are capable of perceiving sounds of the widest frequency range, including ultrasound.

It has been supposed [Konstantinov, Movchan, 1985] that an increase in sensitivity to high-frequency sounds during the evolutionary process is related to the natural selection mechanism that provides for the hearing of short or abrupt sounds, which are characterized by high frequency constituents. Another reason is linked with the necessity for locating the source of sound. Elevated sensitivity to high frequencies is necessary for the maximal use of information about the spectral differences of the signals traveling to each ear. These differences arise because the auricles, as well as the head and body of the animal, shield the auditory system. All this provides information on the site of a sound source. In the study of mammalian auditory system tuning, researchers have discovered that, while in the majority of animals the auditory system is tuned for the reception of high-frequency signals, some species are not able to perceive such signals.

We have focused our research on marine mammals – cetaceans and pinnipeds – which belong to the least studied orders of mammals. Cetaceans and pinnipeds are still of great interest due to the echolocation ability of dolphins and due to the impact the aquatic environment renders on the development of adaptive features in the structure of the auditory system in semi-aquatic and aquatic species.

It is well known that sounds can travel longer distances in water than in air. Ultrasound has a higher frequency than the sound perceived by humans (up to 20 kHz) and, therefore, should spread and fade at markedly shorter distances from a sound source. Representatives of the suborder Odontoceti, which use ultrasound location for orientation, show adaptations for strengthening, as well as focusing, an acoustic wave. This is based on the morphological features of the animal's head (the presence of a fatty jut on the forehead, or melon), and also on the ability of the animal to produce ultrasound or short-wave signals. During the process of echolocation, the dolphin probes its ambient space with ultrasound signals, projecting pencil-thin ultrasound waves from its melon. The width of the ultrasound waves is characterized by an orientation diagram of acoustic radiation, which represents a graphic relation of the pressure value of sending the acoustic signal from the angle between the acoustic axis and the direction to the point. The reflected signal comes back to the dolphin and is received by its auditory system.

Low-frequency signals used by species of the suborder Mysticeti usually fade at a distance not exceeding 20 km. However, there is evidence that infrasound can travel to a distance of 1000 km. Such wave guides are formed at great depths as the result of a pressure increase. The sound thereby spreads with maximal speed and minimal fading.

The use of ecological and morphological approaches during our study of mammals with different ecological specializations has allowed us to establish that the organizational features of the peripheral part of the auditory system are determined by animal adaptation to the specific acoustic characteristics of habitats. The main adaptations in the course of evolution are morphological and functional ones in the peripheral part of the auditory system for hearing sensitivity optimization in physically different habitats. Here, the basic trends in structural evolution for all parts of the peripheral auditory system are examined in the following ecological sequence: terrestrial, subterranean, aerial, semi-aquatic and aquatic species.

The peripheral part of the auditory system includes three components — the outer, the middle and the inner ears. The outer and middle ears receive sounds from the environment and transmit them to the inner ear. A sound wave enters the outer ear, causing the tympanic membrane to vibrate. Through a chain of auditory ossicles, these vibrations are transmitted to the inner ear, where the mechanical vibrations are transformed into nervous signals. First, we will examine the structures of the outer, middle and inner ears.

CHAPTER 1.
STRUCTURE OF THE OUTER EAR OF SOME MAMMALS

Despite numerous morphological and experimental works devoted to the study of the outer ear of mammals [Huber, 1934; Clarke, 1948; Yamada, 1951, 1953; Purves, 1955; Reysenbach de Haan, 1057a, b; Fraser, Purves, 1960; King, 1964; Ramprashad et al., 1971, 1973, 1976; Solntseva, 1971, 1972a, e, 1973a; Simkin, 1977], the existing data are highly fragmentary, since there are no sufficiently complete between-order comparative-anatomical descriptions of the outer ear.

This chapter is based on our own data and is supplemented by previous research. However, because the existing morphological descriptions of the outer ear are incomplete, we have abandoned the customary pattern of presentation according to Simpson's [1945] classification.

When describing the outer ear structure, one must examine representatives of several ecological series – terrestrial, aerial, subterranean, semi-aquatic and aquatic mammals – as these series include all typical patterns of outer ear structure. This allows one to establish trends in morphological changes that have occured in the process of adapting to different environments.

The outer ear of most mammals consists of an auricle and an external auditory meatus, which includes cartilaginous and osseous parts. The cartilaginous part is a continuation of the auricle and is attached to the osseous part, which is located in a tympanic bulla (*bulla tympanica*) and ends with a tympanic membrane, which separates the outer ear from the middle ear.

The auricle consists of elastic cartilage covered by skin. The shape and size of the cartilage vary among species (Fig. 1a, b). The cartilage of the auricle is connected to the cartilage of the external auditory meatus, which also differs in shape, size and location relative to the auditory meatus cavity.

Most mammals have horseshoe-shaped cartilage adjacent to the auditory meatus cavity in the form of one or two cartilaginous plates; moles and seals have four flexibly connected cartilaginous laminae which vary in shape, size and location.

In the auricle, the back (outer surface), scaphoid (inner concave surface), two edges, top and base are discernible. The base of the auricle is located under the skin on a pad of fat. Aural glands and hair bulbs are hidden in the skin of the auricle's inner surface. Some mammals have additional valve-shaped enlargements on the inner face of the auricle, which increases the auricle's size and flexibility. Some animals with especially keen hearing have an enlarged auricle base or voluminous sac. This part of the auricle is responsible for tuning the frequency of acoustic signals perceived at the outer ear level [Simkin, 1977].

Ear muscles are attached to the auricle. Their description is given according to Akaevsky [1962], using the horse as an example. All ear muscles are divided into three groups. The first group forms the tensor muscle of the cartilaginous scutellum and is located in front of the auricle. The second one includes the muscles that attach the auricle to the scutellum or directly to the skull; they are the strongest and enable various movements of the auricle. The muscles of the third group are the weakest and are set completely onto the auricle.

(1) The tensor muscle of the scutellum, *m. scutularis*, comprises a thin muscle plate fastened at a crest around the temporal fossa. At the center of this plate is a plain, unevenly shaped cartilaginous scutellum, *scutulum auriculae*. The tensor muscle of the scutellum is divided into three parts: (a) a muscle between the scutella, *m. interscutularis*, that goes from the outer sagittal crest; (b) a frontoscutellar muscle, *m. frontoscutularis*, which is divided into a frontal part, fixed at the outer frontal crest, and a temporal part, attached to the malar arch; and (c) a cervicoscutellar muscle, *m. cervicoscutularis*, that begins at the cervical crest and ends at the cartilaginous scutellum.

(2) Four adductors turn the auricle, pulling it forward and directing the aural slit in the same direction. Three of them begin from the scutellum: (a) a dorsal adductor, *m. adductor dorsalis*; (b) a medium adductor, *m. adductor auris medius*, which is fixed to the auricle in front of the previous adductor; (c) a ventral adductor, *m. adductor auris ventralis*, fastened to the auricle below; and (d) an outer adductor, *m. adductor auris externus*, that goes from the malar arch to the lateral surface of the auricle.

(3) Three erectors turn the auricle medially and set it vertically: (a) a long erector, *m. levator auris longus*, which goes from the cervical crest to the auricle and inclines it back while two others bow it forward; (b) a short erector, *m. levator auris brevis*, which goes from the scutellum to the auricle and is attached to it in front of the long erector; (c) a medium erector, *m. levator auris medius*, that begins from the outer sagittal crest and ends at the back of the auricle under the long erector.

(4) Two abductors direct the auricle laterally: (a) a long abductor, *m. abductor auris longus*, which goes from the nuchal ligament to the base of

the auricle and binds to it behind the ventral aural muscle; (b) a short abductor, *m. abductor auris brevis*, that begins together with the long abductor and ends on the auricle ventrally of it.

(5) A ventral aural muscle, *m. auricularis ventralis*, which begins on the fascia of a parotid gland and fastens on the auricle under the aural commissure; when it contracts, it pulls the auricle down.

(6) Long and short rotators – *m. rotator auris longus et brevis* – which go from the scutellum to the basis of the auricle.

Along with the aural muscles listed above, there is a row of small muscles on the auricle itself. In some mammalian species, some of the aural muscles are well developed; in others, they are less developed or totally absent. For example, the dog has neither a cervicoscutellar muscle nor a ventral adductor, and has only a short erector. The ventral aural muscle is narrow and long. Its own auricle muscles are well developed. Auricle rotation goes only in the front sector of the circle. In the pig, the ventral and outer adductors can join into a single one or are absent altogether. A short erector is absent. The ventral auricle muscle is short and binary. Auricle rotation is possible at the back sector and at half of the front sector of the circle.

The external auditory meatus (*meatus acousticus externus*) consists of cartilaginous and osseous parts. In the wall of the auditory meatus, three layers are usually discernible: the epidermis, the dermis and cartilage or bone.

A plain, multilayer, nonkeratinizing epithelium covers the auditory meatus cavity, uniformly in some mammals and unevenly in other mammals. The dermal layer is divided into an inner layer, formed by dense connective tissue, and an outer layer, composed of hypoderma. Dense connective tissue is composed of longitudinal fascicles of collagen fibers twisted into thin elastic fibers. The dermal layer contains multiple blood vessels, arteries and veins. Arteries are composed of three membranes: the inner (*tunica intima*), the middle (*tunica media*) and the outer (*tunica externa*). The *tunica intima* consists of endothelial cells of the subendothelial layer. The *tunica media* is composed of separate fascicles of unstriped muscle cells arranged in a circular order. The *tunica externa* consists of adventitial cells and individual argyrophilic fibers.

The structure of a vein is very similar to that of a capillary. The vein wall is composed of two membranes: endothelial and adventitial membranes. The endothelial membrane is a layer of elongated polygonal cells. The adventitial membrane encircles the vessel from the outer side and contains many argyrophilic fibers and adventitial cells.

From the auricle to the osseous part, the auditory meatus is surrounded by cartilage, which takes markedly different forms in different mammals.

In most mammals, the cartilage is arranged around the auditory meatus in a horseshoe-shaped manner; in others, it has the appearance of an open-ended tube, while in some other species it consists of several cartilaginous plates flexibly bound to each other by cellular membranes. Such cartilaginous plates are distinguished from each other by size, shape and location. Auricular glands between the epidermal layer and the cartilaginous plate surround the auditory meatus cavity in most mammals. They are compound tubular and alveolar glands in their structure and holocrine glands in their secretion type. The glands are multilobular and open into the auditory meatus cavity. The cells forming a lobule of the gland have granular cytoplasm and large, round nuclei.

The aural muscles are bound to the external auditory meatus. Its description is given using the Greenland seal as an example, as it is a typical representative of phocids with a reduced auricle but with a more strongly developed auditory meatus.

The cartilage of the auditory meatus of the Greenland seal consists of four plates that are flexibly connected to each other and incompletely surround the meatus. There are spiral, posterior-ventral, dorsal-ventral and medial cartilaginous plates [Ramprashad et al., 1973]. The spiral plate is considered to be the inner auricle. There are eight aural muscles bound to the auditory meatus.

(1) The *m. auricularis anterior* begins around the eyehole and mid-frontal area and binds at the leading edge of the spiral plate. The function of this muscle is to press the membrane wall of the auditory meatus in the spiral plate area.

(2) The *m. auricularis superior* begins at the leading edge of the temporal muscle and ends at the aponeurosis of the meatus membrane part. It contracts the inner auricle and closes the membrane part of the meatus.

(3) The *m. auricularis posterior* begins at the medial-occipital area and cervical muscles and attaches to the inner auricle. This muscle contracts the inner auricle postero-cranially and opens the membrane part of the meatus.

(4) The *m. zygomaticus* begins around the lateral-facial area and attaches to the inner auricle. It contracts the auricle and presses the meatus membrane apart.

(5) The *m. mandibuloauricularis* begins at the lower jaw and fastens at the leading and ventral edges of the spiral plate and inner auricle, along the cranial surface of the dorsal-ventral plate. It stretches the spiral plate and the inner auricle ventrally and shortens the cartilaginous part of the auditory meatus.

(6) The *m. tragicus medialis* begins at the back edge of the posterior-ventral plate and attaches at the frontal edge of the spiral plate and its later-

al surface. The muscle stretches the leading edge of the spiral plate and opens the auditory meatus.

(7) The *m. helices* begins at the back edge of the posterior-ventral plate and attaches along the outgrowth of the spiral plate and at the leading edge of the plate itself. It closes the auditory meatus around the inner auricle.

(8) The *m. antitragicus* begins at the cranial surface of the posterior-ventral platc and attaches at the upper edge of the spiral plate. It stretches the spiral plate ventrally and helps to close the auditory meatus.

1.1. Insectivores *(Insectivora)*

For a description of the outer ear of the order Insectivora, two species are of special interest: the shrew (*Sorex araneus*), which has a complex auricle, and the mole (*Talpa europaea*), which has a completely reduced auricle.

In the shrew, the outer ear consists of an auricle and an external auditory meatus. Separate pocket-shaped cells are located on the inner surface of the auricle. The external auditory meatus is a tube that can be up to 4 mm long. The tube diameter varies slightly from 2.0 × 2.5 mm in an inlet area to 2.1 × 1.5 mm at the site of cartilaginous-osseous transition (Table 1). In a cross section, the empty space of the auditory meatus has an uneven shape with offshoots directed into the auditory meatus cavity. The cartilage consists of three isolated plates surrounding the auditory meatus. The epidermal layer covers the auditory meatus cavity uniformly. The aural glands are characterized by a tubular-alveolar type of structure.

In the mole, the auricle is reduced. The external auditory meatus begins as a small hole located at the base of the skull and is about 1 cm long. The cartilaginous part has the shape of a speaking trumpet, the top of which is faced toward the tympanic bulla. The auditory meatus of the mole contains a sebaceous-glandular organ, which is a complex of holocrine glands. These glands begin near the auditory meatus on the lower surface of the skull base and bear against the petrous bone as a plain trapeze more than 6 mm long. The gland complex continues to the skull centerline and joins with the same construction from the other side. Excretory ducts of glands open with several nipples at the site of the cartilaginous-osseous transition of the auditory meatus [Zawitch-Ossenitz, 1933].

The diameter and shape of the empty space of the auditory meatus vary (Table 1). In the inlet area, the auditory meatus cavity is surrounded by four flexibly interconnected cartilaginous plates. At the end part of the cartilagi-

Table 1. Diameter of the external auditory meatus of mammals (in mm)

Species of mammals	length of auditory meatus	Diameter of the auditory meatus in different parts					
		1	2	3	4	5	6
Sorex araneus	0.4	2.0 x 2.5	2.1 x 1.5	-	-	-	-
Talpa europaea	1.0	2.0 x 2.0	1.7 x 1.5	1.7 x 1.3	-	-	-
Rhinolophus ferrumequinum	1.75 -2.0	2.3 x 1.0	1.8 x 1.5	1.0 x 1.0	-	-	-
Vulpes vulpes	1.7 - 2.0	10.0 x 6.0	8.0 x 7.0	7.0 x 7.0	-	-	-
Canis familiaris	2.0 - 2.5	13.0 x 7.0	11.0 x 9.0	9.0 x 9.0	-	-	-
Mustela vison	1.0 - 1.3	4.0 x 2.5	-	-	-	-	-
Myocastor coypus	1.3 - 1.5	4.0 x 2.5	3.0 x 2.5	-	-	-	-
Enhydra lutris	5.5 - 6.0	11.0 x 4.0	5.5 x 5.0	4.0 x 3.0	-	-	-
Eumetopias jubatus	8.0 - 9.0	5.0 x 4.0	5.0 x 3.0	7.0 x 5.0	8.0 x 5.0	6.0 x 5.0	-
Callorchinus ursinus	5.5 – 6.0	5.0 x 0.7	2.0 x 2.0	2.5 x 1.5	-	-	-
Pagophilus groenlandicus	7.0 – 7.5	5.5 x 4.5	4.5 x 0.3	5.0 x 1.7	5.0 x 2.0	4.5 x 2.0	3.5 x 2.0
Pusa caspica	6.0 – 6.5	4.5 x 1.0	6.5 x 0.3	6.0 x 0.7	4.5 x 1.2	3.5 x 1.5	2.5 x 2.0
Phoca vitulina	6.0 – 6.5	6.0 x 2.0	6.0 x 1.0	4.5 x 0.3	7.0 x 2.0	6.0 x 4.0	7.0 x 4.0
Phoca insularis	7.0 – 7.5	6.0 x 3.0	5.0 x 1.5	6.0 x 2.5	5.0 x 3.0	4.0 x 3.5	-
Delphinus delphis	6.0 – 6.5	1.2 x 0.9	-	0.7 x 0.09	1.3 x 0.4	2.2 x 1.3	-
Tursiops truncatus	6.5 – 7.0	1.3 x 1.0	-	0.8 x 0.1	0.9 x 0.4	1.6 x 0.8	-

nous section, the plates join, surrounding the auditory meatus with one cartilaginous horseshoe-shaped lamina.

Blood vessels of the auditory meatus are few in number. A spacious venous sinus is located at close proximity to the tympanic membrane and surrounds the auditory meatus halfway along the diameter.

1.2. Bats *(Chiroptera)*

The bat auricle is large compared to the animal's head surface and body, and has a very complex structure. There is a well-developed trestle, representing a cutaneous outgrowth lanceolate or clavate in shape. *Rhinolophus*, which has no trestles, is an exception [Airapetianz, Konstantinov, 1965; 1970; 1974]. The size, shape and structure of the auricle vary in different bat species. For example, in *Miniopterus,* the auricle protrudes slightly above the animal head, but in *Plecotus,* its size is greater than the length of the animal body. The cartilaginous part of the auditory meatus has a flared shape with a top faced toward the tympanic bulla. Its length in different species varies insignificantly from 1.75 to 2.5 mm. The diameter and shape of the auditory meatus also vary (Table 1). In the inlet area, the auditory meatus lumen is oval in shape, but in the end part, it becomes round in shape. Aural glands of the auditory meatus are few in number and surround the cavity as a ring. Venous sinuses are located at the end of the cartilaginous part.

1.3. Rodents *(Rodentia)*

In *Myocastor coypus,* the auricle is small and round in shape. The upper edge of the auricle is turned toward the inlet. The lower part of the auditory meatus contains a bundle of long and coarse hair, the top of which is directed caudally.

The external auditory meatus is a straight, short and wide tube about 1.5 cm long. The diameter and shape of the auditory meatus lumen varies from oval in the inlet area to round at its end (Table 1). A horseshoe-shaped cartilage lamina surrounds the auditory meatus cavity as an open-ended tube. The aural glands are not numerous and form racemose gatherings along the hair stem. The gland ducts open into the cavity of the auditory meatus.

1.4. Cetaceans *(Cetacea)*

The outer ear of representatives of the suborders Odontoceti and Mysticeti was studied by many researchers [Home, 1812; Murie, 1873; Weber, 1886; Hunter, 1887; Lillie, 1915; Yamada, 1951, 1953; Yamada, Yoshizaki, 1959; Fraser, 1959; Fraser, Purves, 1954, 1959, 1960; Purves, 1955, 1958, 1966, 1967; Reysenbach de Haan, 1957 a, b; Purves, Utrecht, 1963; Ichichara, 1964; Norris, 1968]. However, the existing data were scanty and, in most cases, concerned a macro-anatomical description of the outer ear.

Odontocetes *(Odotoceti)*

In *Tursiops truncatus*, as in other members of the family Delphinidae, the auricle is completely reduced. The external auditory meatus begins with a small hole located at a distance of several centimeters (up to 4 cm) behind the eye. It has the form of a double-bent tube, in which the lumen diameter and shape change sharply. Curves divide the auditory meatus into three parts. The first and third parts are located horizontally relative to the tympanic bulla, while the second part goes down almost vertically and deviates. On going out of the hypoderm in its distal part, the auditory meatus cavity is overgrown with epithelial cells. When stretched, the proximal part represents an irregularly shaped cone with an oval base (Fig. 2a-d). This part of the auditory meatus is surrounded by an open-ended, horseshoe-shaped cartilaginous tube which is homologous to the cartilaginous curl of the auricle in terrestrial mammals. Well-developed muscles of the outer ear are attached to the proximal part of the auditory meatus; the *m. auricularis externus, m. zygomaticoauricularis* and *m. occipitoauricularis* are located along the auditory meatus [Boenninghaus, 1903].

The auditory meatus of the bottlenose dolphin is about 5.7 to 6 cm long. The length of the distal part is about 0.8 to 0.9 cm, and the length of the proximal part is about 5.6 to 5.9 cm. The diameter and shape of the auditory meatus lumen vary (Fig. 3; Table 1). An obliterated auditory meatus is noted in all studied representatives of the family Delphinidae.

In the common dolphin (*Delphinus delphis*), the auditory meatus is about 6 to 6.5 cm long. In the inlet area, the auditory meatus lumen is narrow. Then, for about 1 to 2 mm, the auditory meatus is overgrown with epithelial cells, after which the lumen reappears in an oval shape with outgrowths directed into the auditory meatus cavity. At the end of the proximal part, the auditory meatus lumen grows sharply in size, while the lumen shape becomes oval (Table 1).

In dolphins, the auditory meatus at the beginning of the proximal part is surrounded by simple auricular glands that are alveolar-tubular in structure. Secretions pass into the alveolar tube, functioning similar to the holomerocrine type.

The auricular glands of the bottlenose dolphin's auditory meatus are more complicated in structure than those of the common dolphin. The beginning of the proximal part of the auditory meatus is surrounded by a glandular complex. It attains a substantial size and is composed of tubular and alveolar glands of a complicated structure, in which the ducts open into an epithelium fold. The gland duct passes into an intercalated part usually located in the middle of a gland lobe, the latter not observed in the common dolphin. The walls of the intercalated part are covered by a nonsecreting, one-layer epithelium, whose cells are cubic in form. The nucleus is irregularly shaped, large and light. The lumen of the intercalated part has secretion cells with basophilic granules and degenerated nuclei. Secretory apinuses are open and consist of cylindrical cells. The apical part of the cell is oxyphilic, and the basal part is basophilic. The apical part of the cell is swollen and does not indicate for apical secretion; that is very similar to a ceruminous gland structure. The shape, size and location of nuclei can vary. The nucleus can be small, oval-shaped and dark and can be located at the basal part of the cell. However, larger nuclei and lesser pyknotic ones can also occur. They are holomerocrine glands according to their secretion type. Secretion cells have monochromatic color [Solntseva, Chernova, 1978, 1980].

The dermal layer contains numerous blood vessels, and large venous sinuses are located at the end of the cartilaginous part. An epidermal layer covers the auditory meatus cavity non-uniformly.

Mysticetes *(Mysticeti)*

The outer ear of mysticetes is composed only of the external auditory meatus due to a complete reduction of the auricle. The external auditory meatus begins with a lentiform deepening behind the eye. In rorquals, the auditory meatus lumen narrows from an inlet toward the tympanic membrane and, after passing through a fatty layer, disappears. After several centimeters, the lumen appears again and leads up to a tympanic membrane [Carte, Macalister, 1868].

One of the peculiarities of the external auditory meatus of mysticetes is the presence of cerumen (earwax), which conforms to the shape of the auditory meatus lumen (conic). The base of this cone is curved according to the shape of a "glove finger" (i.e., the tympanic membrane). The cerumen is created due to the work of modified gland ducts of the "glove finger" and a desquamating epithelium [Purves, 1955]. In right whales, the typical cerumen is not formed; instead, it is small and has a gray-blue color [Yablokov et al., 1972].

1.5. Carnivores *(Carnivora)*

The outer ear of the species of the order Carnivora has been studied in both morphological and experimental works. Research of the outer ear has been conducted using typical representatives of the terrestrial (fox, dog) and semi-aquatic carnivores. In both of the studied species (fox, dog), the outer ear shows great structural similarity. The auricle has the shape of a leathery, rolled-up speaking trumpet, as do most terrestrial mammals. The base of the auricle lies on a fatty pad, which allows for great mobility of the auricle. Crests are located on the inner surface of the auricle, which is equipped with long hair. The external auditory meatus has the shape of a straight, short and wide tube (Fig. 4a, b). In a fox, its length constitutes about 1.7 to 2.0 cm; in a dog, it is about 2.0 to 2.5 cm.

The auditory meatus cavity is surrounded by two cartilaginous plates, forming an open-ended tube. The cartilaginous plates are notable for their size and configuration. The outer plate is 2 to 3 times thinner than the inner plate and has a horseshoe-shaped form. The inner plate, which resembles an open-ended ring, surrounds the auditory meatus cavity. In the inlet area, the auditory meatus lumen has an oval form, and at the end, it has a circular form (Table 1).

The auditory meatus cavity is surrounded by a glandular complex. Small alveolar and holocrine glands of complicated structure open into a nipple composed of an epidermal outgrowth. Above their ducts, the ducts of a glomerular ceruminous gland open into a hair bursa (pouch). A distinctive feature of the tubular glands is the forming of a secretory reservoir. Every gland is composed of several lobes, which vary in size. The glands are formed by cubic cells that have granular cytoplasm and round nuclei [Solntseva, Chernova, 1980].

The auditory meatus cavity is covered by a stratified, nonkeratinizing pavement epithelium. Blood vessels of the auditory meatus are not numerous.

In the outer ear of a mink (*Mustela vison*), the auricle is a small, round and thin leathery fold, on the inner surface of which, near the inlet, the skin outgrowths are located. The inner surface of the auricle is covered by thin hair (Fig. 1a). The inner auditory meatus has the form of a straight, short tube, the diameter and shape of which vary insignificantly (Table 1). The length of the auditory meatus is about 1 to 1.3 cm. The auditory meatus cavity is surrounded by two cartilaginous plates — outer and inner — as inherent for terrestrial carnivores. The inner plate is 2 to 3 times thicker than the outer one and has a horseshoe-shaped form. The outer plate is smaller in size and is located closer to the surface of the epidermis (Fig. 4a, b).

Auricular glands are few in number and are concentrated at the middle part of a hair follicle. The excretory duct of the gland forms a wide cavity that contains hair. The aural glands are similar to those of other carnivores in their structure and secretion type.

In a sea otter (*Enhydra lutris*), the auricle is formed by a thick leathery fold, the edges of which are twisted and tightly adjoined. As a result, the auricle takes on a conical shape. The most developed muscle of the auricle is the *musculus auricularis anterior*.

The external auditory meatus has the form of a straight tube about 5.5 to 6 cm long. As in most mammals, the middle part of the auditory meatus has a trumpet-like form with a top directed toward the tympanic bulla. The auditory meatus cavity is surrounded by four cartilaginous plates, forming an open-ended tube. The location of the plates is similar to that in terrestrial carnivores.

The diameter of the auditory meatus lumen changes from the inlet to the tympanic membrane, although its shape remains oval (Table 1).

The dermal layer contains multiple blood vessels. Aural glands are not numerous and are structured similarly to those of a mink. An epidermal layer covers the auditory meatus cavity uniformly.

1.6. Pinnipeds *(Pinnipedia)*

The outer ear of pinnipeds has been studied in different otariids and phocids [Rosenthal, 1825; Huber, 1934; King, 1964; Ramprashad et al., 1971; Solntseva, 1973a, c, g, 1974, 1975a].

Otariids *(Otariidae)*

For a description of the outer ear of otariids, two representatives are of special interest, namely, the Northern fur seal and the Steller sea lion, which both exhibit different outer ear structures.

The auricle of the Northern fur seal (*Callorchinus ursinus*) is a thick leathery fold that can extend up to 5 cm along the long axis. The edges of the auricle adjoin each other tightly because of the bending of auricle along the long axis. The outer surface of the auricle is covered by sparse and wiry hair. Its inner surface is smooth and without protrusions. Two aural muscles of the outer ear, the *m. auricularis anterior* and the *m. auricularis posterior*, are the most developed.

The external auditory meatus curves at some distance from the inlet and has a length of about 5.5 to 6.0 cm. The auditory meatus is surrounded by cartilaginous plates. The largest plate is horseshoe-shaped, and the other is small and is adjacent to the meatus cavity. In its initial part (i.e., in the inlet area), the auditory meatus lumen has a slit-like shape. In the proximal part, the lumen increases in size and acquires a triangular form, but at the point of the cartilaginous-osseous transition, its shape becomes oval (Fig. 5a-d; Table 1).

The aural glands do not form a continuous glandular ring. These are holocrine glands, composed of four to five lobes. The gland ducts are extensive. The glands are simple alveolar with forked ends according to their structural type. Cerumenal glands are not found (Fig. 6 a, b, c).

There are numerous blood vessels in the connective tissue. The large venous sinuses are located between the cartilaginous plate and the epidermal layer. The epidermal layer covers the auditory meatus uniformly; its thickness remains constant along the whole length.

In the Steller sea lion (*Eumetopias jubatus*), the auricle structure is similar to that of the sea otter. The auricle is twisted into a cone, the top of which is turned caudally (Fig. 1b). The most developed muscles of the outer ear are the *m. auriculus anterior* and the *m. auriculus posterior*.

The external auditory meatus has the shape of a curved tube 9 cm long and is surrounded by horseshoe-shaped cartilage. Although the diameter of the auditory meatus lumen changes from the inlet to the tympanic membrane (Table 1), its form remains oval.

The aural glands are not numerous and surround the auditory meatus cavity like a ring. In their structure, they are simple alveolar glands, not forming a reservoir for a secretion. There are cerumenal glands, which are simple-tubular in structure. The tube is twisted spirally; its long axis goes transversely to the surface of the epidermal layer. The gland tangle is loose. The gland duct opens into a hair follicle a little lower than the holocrine glands. A strong basophilia is typical of the secretory epithelium. The nucleus is located in the center of the cell [Solntseva, Chernova, 1980]. The venous sinuses are located in the connective tissue. The epidermal layer covers the auditory meatus uniformly.

Phocids *(Phocidae)*

The description of the outer ear of phocids is performed in the typical representatives of this suborder: Greenland, Caspian, Island seals, largha and bearded seals.

The outer ear of different phocids shows similar features. Only the external auditory meatus forms the outer ear, as there is a complete reduction of the auricle. At some distance from the inlet, the auditory meatus, located directly under the skin, bends in a knee-like way (double-bent tube) and descends diagonally to the tympanic bulla.

In the largha seal (*Phoca vitulina*), four auricular cartilages of the auditory meatus are described [Böas, 1912], and the same is done for the Greenland seal [Solntseva, 1973a, 1975a; Ramprashad et al., 1973]. All four cartilaginous plates are flexibly connected and have different forms and relative orientation. Lateral, posterior-ventral, dorso-ventral and medial cartilaginous plates are discernible [Ramprashad et al., 1971].

Eight aural muscles are attached to the external auditory meatus in the inlet area before a knee-like bending of the auditory meatus.

The auditory meatus is covered by the epidermis uniformly, the thickness of which varies insignificantly in different species. Numerous holocrine glands are located in derma clustered around the middle part of a hair follicle. Every gland lobe bears a spacious duct, which opens at the top part of the hair follicle. The gland is formed by cubic cells with oval nuclei and granular cytoplasm. The apocrine cerumenal glands, located in the inner dermal layer, encircle the auditory meatus. The walls of these glands are formed by one layer of cubic cells. The cellular cytoplasm contains acidophilic granules [Ramprashad et al., 1971].

The dermal layer is penetrated by numerous blood vessels. Venous sinuses are located at the proximal end of the cartilaginous part and in the osseous part of the auditory meatus.

In the Greenland seal (*Pagophilus groenlandicus*), the length of the auditory meatus constitutes about 7 to 7.5 cm (Table 1). At the distal part, the auditory meatus is circular and has protrusions, directed medially. At the point of a knee-like bending, the lumen acquires a horseshoe-shaped form and is very narrow. Then the lumen narrows at the center and widens peripherally, after which it becomes triangular in shape. At the point of cartilaginous-osseous transition, its shape becomes rounded.

In the Island seal (*Phoca insularis*), the external auditory meatus begins with a small foramen of a triangular shape. The length of the auditory meatus constitutes about 7 to 7.5 cm. At the point of a knee-like bending, the auditory meatus lumen has a shape of a channel with protrusions, directed into the cavity of the meatus. Then the auditory meatus lumen increases in size and its form becomes more rounded (Table 1).

In the Caspian seal (*Pusa caspica*) and the largha seal, the external auditory meatus begins with a small foramen of an oval shape. Its length constitutes about 6.0 to 6.5 cm. At the point of a knee-like bending, the auditory meatus lumen decreases until it is a slit-like aperture with the shape of a horseshoe. The long axis of the lumen is about 6.5 mm; the short axis is about 0.3 mm. Then the lumen acquires a pyriform shape, and at the point of the cartilaginous-osseous transition, its shape becomes rounded (Fig. 7a-g; Table 1).

In the bearded seal (*Erignathus barbatus*), the external auditory meatus begins with a narrow, irregularly shaped foramen. The length of the meatus constitutes about 7.0 to 7.5 cm. At the point of a knee-like bending, the lumen becomes narrower. At the proximal part, the size of the lumen increases and its shape becomes oval (Table 1).

1.7. Features of outer ear structure in representatives of various ecological groups of mammals

The main pattern of outer ear structure in members of various orders of mammals has been described above in taxonomic order, with emphasis on lifestyles associated with terrestrial, subterranean, aerial, semi-aquatic or aquatic habitat conditions.

In terrestrial mammals, the outer ear consists of an auricle curled up like a speaking-trumpet and an external auditory meatus, which has the shape of a straight, short and wide tube.

In subterranean forms (mole), the auricle is completely reduced and the external auditory meatus acquires the shape of an elongated tube, located superficially on the skull base.

Among chiroptera (aerial forms), the size and shape of the auricle vary significantly. As in most terrestrial forms, the auditory meatus of members of the order Chiroptera resembles a funnel-shaped tube.

In semi-aquatic forms, a morphological reconstruction of the auricle broaching its outer parts is evident (Diagram 1). For example, in the auricle of the mink, a cutaneous outgrowth appears near the inlet. In the nutria, the upper edge of the auricle forms a fold located above the upper part of the auditory foramen. From below and adjacent to it, there is a dense hair bundle of long, coarse, almost waterproof hairs. In the sea otter and the Steller sea lion, a significantly greater morphological reconstruction of the auricle is evident. The auricle acquires the shape of a thick leathery fold, in which the edges are tightly adjoined. This type of auricle is in the shape of a cone with the top directed caudally. In the fur seal, the auricle is bent along the main axis, and, consequently, its side surfaces are tightly adjoined.

In phocids and cetaceans, the auricle is reduced, similar to that observed in subterranean forms. The external auditory meatus begins with a small foramen, whereas that in cetaceans shows an inlet significantly smaller than in phocids.

In terrestrial, aerial, subterranean and semi-aquatic forms, the external auditory meatus bears the shape of a straight tube, the diameter of which varies insignificantly along the tube, while in aquatic mammals the auditory meatus is more complicated.

In the Steller sea lion (*Eumetopias jubatus*), the auditory meatus bends, but the diameter and shape of its lumen maintain features inherent to terrestrial forms. In a fur seal, the auditory meatus curves more extensively and attains a double-bent form.

In contrast to otariids, in which the auditory meatus is directed straight to the tympanic bulla, the auditory meatus in phocids is located directly under the skin and is bent.

In most mammals, the external auditory meatus is surrounded by horseshoe-shaped cartilage. In species with a reduced auricle (moles, phocids), this cartilage is composed of four plates flexibly connected to each other. For phocids and some otariids, a change in diameter and shape of the auditory meatus lumen is typical.

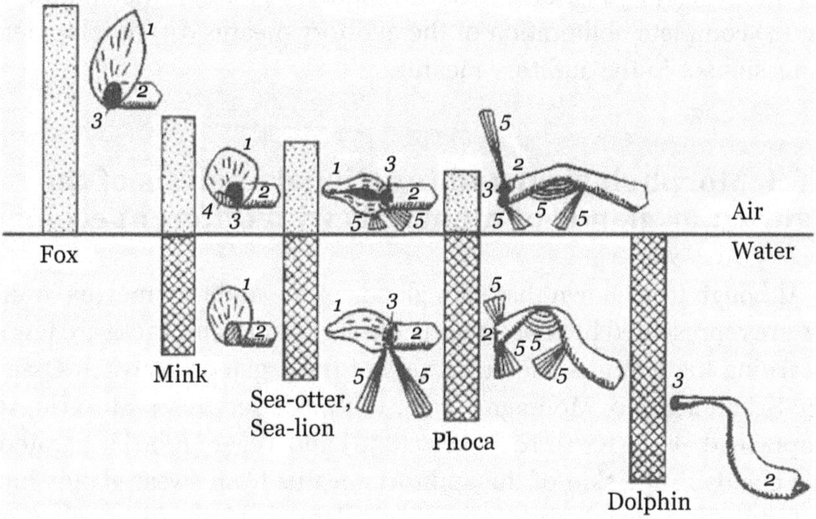

Diagram 1. Structure of the outer ear of terrestrial, semi-aquatic and aquatic mammals. Mechanisms opening the meatus in air and closing it under water. Vertical columns are an average specific duration of stay in air and under water: 1 — pinnal helix; 2 — auditory meatus; 3 — orifice; 4 — skin valve; 5 — ear muscles; 6 — tympanic membrane ligament.

In odontocetes (dolphins), the construction of the auditory meatus differs significantly from that of all other analyzed species. The bends divide the auditory meatus into three sections (two horizontal and one vertical); as a result, the auditory meatus has a pronounced double-bent form. At some distance from the inlet, the lumen of the auditory meatus is obliterated. Hence, two sections are formed: distal and proximal ones. Similar obliteration of the auditory meatus is noticed in a sperm whale [Clarke, 1948; Yamada, 1953a, b]. Nevertheless, there are some data that in some species of odontocetes, the auditory meatus is not obliterated [Reysenbach de Haan, 1957a, b; Fraser, Purves, 1960].

In animals whose way of life is connected with subterranean and aquatic habitat conditions (moles, sea otters, pinnipeds, cetaceans), well-developed venous sinuses are located in the proximal end of the cartilaginous part of the auditory meatus, not far from the tympanic membrane, and also in the osseous part of the auditory meatus.

Thus, the outer ear of different ecological groups of mammals is characterized by a diversity of adaptations [Solntseva, 1971, 1972a, c, 1973a, c, e, g, 1974, 1975a, 2006] that could have appeared on the way to dwelling in subterranean and aquatic environments. These modifications concern the following structures: (1) reduction of the auricle; (2) change in shape of the auditory meatus, as well as in its diameter and lumen form; (3) appearance of flexibility at the junction of the cartilaginous plates of the auditory meatus; (4) complete obliteration of the auditory meatus; (5) development of venous sinuses in the auditory meatus.

1.8. Morphological and functional analysis of the auricular glands of mammals with different ecologies

Although it is known that skin glands of the auditory meatus in mammals are represented by tubular and alveolar formations, there are few data concerning the features and structures of these glands [Zawitch-Ossenitz, 1933; Schaffer, 1940; Montagna et al., 1948; Nielsen, 1953; Murariu, 1976; Ramprashad et al., 1971]. Köllicker [1850] and Teger [1906] separated tubular glands in the skin of the auditory meatus from sweat glands for the first time and named them "cerumenical" glands. Geinold [cited after Schaffer, 1940] also considered these glomerate glands as different from ordinary sweat glands. The main difference is that the cylindrical glandular epithelium of cerumenical glands has no sharply outlined border refracting light to a great extent, and the cytoplasm of the apical part of a secretory cell is usually turgid. The ontogeny of cerumenical glands shows their relationship to apocrine glands, when the connection of gland ducts to hair follicles can be safe. Thus, long before the determination of apocrine secretion [Schiefferdecker, 1917], its variety was described in cerumenical glands.

In addition to cerumenical glands, the skin of the auditory meatus of mammals contains holocrine glands. Some authors place this type of gland among the cerumenal ones, because they consider cerumen (earwax), the product of secretion of both tubular and alveolar glands [Nielsen, 1953; Murariu, 1976]. However, Schaffer [1940] supposed that the secretion of tubular glands is not a compound of cerumen and, conversely, removes the earwax by dissolving it.

Holocrine glands are usually connected by ducts with hair and secreted lipids of cerumen: triglycerides, cholesterol, some plasmols and, possibly, phospholipids [Montagna et al., 1948]. Tubular glands produce proteins. The basophilia of their secretory epithelium is the result of a protein synthesis [Caspersson, 1947]. The secretory activity of tubular glands produces a small quantity of pigment, which in its histochemical indices in animals differs from the pigment of human earwax. In humans, both tubular and alveolar glands in the skin of the auditory meatus secrete lipids [Montagna et al., 1948].

Thus far, I have compared the structures and features of the skin glands of the external auditory meatus in representatives of phylogenetically close and distant orders of mammals, including mammals of aerial, terrestrial, subterranean, semi-aquatic and aquatic habitats. This analysis makes it possible to determine the functional importance of both types of auricular glands and to follow a trend in the morphological changes which appear in the processes of mammalian adaptation to different habitats.

1.8.1. Insectivores *(Insectivora)*

In all representatives studied, auricular glands are typically present (Diagram 2; Tables 2, 3). Similarly, the glands in hedgehogs are holocrine formations of a small size (113 × 113 μm) and rare, simple tubular glands. The diameter of the secretory part of these tubular glands reaches 45 μm (Table 3). Holocrine glands are found two by two near every hair, extending into the auditory meatus cavity. The duct of every gland opens into a hair follicle. The cytoplasm of the secretory cells of the tubular glands is basophilic and is characterized by the absence of an apical secretion.

In shrews and moles, there are thickened strips of skin, covered in hair, in the auditory meatus. They are gaunt along the auditory meatus and consist of small (moles) or enlarged (shrews) holocrine glands and simple tubular glands beneath them (Diagram 2). Sometimes the secretory part of the tubular gland may be twirled into a tangle (common shrew). The skin of the auditory meatus in a pigmy shrew (*Sorex minutus*) does not contain tubular glands. In different species, the sizes, shapes and locations both for the separate glands and for glandular complexes formed by them vary (Tables 2, 3). Glands reach maximum size in the muskrat (*Desmana moschata*), Eurasian shrew (*Neomys sp.*) and common shrew (*Sorex sp.*)

The holocrine glands in moles and shrews, as in hedgehogs, are usually simple alveolar, have a bag-shaped form and are located by two or by four near every hair, with which are bound by ducts. Tubular glands have a cylindrical secretory epithelium, the cytoplasm of which does not show apical secretion. The duct of a tubular gland opens into the common duct with a hair epidermal crater at the depth about 23 μm from the skin surface.

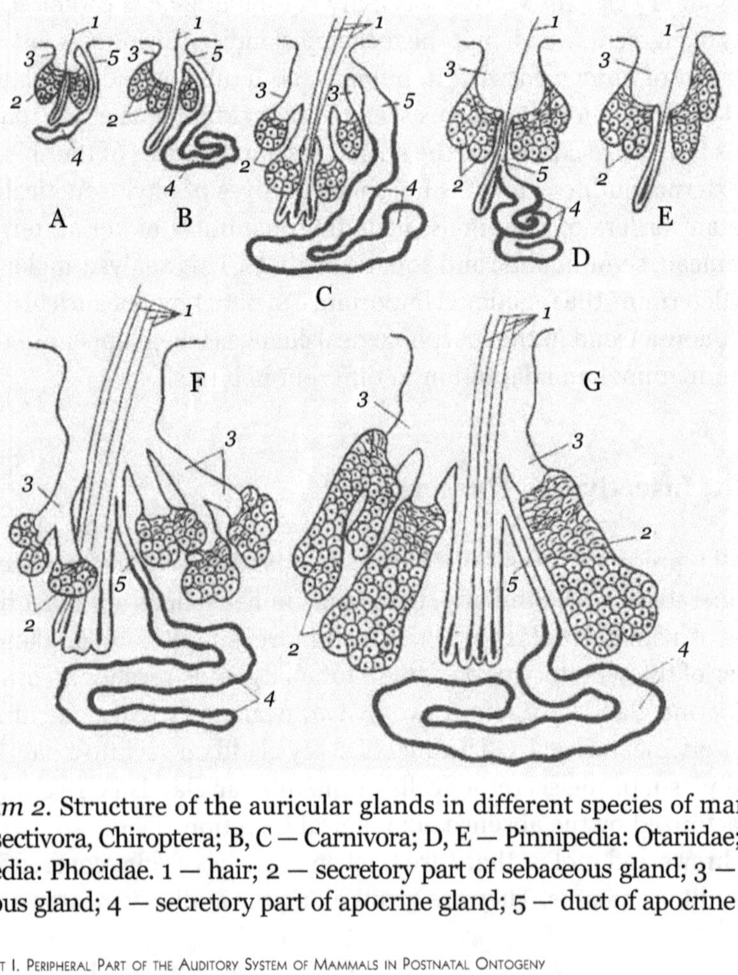

Diagram 2. Structure of the auricular glands in different species of mammals. A — Insectivora, Chiroptera; B, C — Carnivora; D, E — Pinnipedia: Otariidae; F, G — Pinnipedia: Phocidae. 1 — hair; 2 — secretory part of sebaceous gland; 3 — duct of sebaceous gland; 4 — secretory part of apocrine gland; 5 — duct of apocrine gland.

Table 2. Date of topography of the auricular glandes of the external auditory meatus in mammals with different ecologies

Species	Cerumenal glands	Holocrine glands	Complex cerumenal and holocrine glands	Complex merocrine alveolar-tubular glands
Talpa europaea	+	+	+	-
Talpa coeca	+	+	+	-
Talpa altaica	+	+	+	-
Desmana moschata	+	+	+	-
Neomys fodiens	+	+	+	-
Sorex caecutiens	+	+	+	-
Sorex caucasica	+	+	+	-
Sorex minutus	+	+	+	-
Sorex araneus	+	+	+	-
Sorex raddei	+	+	+	-
Crocidura suaveolens	+	+	+	-
Crocidura russula	+	+	+	-
Erinaceus europaeus	+	+	+	-
Rhinolophus ferrumequinum	-	+	+	-
Vespertilio pipistrellus	-	+	-	-
Vespertilio murinus	-	+	-	-
Myotis blythi	-	+	-	-
Plecotus auritus	+	+	+	-
Murina leucogaster	+	+	+	-
Hipposideros commersoni	+	+	+	-
Triaenops persicus	+	+	+	-
Pteropus tonganus	-	+	-	-
Vulpes vulpes	+	+	+	-
Mustela vison	+	+	-	-
Enhydra lutris	+	+	-	-
Callorchinus ursinus	-	+	-	-
Eumetopias jubatus	+	+	-	-
Pagophilus groenlandicus	+	+	+	-
Pusa caspica	+	+	+	-
Phoca vitulina	+	+	+	-
Phoca insularis	+	+	+	-
Erignathus barbatus	+	+	+	-
Tursiops truncatus	-	-	-	+
Delphinus delphis	-	-	-	+

Table 3. Mean sizes (in mm) of auricular glands of some insectivores and bats

Species	Holocrine glands		Cirumenal glands	
	Sizes of a glandular complex	**Sizes of a gland**	**Sizes of glandular glomi**	**Diameter of glandular tubule**
Desmana moschata	2.9x0.2	0.2x0.1	0.2x0.1	0.05
Talpa europaea	0.7x0.7	0.06x0.03	Absence	0.02
T. coeca	1.0x0.3	0.06x0.03	Absence	0.03
T. altaica	1.1x0.5	0.06x0.06	Absence	0.05
Neomys fodiens	1.4x0.1	0.1x0.1	0.1x0.1	0.03
Sorex caecutiens	1.8x0.4	0.6x0.1	0.2x0.1	0.06
S. caucasica	1.4x0.1	0.1x0.06	0.2x0.1	0.05
S. minutus	2.1x0.2	0.1x0.06	Absence	Absence
S. araneus	2.1x0.4	0.2x0.2	Absence	0.03
S. raddei	1.4x0.2	0.1x0.2	Absence	0.05
Crocidura suaveolens	2.1x0.1	0.1x0.1	Absence	Absence
C. russula	0.7x0.1	0.1x0.06	Absence	Absence
Erinaceus europaeus	Absence	0.1x0.1	0.1x0.1	0.05
Rhinolophus ferrumequinum	1.7x0.1	0.1x0.1	Absence	Absence
Plecotus auritus	1.8x0.2	0.06x0.06	Absence	0.09
Murina leucogaster	1.1x0.3	0.1x0.1	Absence	0.06

In addition to glandular formations in the auditory meatus of members of the order Insectivora, there are separate holocrine glands located by twos near every hair. The skin around the glands is thickened and acquires a beadlike texture.

1.8.2. Chiroptera *(Chiroptera)*

The external auditory meatus in members of the order Chiroptera contains little earwax. The skin of the meatus is characterized by a thin epidermis, the absence of an active keratinization process and a poor development of glandular formations. As a rule, the glandular formations are represented by small (up to 22 μm), bag-shaped holocrine glands located two by two near every hair.

In individual representatives of the order, holocrine glands of usual or enlarged size are tightly closed, forming a glandular complex (Table 2). One or several of such thickenings are stretched alongside the auditory meatus to a tympanic membrane. In these skin sections, tubular glands are located under hair bulbs. Alveolar and tubular glands in members of the order Chiroptera reveal great similarity in structure compared to those observed among members of Insectivora (Diagram 2). As regards the latter order, apical secretion is absent from the glandular epithelium of the tubular glands in bats.

The number, size and location of auricular gland complexes vary in different species of this order. For example, in the greater horseshoe bat (*Rhinolophus ferrumequinum*), the glandular complex consists of only holocrine glands. Tubular glands are absent from this species. Auricular glands reach maximum development in Commerson's bat, the greater horseshoe bat and the long-eared bat (*Plecotus auritus*) (Table 2).

1.8.3. Carnivores *(Carnivora)*

In different carnivores, auricular glands can be underdeveloped (sea otter, mink) or form a glandular complex surrounding the external auditory meatus (fox – Diagram 2; Tables 2, 4).

In the mink and the sea otter, the holocrine glands of the auditory meatus are small in size (on average 339 × 445 μm) and are complex alveolar. They are located two by two near every hair and connected to it by ducts. As a whole, these glands are similar to holocrine glands of the Insectivora order in structure, but differ from them by lobulous secretory section and widened excretory duct. Tubular glands are represented in the form of convoluted tubes, the secretory sections of which are twirled into a tangle. The diameter of the secretory section is about 56 μm (Table 4). The structure of the apocrine secretory epithelium is similar to that observed in the orders Insectivora and Chiroptera. The epidermis of the auditory meatus of these species includes multiple pigment granules.

A more complicated structure of holocrine glands is found in the fox than in the mink and the sea otter (Diagram 2; Table 4). Glands in the fox are represented by poly-alveolar formations. Holocrine glands surround hair fascicles and are connected to them by ducts. The structure of the tubular glands is similar to that of the order described above; however, in the proximal part, the duct can become broadened, forming a reservoir (up to 339 μm in diameter) filled with a homogeneous light-yellow mass. The duct of the tubular glands opens into a hair follicle above the point of follicle confluence with holocrine gland ducts. In the fox, the skin epidermis of the auditory meatus, unlike that of the mink or the sea otter, is thick, folded and poorly pigmented.

Table 4. Mean sizes (in mm) of auricular glands of some carnivores, seals and whales

Species	Holocrine glands					Cirumenal glands		
	Thickness of glandular complex	Sizes of a gland	Sizes of a glandular lobe	Diameter of a common duct	Depth of glandular layer	Sizes of a gland	Diameter of a secret tubule	Depth of glandular layer
Vulpes vulpes	0.8	0.5x0.5	0.1x0.06	0.03	0.3	0.6x0.2	0.02	0.7
Mustela vison	Absence	0.3x0.4	0.2x0.1	0.01	0.1	0.2x0.5	0.03	0.5
Enhydra lutris	Absence	0.4x0.4	0.2x0.07	0.07	0.07	0.07x0.2	0.05	1.1
Callorchinus ursinus	Absence	0.1x0.1	0.1x0.03	0.08	0.1	Absence	Absence	Absence
Eumetopias jubatus	Absence	0.1x0.1	0.05x0.03	0.02	0.1	0.3x0.5	0.03	Absence
Pagophilus groenlandicus	1.4	1.7x1.1	0.7x0.4	0.02	1.0	0.1x0.2	0.05	1.0
Pusa caspica	1.0	1.0x0.4	0.3x0.06	0.02	0.7	0.1x0.3	0.06	0.7
Phoca vitulina	1.4	1.4x0.7	0.4x0.1	0.03	1.4	0.2x0.2	0.01	1.4
P. insularis	1.0	1.0x1.0	0.3x0.1	0.01	1.1	0.06x0.2	0.06	1.1
Erignathus barbatus	1.1	1.0x0.9	0.5x0.2	0.02	1.6	0.1x0.2	0.02	1.6
Tursiops truncatus	0.9	Absence	Absence	Absence	Absence	0.4x1.4	0.03	Absence
Delphinus delphis	0.2	Absence	Absence	Absence	Absence	0.1x0.7	0.05	Absence

1.8.4. Pinnipeds: Phocids, Otariids *(Pinnipedia: Phocidae, Otariidae)*

Phocids

The presence of a continuous glandular ring surrounding the auditory meatus cavity and consisting of large holocrine glands is inherent to all species of phocids. The thickness of the ring usually varies in different species, but the greatest thickness is found in a largha seal (Diagram 2; Table 4). In the studied species of phocids, the glandular ring consists of different numbers of hypertrophied holocrine glands, complex poly-alveolar in structure. Three to six such glands open into every hair follicle. The form of the secretory alveolus of a holocrine gland is mainly cylindrical with an alveolar distal part. However, in the Caspian seal, the glands with a round alveolus predominate.

Tubular glands are inherent to the auditory meatus of all species studied (Table 2). Their structure is similar to that of the order Carnivora. Tubular glands lie under hair bulbs; their secretory section is twirled into a tangle. However, unlike Carnivora, in phocids, a narrow duct of a tubular gland does not form widenings and opens into a hair follicle beneath the holocrine gland ducts. The sizes of tubular gland tangles, secretory sections and ducts vary among species (Table 4). The glandular epithelium of the tubular ducts is characterized by basophilia and swollen cytoplasm of a cellular apical part.

Otariids

In the Northern fur seal (*Callorchinus ursinus*) and the Steller sea lion (*Eumetopias jubatus*), the holocrine glands do not form an entire glandular ring around the auditory meatus cavity. Instead, they encircle sparse hairs. Holocrine glands are small in size (on average up to 113 × 113 μm; Table 4) and have a complex alveolar structure. Each gland consists of four to five rounded lobes.

Tubular glands are absent from the fur seal but are present in the Steller sea lion, in which they are simple tubular in structure, and are separated from one another. Each secretory tube is twisted into a ball; its long axis is perpendicular to the skin surface. The ball is usually small (up to 339 × 450 μm; Diagram 2; Tables 2, 4). The duct of a tubular gland, as in phocids, opens into the hair follicle under the ducts of the holocrine glands. A strong basophilia of the cytoplasm, a central location of the nuclei and the absence of apical secretion are typical of the secretory epithelium of the tubular gland. The tubular lumen contains granules of a dark-orange color.

1.8.5. Cetaceans *(Cetacea)*: *Odontoceti-Delphinidae*

In dolphins, unlike all species studied, hairs are completely absent, as are the associated alveolar and tubular glands in the skin of the auditory meatus. The proximal part of the auditory meatus is surrounded by a ring, which is composed of glandular tissue. The latter is a complex of poly-alveolar-tubular glands in a bottlenose dolphin and simple alveolar-tubular glands in a common dolphin (Diagram 3; Tables 2, 4).

In the bottlenose dolphin (*Tursiops truncatus*), a multilayer pavement epithelium of the proximal part of the auditory meatus is notable for its significant thickness and rugosity. The epithelium cells constantly undergo a desquamation process. The nuclei of the epithelium's outer layer are one-half the size of the nuclei of the basal layer. They are usually pycnotic and dense, and have an uneven shape. The glandular ring, which surrounds the proximal part of the auditory meatus, consists of large glands with ducts that open in the folds of the epithelium of the auditory meatus. The excretory duct of a gland continues into the intercalated part, located in the center of a secretory lobe. The lumen of the intercalated part and excretory duct contain a mass of an eosinophilic secretion with basophilic granules and degenerated nuclei. Unlike the ducts, the secretory tubes and alveoli are surrounded by myoepithelial cells. The glandular epithelium is made up of cylindrical cells, which have cytoplasm that is stained polychromatically: its apical part is oxyphilic, and its basal part is basophilic. Apical secretion of the apical part of the cytoplasm in the secretory cells is absent. The secre-

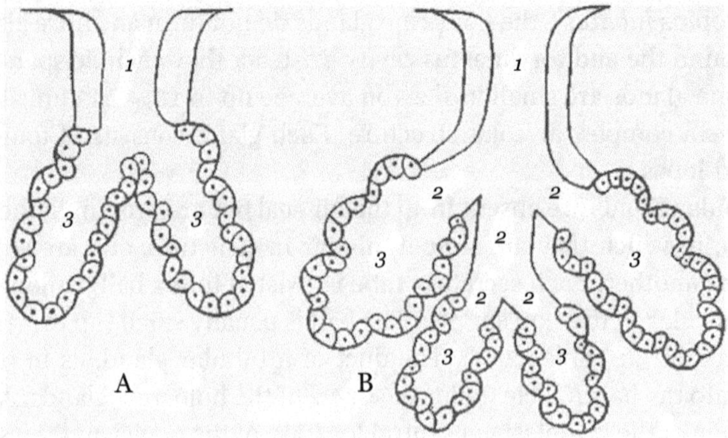

Diagram 3. Structure of the auricular glands in dolphins: A – *Tursiops truncatus*; B – *Delphinus delphis.*

tion type of the glandular epithelium of the auricular gland of the bottlenose dolphin is determined by us as a merocrine one with the elements of a holocrine secretion.

In a common dolphin *(Delphinus delphis)*, a homologous gland is much less developed (Table 4). Its thickness at cross sections does not exceed 0.2 mm. The gland complex does not completely surround the proximal part of the auditory meatus. The structure of a gland in a common dolphin is simpler than that in a bottlenose dolphin (Diagram 3). Intercalated parts of ducts are absent, and a secretory tube directly converts into the excretory duct. As in a bottlenose dolphin, the glandular epithelium of a common dolphin functions according to a holocrine type.

The comparison of auricular glands in the skin of the external auditory meatus of the studied species with their usual glands [Sokolov, 1973] showed that each order has its own type of structure. Holocrine glands of the auditory meatus differ from the ordinary skin glands by the presence of a reservoir for secretion, a widened duct and pigment granules. Tubular skin glands of the auditory meatus differ from the ordinary tubular apocrine glands by the absence of an apical secretion. Despite the fact that the structure of the cytoplasm in the tubular glands of the apical part of the secretory cells allows us to consider them cerumenal tubular glands, it is possible to share Murariu's opinion [Murariu, 1976] that it would be more correct to name holocrine glands as cerumenal ones, because they produce cerumen. Holocrine glands are numerous and more developed compared to tubular apocrine glands, which are rare, isolated from each other and do not form glandular complexes. In species such as a greater horseshoe bat (*Rhinolophus ferrumequinum*), a greater tube-nosed bat (*Murina leucogaster*), a pigmy shrew (*Sorex minutus*) and a Northern fur seal, tubular glands are absent, disproving Schaffer's supposition [Schaffer, 1940] about the lytic role of an apocrine auricular gland secretion. However, direct dependence between the quantity and developmental extent of holocrine and apocrine glands has not been traced. For example, in a fox, tubular glands are more developed than holocrine glands.

The comparison of the holocrine auricular glands in representatives of different systematic groups showed that they have the simplest structure in the orders Insectivora and Chiroptera (simple alveolar). The structure of the glands becomes complicated in the Carnivora order (complex alveolar and complex poly-alveolar). Glands reach maximum development in phocids (poly-alveolar, complex alveolar – Diagram 2). Within the Insectivora order, the structures of auricular glands vary. In a hedgehog, the glands differ from those of a shrew and a mole by their small size and specific location; they do not form entire glandular complexes.

In Chiroptera and Insectivora, the holocrine glands can form glandular complexes. The presence of these glandular complexes in the skin of the auditory meatus is a specific characteristic and is not inherent for all representatives of the order. However, in all studied representatives of the Insectivora order (moles and shrews), auricular glandular complexes are present.

In different representatives of Carnivora, auricular glands are unequally developed. The structure of the holocrine glands becomes complicated in the fox compared to that of the mink or the sea otter.

Among the pinnipeds, the structure of the holocrine glands differs sharply in phocids and otariids. In phocids, the holocrine glands have a compound structure and form an entire glandular ring around the auditory meatus cavity. For the representatives of this suborder, a derated form of the secretory alveoli of the holocrine gland is typical (from cylindrical in the Greenland, Island, largha and bearded seals to oval in the Caspian seal).

The structure of the tubular auricular glands becomes complicated in the orders Carnivora and Pinnipedia compared to that of Insectivora and Chiroptera (Diagram 2). This becomes apparent due to the elongation and folding of the glandular secretory part. The excretory duct of a gland is always connected to a hair follicle. The level of its confluence with the hair bursa is specific to every order and reveals similarities with the level of confluence with the hair bursa of the ordinary apocrine glands. In the orders Insectivora and Chiroptera, the ducts of the tubular glands open into an epidermal crater common with a hair at an insignificant depth from the skin surface. In carnivores, the duct of a tubular gland is shifted in a caudal direction along the hair and lies deeper than in insectivores, opening into the hair bursa above the ducts of holocrine glands. In pinnipeds, the ducts of the tubular glands open below the ducts of the holocrine glands.

Thus, the structure of the alveolar and tubular auricular glands reveals greater variability within five orders of mammals, becoming complicated in the most highly organized and phylogenetically more-recent forms. The structure of holocrine glands changes from simple alveolar to complex polyalveolar. Simple alveolar glands of the Insectivora order can be a primordial form for the transformation into more complicatedly organized glands of the order Carnivora. By sinking into the derma alongside the hair follicle and further hypertrophic development, the alveolar and tubular glands of carnivores can develop into complex poly-alveolar, complex alveolar and simple tubular glomerular glands of pinnipeds. Complication of the structure of the auricular glands in the series Insectivora – Carnivora – Pinnipedia confirms the notion about the phylogenetic kindred of these orders. The similar structure of the glands in Insectivora and Chiroptera can also be explained by their phylogenetic kindred.

The main function of the auricular glands is the production of earwax, which is composed of holocrine and apocrine gland secretions, pigment, produced by glands, and desquamating keratinized scales of the auditory meatus epidermis. The earwax provides a mechanical adipose greasing of the auditory meatus' skin, protecting it from water penetration, and also serves as a barrier against parasites. As water is especially rich in parasites and, thus, very dangerous for the auditory organ, the auricular glands appear to be better developed in semi-aquatic forms, of which an open auditory meatus is characteristic (desman, Eurasian shrew, mink, sea otter, pinnipeds). They are absent from obligate hydrobionts with a constantly closed auditory meatus (dolphins). In pinnipeds, the auricular glands are the best developed of the species, as they are more closely associated with aquatic conditions. For example, in phocids, these glands are better developed than in representatives of otariids, the latter being less strongly related to aquatic habitats. This allows us to suggest the existence of a correlation between an effective functioning of the auricular mechanisms which close the auditory meatus when submerging under water and the degree of auricular gland development.

In dolphins, the auditory meatus is constantly closed, thus naturally protecting it from water and parasite penetration. Alveolar and tubular glands are absent from both the distal and proximal parts of the auditory meatus (Diagram 3). A specific protein-mucoid gland, the ducts of which open in the proximal part of the auditory meatus, attains a significant development in dolphins. The secretion of these glands is produced in the closed part of the auditory meatus and can help level the pressure in the hearing organ when the animal is submersed at a greater depth. There is also a possibility that this specific gland participates in the production of a protein emulsion that fills the peribullar sinuses surrounding the tympanic bulla.

Aerial and terrestrial habitats are relatively safe in regard to the auditory organs; therefore, the extent of auricular gland development in Insectivora and Chiroptera is insignificant.

In subterranean forms (moles), or in forms that move in subterranean tunnels to find food in the soil and in forest (shrews), the auditory meatus is protected with lavish adipose greasing produced by the auricular glands.

Thus, the degree of auricular gland development and the intensity of their secretory function can be viewed as an adaptation of the outer ear of mammals to different habitat conditions, providing mechanic and likely also bacteriostatic protection to the auditory meatus [Solntseva, Chernova, 1978, 1980]. The structure and features of the auricular glands are species-specific and can serve as taxonomic characteristics.

Chapter 2. Structure of the middle ear in mammals

The middle ear structure of mammals has been attracting the attention of various scientists for a long time, so at present there are numerous fundamental works devoted to a comparative and anatomical study of middle ear elements in the representatives of different orders [Hyrtl, 1844, 1845, 1848; Claudius, 1867; Doran, 1878; Denker, 1899; Bondy, 1907; Frey, 1911; Crowe et al., 1931; Wassif, 1948, 1950; Yamada, 1953 a, b; Webster, 1960, 1966; Werner, 1960; Kobrak, 1963; Hentzen, 1970; Solntseva, 1972b, 1973b, d, f].

Based on the available literature and his own data, German researcher G. Fleischer [1973 a, b] provided a comparative anatomical study of the tympanic bulla in representatives of different orders. Fleischer's data, along with the evidence obtained by other researchers mentioned above as well as myself [Solntseva, 1975a, b], form the basis of this study.

The *bulla tympanica* of mammals represents a periotic-tympanic complex (Fig. 8a-i). The tympanic bone is composed of thin osseous walls, forming the tympanum, in which the elements of the middle ear are localized. The inner ear is situated in periotic bone.

The middle ear of mammals contains a tympanum (*cavum tympani*), which by means of an auditory tube is connected with the gullet cavity, tympanic membrane and a chain of the auditory ossicles (*ossicula auditus*): malleus, incus and stapes. Ligaments and two muscles of the middle ear (*m. tensor tympani, m. stapedius*) are bound to the auditory ossicles. Muscles of the tympanum regulate the transition of auditory energy and preserve the inner ear from super-intensive sounds. Ligaments keep auditory ossicles in a definite position.

The middle ear is separated from the outer ear by a tympanic membrane, stretched onto a tympanic ring (*anulus tympanicus*) (Fig. 9a-c). The tympanic membrane is formed by three layers. Its basis is composed of radially and circularly directed connective-tissue fibers, which grow together with the periosteum of the malleus handle (*manubrium mallei*), set into the tympanic membrane. From inside, the tympanic membrane is covered by a mucous membrane (*stratum mucosum*), formed by a pavement epithelium; from the surface, the tympanic membrane is covered by a lacking hair, glands and a papillary layer tegument (*stratum cutaneum*). The tympanic membrane has the shape of a cone, with the top directed into the tympanic cavity. A stretched part (*pars tensa*) and a small, unstretched part (*pars flacida*), which is more flexible due to the lack of the basis, are discernible on the tympanic membrane (Fig. 10a, b).

The tympanum is underlaid by a mucous membrane, in the middle layer of which there are blood vessels which regulate the degree of the stratum's thickness. The tympanum is located between the external auditory meatus and the inner ear labyrinth, and is separated from these formations by thin membranes. Six walls are discernible in the tympanum: an anterior wall, where the orifice of the Eustachian tube is situated; a posterior wall, having a foramen leading into the antrum mastoideum; medial, upper, and lateral walls, formed by the tympanic membrane; and a bottom wall. The medial wall is formed by a labyrinth capsule and separates the middle ear from the inner ear. This wall contains oval and round windows. The upper wall has a form of osseous partition [Kobrak, 1963].

The auditory ossicles of mammals are more complicated than in representatives of other classes of terrestrial vertebrates. The malleus is differentiated into the head of the malleus (*capitulum mallei*), neck of the malleus (*collum mallei*) and handle of the malleus. The head of the malleus bears an articular surface for a junction with the incus. At the medial side of the handle of the malleus, near the neck of the malleus, a small muscular arm (*pr. muscularis*) is located, which serves for the fastening of a tensor muscle of the tympanic membrane. At the dorsal edge of the handle of the malleus, a short arm (*pr. brevis*) is fastened; and at the dorsal edge of the neck of the malleus, a long arm (*pr. longus*), which is reduced in most mammals to a small, sharp *pr. gracilis*, is connected with the help of a ligament within the tympanum's wall. In other species, the long arm is large in size and connects rigidly with the wall of the tympanum. The *m. tensor tympani* is attached to the handle of the malleus, under whose contraction the tympanic membrane is pulled inside and, through a system of auditory ossicles, presses the stapes into an oval window. The malleus is fixed in a definite position with the help of three ligaments: a fore bundle that reaches the *fissura Glasseri* and gets attached to the *pr. anterior* of the malleus; a ligament located in the area opposite the neck of the malleus; and an axial ligament representing the axis, around which the malleus rotates.

The incus has a body and two components: the long arm (*crus longum*) and the short arm (*crus breve*). The long arm is stretched parallel to the handle of the malleus. Its lower end is curved, forming a junction with the stapes with the help of a lenticular arm. The short arm is located in the deepening of the osseous tympanum, where it is bound by a ligament.

The stapes is composed of a body formed by two crura of the head (*capitulum stapedis*) and a footplate (*basis stapedis*). The head of the stapes is connected with the long arm of the incus. The crura form an inter-ear area. The footplate of the stapes is oval and is kept in an oval window of the inner ear with the help of a ring-shaped ligament. The *m. stapedius* is attached to the head of the stapes, lying in the area of the posterior wall of the tympanum.

Below, different versions of the auditory ossicle structure and the ways of their junction with the tympanum will be examined.

2.1. Monotremats *(Monotremata)*

The study of middle ear structure in monotremats has been performed using two typical representatives of this order, the echidna and the platypus [Hyrtl, 1844; Doran, 1878; Eschweiler, 1899; Denker, 1901; Bondy, 1907; Frey, 1911; Hopson, 1966]

In the echidna (*Tachyglossus*) and the platypus (*Ornithorynchus*), the malleus is large and is in the form of a flat plate. The head of the malleus is comparatively small. The long arm (thin arm) of the malleus is incompletely fused to the wall of the tympanum. The incus is small compared to the malleus. The stapes is not differentiated into crura and is similar in structure to the column of birds. The base of the stapes is held in an oval window by a wide, ring-shaped ligament. Of the muscles of the middle ear, only the *m. tensor tympani* is present. It is significant that only the *m. tensor tympani* is also seen in birds. The tympanic bone is located at the skull base.

2.2. Marsupials *(Marsupialia)*

The middle ear of marsupials has been studied in several representatives belonging to different families [Hyrtl, 1844; Doran, 1878; Denker, 1899; Van Kampen, 1905; Bondy, 1907; Segall, 1969].

In the brown four-eyed opossum (*Metachirops*), the *pars transversalis* of the malleus is well developed. The long arm is rigidly joined to the wall of the tympanum. In incus structure, the long arm deserves attention due to the shape of being as if gouged from inside. The stapes is located in such a way that its crura face one another dorsoventrally. The tympanic bone is circular in shape and is bound to the skull by connective tissue.

In the marsupial mole (*Notoryctes*), a slightly different structure of middle ear elements is seen. A *pars transversalis* is absent from the malleus, and its long arm is only rudimentary. While in *Metachirops,* the articular surfaces of the malleus are well defined and are located at an angle to one another, in *Notoryctes* they are markedly less defined and are noticeably smoothed. The malleus also has a smooth head and a broadened handle. The incus is significantly greater in size than the malleus. Its long arm is quite reduced, and the lenticular arm is situated at the ventral edge of the incus body. The short arm is located dorsally and is rigidly bound to the wall of the tympanum. The stapes repeats the shape of the auditory column of birds. Its base fits tightly to the oval window of the inner ear.

In *Petaurus*, the malleus shows a well-defined head. The handle of the malleus is fastened not in the middle of the tympanic membrane, but is shifted a bit rostrally. The long arm of the malleus is pronounced and knits together with the wall of the tympanum. The incus is larger than the malleus. Its short arm is located perpendicularly toward the handle of the malleus. The stapes has the shape of a hollow skittle. The *m. tensor tympani* is fastened to the muscular arm of the malleus. The tympanic bone is round in shape and is bound to the other bones by connective tissue.

Thylogale kangaroos show a well-developed long arm of the malleus, which is widened ventrally and fused together with the tympanic bone. The stapes is asymmetrical. The tympanic bone has the shape of a broadened circle.

In marsupials, two structural types of the sound-transferring apparatus are distinguishable. There are forms with a large long arm of the malleus virtually fused to the tympanic bone, with small articular surfaces, a well-developed *pars transversalis* and a small incus, the short arm of which lies parallel to the handle of the malleus. In other species, the long arm of the malleus is crooked; the short arm is located perpendicularly to the axis of the handle of the malleus. Both of these complexes distinguished are interconnected with transitions. The stapes can be differentiated into crura, forming an area between them and showing the shape of a column. The tympanic bone is large.

2.3. Insectivores *(Insectivora)*

The works of Denker [1899]; Bondy [1907]; Frey [1911]; and Doran [1878] concern a description of middle ear structure.

In *Crocidura,* a well-developed long arm knits with tympanic bone. There is a hollow-shaped deepening between the articular surface of the malleus and the handle of the malleus. The incus is small; its short arm repeats the direction of the long arm of the malleus. The lenticular arm of the incus is well developed. The stapes is very fragile, with an oval base and a rounded head. The tympanic bone is located on the ventral side of the skull. It is fastened to the skull bones in the dorsal and occipital parts by means of connective tissue.

The mole family (*Talpa*) is represented by forms in which the long arm of the malleus does not grow together with tympanic bone; they are joined, instead, by a short ligament. The muscular arm of the malleus is absent. Contrary to species of the order Insectivora examined above, the incus in moles is strongly enlarged compared to the malleus. The long arm of the incus is greatly gouged on the lateral side. The short arm is noticeably shortened. The lenticular arm of the incus is well developed. The base of the stapes is markedly enlarged. The tympanum is small in size and is penetrated by osseous beams. The tympanic membrane lies at the skull base.

Thus, the middle ear of different members of Insectivora fails to represent a single structure. In most forms, as in shrews, the long arm of the malleus is rigidly knitted with the wall of the tympanum, the *pars transversalis* of the malleus is well developed and the muscular arm is present on the handle of the malleus.

In the mole, the long arm of the malleus is flexibly connected to the wall of the tympanum, the muscular arm is absent and the tympanic membrane rests on the skull base.

2.4. Flying lemurs *(Dermoptera)*

The middle ear in flying lemurs has been studied using a few species [Van Kampen, 1905; Klaauw, 1922].

In *Cynocephalus*, the malleus has a barb-shaped, thin arm. The *pars transversalis* of the malleus is absent. The handle and head of the malleus are well developed. The incus is large compared to the malleus, in that its body is markedly greater than the head of malleus. The muscular arm is absent. The stapes is large, with pronounced crura. The head of the stapes is smooth, as are the inner walls of the tympanum.

Thus, the middle ear of flying lemurs reveals the features typical of most mammals. In this way, the long arm of the malleus in the Dermoptera order is already transforming into the long arm, which does not grow together with the wall of the tympanum.

2.5. Chiroptera *(Chiroptera)*

The structure of the middle ear in the order Chiroptera has been described using several species [Hagenbach, 1835; Bondy, 1907; Frey, 1911; Gray, 1913; Klaauw, 1922; Wassif, 1950; Henson, 1961, 1965; Airapetiantz, Konstantinov, 1965, 1974].

Compared to the size of the cochlea, the tympanum of bats is small. The inner surface of tympanum has smooth walls. The auditory ossicles are shortened in size, are thin and have deep grooves. A thin osseous plate is located in the joint area between the malleus and the incus. Among the representatives of the Macrochiroptera suborder, the structure of auditory ossicles does not reveal species diversity.

In the genus *Pteropus,* a well-developed long arm of the malleus grows together with the wall of the tympanum. The muscular arm is absent. The incus is notable for its small size. The stapes has a concave base and two thin crura.

The malleus of *Rousettus* has a spacious *pars transversalis* on its dorsal side and a well-developed muscular arm. A column-shaped formation is found where the long arm is connected to the wall of the tympanum.

Rhinolophus species show the same type of structure of the middle ear elements as observed in the suborder Macrochiroptera. The long arm of the malleus is well developed and rigidly knitted with the wall of the tympanum. The muscular arm is located on the dorsal wall of *pars transversalis.* The articular surface of the malleus is not large; its two parts are located at right angles toward each other. The incus is smaller than the malleus. The stapes is differentiated into two flat crura – the head and muscular arm. The tympanicum has the form of an unevenly broadened circle. The periotic-tympanic complex is joined with the skull by means of connective tissue.

In *Nyctalus,* the long arm of the malleus grows together with the tympanic bone. There is an impression/deepening on the small articular surface of the malleus, as noted in *Pteropus.* Both articular surfaces of the malleus are located at an angle to each other. The muscular arm of the malleus is well defined and is thickened dorsally. The incus is small in size and has an enlarged long arm. The stapes has a small head and thin crura. The tympanic bone is broadened ventrally.

In *Pteropus,* the tympanic bone is ring-shaped; in different *Pteropus* species its thickening appears at the ventral and medial edges. The ear muscles are very well developed and are remarkable for their great size compared to the animal's body size. The muscles are located in the osseous deepenings.

Thus, the structures of the middle ear in the suborders Macro- and Microchiroptera reveal significant differences. A muscular arm of the malleus is absent from most members of Macrochiroptera, whereas in all representatives of Microchiroptera, this arm is well defined. In both suborders, the long arm of the malleus grows together with the tympanic bone. The stapes is differentiated into crura, forming an area between them.

2.6. Primates *(Primates)*

The structure of the middle ear among primates has been studied in species of tree shrews (*Tupaia*), galagos (*Galago*), macaques (*Macaca*), chimpanzees (*Pan*) and some other genera [Denker, 1899; Bondy, 1907; Frey, 1911; Holz, 1931; Gray, 1953; Werner, 1960; Segall, 1969].

In tree shrews, a peculiar structural type of the auditory ossicles is evident, which has not been found in the species described above. The long arm of the malleus in the form of a thickened vertical plate is connected with the wall of the tympanum with the help of particular arms. The head of the malleus is not large, and the arms diverge sharply from it. The neck of the malleus and the muscular arm are absent. The handle of the malleus is slightly rounded and widens at the end. The *pars transversalis* of the malleus is absent. The incudomalleal articulation shows a plain surface. The incus is notable for its small size. The long arm of the incus is parallel to the handle of the malleus, and the short arm is perpendicular to the handle of the malleus. The lenticular arm is well defined. The stapes is large, and its crura are hollow. The base of the stapes is shaped like a kidney. The muscular arm of the stapes is not defined. The tympanic bone has a round shape.

In galagos, the structure of the auditory ossicles reveals a lot of features similar to those of *Tupaia*. The long arm of the malleus grows slightly together with the tympanic bone. The malleus has a small head. The incudomalleal articulation has a deep, rounded groove. The muscular arm of the malleus is located at the middle of the handle of the malleus, which broadens at the end like a spatula. The incus is larger than the malleus. Its long and short arms are of the same length. The lenticular arm is well defined. The stapes is shaped like a top and is hollow. The hole formed by the crura is small. The head of the stapes and the muscular arm are well defined.

In macaques, the malleus has a long arm (thin arm) that is not fastened to the wall of the tympanum. The head of the malleus is massive. The muscular arm is located in the middle part of the handle of the malleus. The incus is large compared to the malleus. The stapes is shaped like a top and is hollow. The tympanic membrane is round. The tympanic bone encloses the cavity of the middle ear ventrally and forms an osseous part of the auditory meatus. The tympanum is narrow.

In chimpanzees, the structure of the middle ear is similar to that of humans. The malleus preserves the still-crooked long arm, but already reveals classical partition into the head, neck and handle of the malleus. The muscular arm is completely reduced. The incus is heavy outwardly, but the mass of its body is greater than the mass of the head of the malleus. The long arm of the incus at the distal part is wider than at the proximal part and is peculiarly crooked. The lenticular arm is pronounced but is very short. The stapes has two hollow crura, and its base fits easily into the oval window of the inner ear. There is a small muscular arm on the stapes. The tympanic bone forms the osseous part of the auditory meatus and confines the tympanum. It grows together with the adjacent bones of the skull along its whole extent.

Thus, among members of the order Primates, a pattern of similarity is revealed in middle ear structure. The long arm of the malleus is thin, transitioning to a gracile arm, which is not joined to the wall of the tympanum. The muscular arm is located on the handle of the malleus. The incus is large, and the stapes is differentiated into crura. It is worth mentioning, though, that in tree shrews and semi-monkeys, the tympanic bulla is large and spacious, while in true monkeys it is in most cases narrow and small.

2.7. Edentates *(Edentata)*

A description of the middle ear has been made using representatives of four orders [Hyrtl, 1844; Doran, 1878; Van Kampen, 1905; Bondy, 1907; Frey, 1911; Gutha, 1961], with significant differences revealed in middle ear structure.

In *Bradypus,* the malleus has a barb-shaped thin arm which is not grown together with the tympanic bone. The head of the malleus is great, and the neck is not defined. A muscular arm is absent. The handle of the malleus is shaped like a spatula. A deep groove is inherent to the malleus-incus joint. The incus is small; its long arm is parallel to the handle of the malleus, and the short arm is perpendicular to it. The lenticular arm is well defined. The stapes is not differentiated into crura. There is a ribbon-like osseous plate between the head and the base of the stapes. The tympanic bone forms most of the tympanic bulla.

In the giant, or three-fingered, ant-eater (*Myrmecophaga*), the auditory ossicles are similar in structure with those of *Bradypus*. The long arm of the malleus grows together with the tympanic bone, forming a thin osseous plate. The malleus is divided into the head, neck and handle. The base of the handle, which is the place of the *m. tensor tympani's* attachment, is rough. Incudomalleal articulation forms two articular surfaces. The incus is large and has processes of different lengths. The stapes includes developed crura, although the front of them is more developed than the back one. The base of the stapes is slightly prominent. The tympanic bulla is spacious.

In the three-belted armadillos (*Tolypeutes*), the long arm of the malleus is developed and is thickened ventrally. Its junction with the tympanic bone is not complete, and there is a small slot in-between. The *pars transversalis* of the malleus is wide. The incus is smaller than the malleus. A small head of the stapes is delimited from the body. The tympanic bone is unevenly thickened. The tympanum is not closed ventrally by a bone.

Pichiciegos (*Chlamyphorus*) differ from the previous genus in their middle ear structure. The long arm of the malleus is barb-shaped and enlarged in size. The head of the malleus is well defined, and its handle is broadened. The incus is large; the mass of its body is greater than that of the head of the malleus. The long arm of the incus carries a lenticular arm. The stapes is shaped like a kidney and has a large base. Flat crura of the stapes are transferred by the middle of its base; a similar construction is noted in *Chinchilla* and *Giraffa* species and is determined as a secondary columella, which is typical of the genera *Notoryctes* and *Manis*. The tympanic bone forms most of the tympanic bulla.

Thus, representatives of edentates differ significantly from each another in middle ear structure. Along with the forms in which the long arm of the malleus and the *pars transversalis* are well developed, there are species in which the long arm of the malleus is shortened down to a small process. In the former case, the articular surface of the malleus is small, while in the latter it is enlarged. The head of the malleus is defined only in the forms with a short long arm. Very different variations are found in stapes structure. The tympanic bone in the former case is rounded, as opposed to its being markedly broadened in the latter case.

2.8. Pangolins (*Pholidota*)

There is very little information in the available literature concerning the middle ear in pangolins [Denker, 1899; Eschweiler, 1899].

In the genus *Manis*, the malleus carries a sharp long arm. The head of the malleus is massive. Articular surfaces of the malleus are located at 80- to 90-

degree angles to each other. The surface in the area of incudomalleal articulation forms a deep groove. The incus is strongly enlarged. The muscular arm of the stapes is poorly defined, and the head of the stapes is not bordered from the crura. A trough is located on the inner surface of the tympanic bone.

Thus, based on numerous morphological structures of the middle ear of Pholidota, this order represents highly specialized mammals.

2.9. Lagomorphs *(Lagomorpha)*

The middle ear of Lagomorpha was studied by numerous researchers [Hagenbach, 1835; Bondy, 1907; Frey, 1911; Gray, 1913; Cockerell et al., 1914; Holz, 1931]. Here, we will examine the peculiarities of the middle ear structure of the rabbit (*Oryctolagus)* as a typical representative of this order.

The head and neck of the malleus are well defined. The long arm is represented by an osseous plate. A muscle process is located at the distal part of the handle, and has a shape of a sharp skittle. Incudomalleal articulation is characterized by a deep relief. The incus is comparatively large; its size is greater than that of the malleus. The lenticular arm is well defined. The base of the *manubrium mallei* lies asymmetrically on the tympanic membrane. The stapes is shaped like a top, with a flat base and hollow crura. The head of the stapes and its the muscular arm are well defined. The tympanic bone participates in the formation of the osseous part of the auditory meatus.

2.10. Rodents *(Rodentia)*

The features of the middle ear structure of rodents are among the best studied in mammals [Hagenbach, 1835; Denker, 1899; Bondy, 1907; Frey, 1911; Cockerell et al., 1914; Holz, 1931; Webster, 1960, 1961, 1966].

In *Sciurus,* the long arm of the malleus is thin and grows rostrally together with the tympanic bone; its head is enlarged. The muscular arm is located in the middle part of the handle. The incus is larger than the malleus; its long arm lies parallel to the manubrium. The stapes is differentiated into crura, which are rostrally widened.

In *Spalax,* the long arm of the malleus grows incompletely together with the tympanic bone, providing for easy movement of the auditory ossicles. The head of the malleus is flat, and the handle broadens at the end like a spatula. The muscular arm is absent. Incudomalleal articulation is quite flat.

The incus is larger than the malleus. The stapes is differentiated into two thin and round crura; its base is plain. From the anterior crus, the osseous plate is spread, shortening the space between the crura.

A similar structure has been noted in the mole and in the genus *Heliophobius.* The head and muscular arm of the stapes are well defined. The tympanic ring is round. The tympanic bone forms the tympanum, which has smooth walls but lacks osseous beams and columns.

In *Hystrix,* the long arm of the malleus is shortened down to a small, weakly defined process at the ventral edge of the malleus head. A muscular arm is located in the middle part of the handle. The long arm of the incus is elevated ventrally and slightly occipitally. Compared to the length of crura, the stapes has a large base. The crura are hollow and are located at the middle of the base. From the posterior crus, an osseous plate diverges, which confines the space between the crura. A tympanic ring has the shape of a circle.

In the middle ear of *Heliophobius,* the malleus grows together with the body and the long arm of the incus. The malleus is preserves the diminutive long arm. The head of the malleus and the incus body are flat. The handle is crooked and forms a spatula-shaped widening. Both crura of the stapes are formed by thin osseous plates, which confine a small area between them. The base of the stapes is plain. The head of the stapes is sharply delimited from the stapes body. In some parts, the tympanic bone is of an uneven thickness and forms the osseous part of the auditory meatus.

In hamsters (*Cricetus*), the long arm of the malleus is rigidly knitted with the tympanic bone. There is *pars transversalis* of malleus. On the handle, broadened in its proximal part, a top-shaped muscular arm is lcoated. The stapes is differentiated into crura – a small head and a plain base. The tympanic ring has an oval shape.

The middle ear of gray voles (*Microtus*) reveals a lot of structural patterns similar to those in *Sciurus*. The long arm of the malleus is thin and grows together with the tympanic bone. The incus is not large; osseous columns diverge from it. The whole tympanum is penetrated by small osseous beams. The tympanic bulla is small and slightly flat.

In rats (*Rattus*), the long arm of the malleus is rigidly knitted with the tympanic bone. The *pars transversalis* of the malleus and also the muscular arm, located on the handle, with the shape of a short horn, are well defined. The handle ends with a narrow blade. The articular surface of the malleus is small. The incus is not large; its long arm is parallel to the *pars transversalis* of the malleus, and its short arm is directed toward the long arm of the malleus. The lenticular arm is well defined. The stapes is differentiated into two hollow crura. The periotic and tympanic bones are weakly connected with the skull.

In *Hydrochoerus,* the malleus and incus are poorly connected with each other. A crooked long arm of the malleus is located ventrally to the head of the malleus, which has a thin neck.

In *Chinchilla,* a rudimentary long arm of the malleus is located on the ventral edge of the head of the malleus. Incudomalleal articulation is flat. The fastening between the malleus and the incus occurs at the later phases of ontogenesis [Fleischer, 1973 a, b]. The stapes is differentiated into crura, which are located close to the middle of the base. A large tympanum is interlaced with osseous plates. Radial beams are radiated from the tympanic ring.

In the nutria (*Myocastor coypus*), the long arm of the malleus does not join to the wall of the tympanum. The malleus is rigidly bound with the incus. The handle of the malleus is thin and long, and attaches to the tympanic membrane along the full length (Fig. 9a). Both processes of the incus have about the same length. The stapes is differentiated into crura, forming a large area between them. The crura are hollow and are very fragile. The base of the stapes is oval in shape (Table 5).

Thus, in species of some rodent genera, different types of middle ear structure are evident. Some forms have a well-developed long arm of the malleus, which is rigidly grown together with the wall of the tympanum, well-developed *pars transversalis* of the malleus and a small incus. In other forms, the incus is large, and the long arm of the malleus is crooked and is not grown together with the wall of tympanum; in these forms, the auditory ossicles may perform free movements.

2.11. Cetaceans (*Cetacea)*

At present, quite detailed information on middle ear morphology in the suborders Odontoceti and Mysticeti is available [Hyrtl, 1844; Claudius, 1858; Doran, 1878; Beauregard, 1894; Denker, 1902; Boenninghaus, 1903; Hanke, 1914; Kernan, 1919; Yamada, 1953a, b; Edinger, 1955; Fraser, Purves, 1954, 1959, 1960; Reysenbach de Haan, 1957a, b; Fraser, 1959; Purves, Utrecht, 1963; Purves, 1966; Giraud-Sauveur, 1969; Solntseva, 1972b; 1973b, d, f, g; 1978; Yablokov et al., 1972; Fleischer, 1973a, b].

Odontocetes can develop two types of acoustic signal reception. The first type is typical of the species which have a tympanic membrane ligament (Delphinidae).

The other pattern is principally a new type of vibrations developed in the middle ear (*Kogia, Phiseter catodon, Berardius*) [Fleischer, 1975].

The tympanic bulla of cetaceans is the most unique in its structural organization compared to other species of mammals (Fig. 8f, g, i). It consists of a periotic-tympanic complex. The tympanic bone is composed of thin osseous walls, forming the tympanum, in which the elements of the middle ear are located. The inner ear is situated in periotic bone.

In the family Delphinidae, the tympanic bulla is organized according to a common scheme. The tympanic and periotic bones are partially knitted with each other in the area of the posterior, sigmoideus and tubarius (additional ossicle) arms [Kasuya, 1973; Pilleri, 1987]. In the Amazon dolphin (*Inia geofrensis*), unlike the marine dolphins (*Tursiops truncatus, Delphinus delphis, Phocaena phocaena, Stenella attenuata),* the tympanic bulla is formed by thick osseous walls. The tympanic and periotic bones in river dolphins relatively bear against each other, giving the impression of an entire structure. The tympanic bone is massive, twice the size of the periotic bone. It is pear-shaped and is divided into medial and lateral lobes. The lateral lobe is more prominent compared to the medial one. A cross furrow extends along the medial part of the lateral lobe, dividing this lobe into two equal parts. This feature is not found in any marine dolphin species (Fig. 28).

In marine species of dolphins, the tympanic bulla is formed by more thin and rather fragile osseous walls. The tympanic and periotic bones appear to be isolated from each other and are approximately equal in size. The tympanic bone has a lengthened form with well-defined medial and lateral lobes. This is especially evident when comparing the dorsal surface of the tympanic bulla in river and marine species of dolphins.

A comparison of the ventral surface of tympanic bulla in river and marine species of dolphins shows that in river dolphins, the periotic bone is more wide and short and has a smoothed surface. In marine species, the *os perioticum* is lengthened and has a ribbed surface, which is stretched along the whole length of the periotic bone.

In river dolphins, a sigmoideus process is well developed; it is thickened, lengthened, separated from the other structures, and it is twice the size of this process in marine species of dolphins. Anterior and posterior processes of *os perioticum* in the river dolphin are less developed than in marine species, in which these processes are massive and broad. The anterior process in a river dolphin is sharp, while in marine species it has a spade-shaped form with smooth and rounded edges. The posterior process of the *os perioticum* in the river dolphin is smooth and is not defined, while in the marine dolphins this process is hypertrophied, and has a few small and thin thorns on the top. A deepening for the stapedius muscle of the middle ear in the river dolphin is oval and profound; in marine species, it is wider, not profound and has no distinct borders.

In all species of Delphinidae, a tubarius arm (additional ossicle) is located in the area of the anterior surfaces of the tympanic bulla. This arm is equally well developed in the river and marine species of dolphins.

The auditory ossicles in dolphins have specific structural features, join location and the way of fixation in the tympanum.

In the common dolphin (*Delphinus delphis*), the bottlenose dolphin (*Tursiops truncatus*) and the common porpoise (*Phocaena phocaena*), the long arm of the malleus is thick and rigidly joined to the wall of the tympanum. In all three types, the structure of the auditory ossicles reveals many similar traits. The whole malleus is composed of a compact and cylindrical *pars transversalis*. Incudomalleal articulation has two completely smooth articular surfaces, which are slightly prominent in the malleus and a bit concave in the incus. Articular surfaces are located at right angles to each other. The handle is reduced to a small prominence. The incus is compact. Its long arm is thickened and ends with a lenticular arm. The short arm is well defined, significantly prolonged and elastically connected with the wall of tympanum. Both processes are located 45-degree angles to each other. The stapes in a bottlenose dolphin and a common dolphin is short, not differentiated into crura and has the shape of a cone with a smoothed top (Table 5). In porpoises, the thickened crura of the stapes confine a small area between them. The point of the *m. stapedius* attachment is a wide side of the stapes contrary to most mammals. The base of the stapes fits in the oval window of the inner ear tightly. While there are data indicating that the base of the stapes knits with the oval window [Hyrtl, 1844; Beauregard, 1894], a number of researchers have demonstrated otherwise [Claudius, 1858; Denker, 1902; Yamada, 1953a, b; Fraser, Purves, 1954; Reysenbach de Haan, 1957; Fleischer, 1973a, b; Solntseva, 1969, 1973b, 1975a, b]. The head of the stapes is not delimited, and the muscular arm is not defined. Auricular muscles are enlarged and have a fan-shaped form. The auditory ossicles are rigidly bound with each other, especially in the area of the incudomalleal joint. The tympanum is formed due to the partial growing together of tympanic and periotic bones.

In dolphins, the joint of the long arm of the malleus and tympanic bone is rigid because this arm is knitted with tympanic bone. As a result, the malleus and the incus can function only as a whole structure.

The tympanic membrane of dolphins is strongly modified (Fig. 9c); it is rounded, markedly thickened, resilient and is not connected directly to the handle of the malleus. Instead, a triangular ligament is asymmetrically bound to the tympanic membrane, forming an additional lever in the line of auditory ossicles that significantly enhances the efficiency of the sound-pressure transition under water.

The tympanic bulla consists of the periotic-tympanic complex. The tympanic bone is formed by thin osseous walls, creating a tympanum, in which the elements of the middle ear are located. The inner ear is located in periotic bone. The tympanic bulla in dolphins does not grow together with the bones of the skull and is instead connected to them by a tendinous ligament. In odontocetes and mysticetes, among the elements connecting the periotic bone with the skull, there are numerous small sesamoid bones, which provide some mobility of the tympanic bulla relative to the skull [Yamada, 1953 a, b].

The tympanic bulla is surrounded by a system of sinuses, which have an osseous base and are covered by a mucous membrane, which includes multiple protein glands. There are some suppositions that sinuses are the result of the tympanum overgrowth. [Boenninghaus, 1903].

Through the sinus system, an adipose tension is stretched, down to the lower mandible, through the pterygoidal sinus and reaching the lateral side of the tympanic bulla. This connection of tympanic bulla with the lower mandible is expressed at a different degree in various species and is named as a sound crater or canal [Boenninghaus, 1903]. The roles of the specific gravity [Denker, 1902] and pressure regulators in the middle ear [Kellogg, 1928] are also prescribed to the sinuses of odontocetes.

The tympanum is filled not only with the air, but also with an emulsion [Fraser, Purves, 1960], which we failed to reveal [Solntseva, 1975a]. The tympanum is covered with a mucous membrane, which includes numerous blood vessels. Most of the cavity is filled with a cavernous plexus, consisting of a dense net of blood vessels. The venous sinuses are located in the area of the tympanic membrane.

One of the characteristic traits of the *Kogia* is the presence of the vigorous and pneumatized *pars mastoidea* of tympanic bulla, which serves as a place of attachment of the latter to the skull. This peculiarity is found also in other species of cetaceans. The functional importance of this type of connection with the skull contrary to ligament hanger is still unknown.

The important traits of the middle ear specialization in the *Kogia* are the absence of a tympanic membrane, tympanic ligament and *apes transversalis* of the malleus, and the presence of a thin osseous lamina in the lateral wall of the tympanum. It has been suggested that a sonar system and an ultrasound orientation exist in representatives of the *Kogia*, similar to what is found in other odontocetes. Due to the absence of a tympanic membrane, the lateral wall is a specialized sound-receiving surface. This wall includes an adipose body, which, as shown by Norris [1968], conducts auditory signals to the auditory organ. In its function, the lateral wall of the tympanic membrane is an osseous membrane [Fleischer, 1975].

The structure of the auditory ossicles is such that the incudomalleal complex is not able to conduct any sound vibrations. The vibrations of the lateral wall are transmitted through the area where it is joined to the malleus. However, the thickness and the position of the incus promote the acceptance of the vibrations between periotic and tympanic bones and their transmission to the chain of the auditory ossicles [Fleischer, 1975].

The principle of the auditory ossicles' structure is similar in odontocetes and mysticetes.

In minke whales (*Balaenoptera*), the long arm of the malleus is long and wide, and its proximal end is notched. The joint between it and the wall of the tympanum is rigid. The articular surface of the malleus is spherically enlarged. The handle of the malleus is shortened up to the little cylindrical process. A ligament of the tympanic membrane is attached to the handle along the full length. [Fraser, Purves, 1960]. The handle enlargement occurs in mysticetes due to the muscular arm [Yamada, 1953 a, b]. The incus is wide and short. Its short arm is quite reduced. The long arm is strongly thickened, lengthened and ends with a special cylindrical lenticular arm. The stapes is differentiated into crura. The area between the crura is covered in an osseous plate. The base of the stapes is oval. The *m. stapedius* is attached to the narrow side of the stapes (Fig. 11 a, b).

The tympanic bone in mysticetes forms a heavy, compact tympanic bulla, which knits with periotic bone at only two points. In mysticetes, the tympanic bulla does not participate in the formation of the skull wall, from which it is separated by the system of sinuses. The sinuses of mysticetes are constructed in a simpler way than in odontocetes. The sinus system is composed of the pterygoidal and peribullar sinuses. The tympanum is covered by a folded mucous membrane with a cavernous plexus [Boenninghaus, 1903].

Thus, both in odontocetes and mysticetes, the long arm grows together with the tympanic bone, and in odontocetes, it is especially massive. The handle of the malleus is reduced in representatives of both suborders. However, both of them have massive *pars transversalis* of the malleus. The *musculus tensor tympani* begins in a deepening at the occipital end of the *pars transversalis* in dolphins, while in mysticetes it is at its rostral end. In all cetaceans, the tympanic membrane is drastically changed. It is fastened at the lateral part of the *pars transversalis*. The incus is enlarged in both forms. In odontocetes, the short arm of the incus is lengthened; in mysticetes, it is quite reduced. The long arm of the incus is well defined in the representatives of both suborders, Odontoceti and Mysticeti.

The stapes in dolphins can be differentiated into crura, and it can also have the shape of a smoothed cone without visible differentiation into crura. In mysticetes, the area between the crura is covered in an osseous plate. The base of the stapes has a round shape in the Odontoceti suborder and an oval shape in the Mysticeti suborder.

2.12. Aardvarks *(Tubulidentata)*

The study of middle ear structure in this group has been recorded in the works of Hyrtl [1848] and Doran [1878].

In the aardvark (*Orycteropus afer*), the long arm of the malleus is rigidly fastened with a tympanic bone. The articular surface is greatly enlarged. Between the articular surface of the handle of the malleus and the incus, a step-like area is located, which is the place of the tympanic muscle's attachment. The handle widens at the end. The incus is small. Its short arm continues in the direction of the long arm of the malleus. The long arm of the incus is perpendicular to the long arm of the malleus. The lenticular arm is well defined. The stapes is small and has round crura. On the back ear the muscular arm is located. The head of the stapes is not defined. The tympanic bone is rounded.

2.13. Proboscides *(Proboscidea)*

Descriptions of the middle ear of Proboscidea have been made using several typical representatives of this order [Claudius, 1865; Kolmer, Eisinger, 1923; Holz, 1931; Keen, Grobblaar, 1941].

In elephants from the family Loxodontida, the malleus has a short *pr. gracilis*, which does not grow together with the tympanic bone. The head of the malleus is quite rectangular in form and is not very large. A *pars transversalis* is barely visible. A muscular arm is absent, and the handle of the malleus is not enlarged. The surface in the area of the incudomalleal articulation is plain. The incus is of a greater size; its processes are comparatively long and thin. The long arm has a small eminence for a lenticular arm. The malleus and incus have small foramens in different places, leading into the cavity of ossicles. The stapes is differentiated into thin crura. The area between the crura is markedly defined. The base of the stapes is slightly prominent. The head of the stapes is not delimited from the crura. The tympanic ring is narrow in the rostral part and wide in the occipital part. The tympanic bone forms most of the tympanic bulla, the ventral part of which is penetrated by osseous beams. From the lateral side the tympanic bone is completely covered by the vault of squamosal, which forms a false auditory meatus.

2.14. Hyracoideans *(Hyracoidea)*

Only few and fragmentary literature data are available on the middle ear of the order Hyracoidea [Doran, 1878; Klaauw, 1922, 1930; Keen, Grobblaar, 1941].

In Procavia, the long arm of the malleus is reduced. The head of the malleus is rectangular. The manubrium possesses a small muscular arm and ends in a spatula-shaped enlargement. The incudomalleal articulation has two articular surfaces. The body of the stapes is small and has unusually long processes. The lenticular arm is delimited from long arm of the incus. The stapes is differentiated into two hollow crura; the base of the stapes is also hollow and prominent. The head of the stapes is not defined. The tympanic bone, together with the entotympanicum, form the tympanum, whose inner walls are smooth.

2.15. Perissodactyls *(Perissodactyla)*

Among species of the order Perissodactyla, middle ear morphology has been examined in the tapir and the zebra [Denker, 1899; Bondy, 1907; Frey, 1911; Holz, 1931; Keen, Grobblaar, 1941].

In *Tapirus*, the malleus has a massive long arm and a well-developed *pars transversalis*, on whose occipital part the muscular arm is located. The manubrium is not widened in its proximal part. The stapes is constructed asymmetrically and is differentiated into crura, the back part of which is the most thickened. The base of the stapes is slightly prominent. The head of the stapes and muscular arm are well defined. The tympanic bone has a semicircular shape, which is ventrally widened.

The structure of the middle ear of a zebra is sharply different from that of a tapir. The long arm of the malleus is thin and shows the form of a trough. A muscular arm, located at the base of the manubrium, is directed ventrally and rostrally. The manubrium is located in the middle of the tympanic membrane. The incus is massive. Its short arm is located along the same line with the proximal part of the long arm of the malleus. The stapes is hollow with a flat base. The head is insignificantly detached from the body of the stapes, and the muscular arm is poorly defined. The radial osseous septa diverge from the tympanic ring and penetrate the whole tympanum.

Thus, the middle ear of the examined perissodactyls reveals different types of structure.

2.16. Artiodactyls *(Artiodactyla)*

In the order Artiodactyla, the structure of the sound-transmission apparatus has been studied in species representing different families [Hagenbach, 1853; Denker, 1899; Bondy, 1907; Frey, 1911; Holz, 1931; Keen, Grobblaar, 1941; Gray, 1953] .

In the wild boar (family Suidae), the long arm of the malleus grows together with the tympanic bone and is strongly enlarged. The handle of the malleus is located asymmetrically on the tympanic membrane. The base of the manubrium is connected by means of a neck with the articular surface of the malleus. The muscular arm of the malleus is well developed. The incus is small; its short arm has the same direction as the part of the long arm of the malleus knitted with the tympanic bone. The stapes is differentiated into two hollow crura and a hollow base. The head is not separated from the body of the stapes. The tympanic ring is asymmetrical. The tympanic bone forms a long tympanum.

In *Hippopotamus,* the malleus with its thin long process grows together with the lateral wall of *fissura petrotympanica.* The articular surface is enlarged on the large head of the malleus. The muscular arm is located quite rostrally, and its distal part reveals a slight thickening. The handle of the malleus is crooked and located asymmetrically on the tympanic membrane, as in the genus *Sus.* The body of the incus is enlarged. The stapes is triangular. Its posterior crus is markedly wider than the anterior one. The base of the stapes is not prominent. A head is absent. The tympanum is separated by septa, located radially. The ventrorostral part of the tympanic bone, where the cochlea is located, is not fused to the other parts of the skull, but is joined with them by means of connective tissue.

In the middle ear of camels (*Camelus*), the malleus has a poorly defined long arm. The head of the malleus is enlarged. The muscular arm is reduced. The handle of the malleus ends with a long spatula-like enlargement. The incudomalleal articulation is faced occipitally and has a smooth surface. The incus is enlarged. The stapes is large, has a rectangular form and is hollow. Its base is prominent. The posterior ear is wider than the anterior ear. The tympanic ring is asymmetrical. The position of the manubrium is shifted from the center of the tympanic membrane. The tympanum is filled with spongy tissue.

The structure of the middle ear of roes (*Capreolus*) is sharply different from that of other representatives of this order. The long arm of the malleus is massive, wide and is rigidly grown together with the tympanic bone.

The articular surface of the malleus is small, and the *pars transversalis* is not divided. The muscular arm is located on the occipital end of the *pars transversalis*. The base of the handle of the malleus is located in the center of the tympanic membrane. The incus is small compared to the malleus. The lenticular arm is well defined. The stapes is differentiated into two hollow crura, which form a narrow space between them. The head of the stapes is delimited from the crura. The base is quite plain. The crura are fixed closer to the middle of the base. The tympanum is not pneumatized.

In the genus *Giraffa,* the long arm of the malleus is connected with the tympanic bone in the same way as in the previous genus. However, it is significantly smaller than in *Capreolus*. The articular surface of the malleus is large. The muscular arm is well defined. The handle of the malleus is located in the center of the tympanic membrane. The incus is larger than in the genus *Capreolus*. The stapes is differentiated into the highly unusual in structure crura, which are located closer to the middle of the base. The head and muscular arm of the stapes are well defined. The tympanic bulla is spacious.

Thus, the structure of the middle ear of the order Artiodactyla reveals significant variations in the representatives of different families. In some species, the long arm, which is rigidly grown together with the tympanic bone, a small articular surface in the area of the incudomalleal articulation, and a well-defined *pars transversalis* of the malleus are inherent. On the other hand, some forms have a poorly defined long arm of the malleus. The tympanum may have smooth walls inside and also may be pneumatized.

2.17. Sirens *(Sirenia)*

The middle ear of the order Sirenia has been studied in manatees (*Monatus* genus) [Hyrtl, 1844; Claudius, 1867; Doran, 1878; Van Kampen, 1905; Fleischer, 1971].

In manatees, the auditory ossicles are large and heavy. The articular surface of the malleus is rostrally elevated. The head of the turns into a round handle. The long arm of the malleus is located in line with the short arm of the incus. The manubrium has a bend, which provides a tension of the tympanic membrane. The whole manubrium is composed of a compact bone and is deprived of the offshoots. The malleus has two smooth articular surfaces, located at an angle to each other. The incus is massive and heavy. Its long arm forms a bend. The short arm grows together with the tympanic bone. The articular surface of the malleus and the *cr. breve* of the incus form an axis, whose points of rotation are located in both places of accretion. The stapes is large and heavy. Its crura are distorted relative to each other. The base of the stapes is not large. The tympanic bone forms a semicircle, which in the occipital and rostral parts fuses with the periotic bone. The periotic and tympanic bones are not grown together with the surrounding bones of the skull, but are located free in the smooth cavity of the squamosal.

2.18. Carnivores *(Carnivora)*

Descriptions of middle ear structure have been made using typical representatives of terrestrial and semi-aquatic carnivores [Hagenbach, 1835; Denker, 1899; Bondy, 1907; Frey, 1911; Holz, 1931; Keen, Grobblaar, 1941; Gray, 1953; Solntseva, 1975a].

In foxes (*Vulpes*), the long arm of the malleus is rigidly fastened with the tympanic bone. A *pars transversalis* of the malleus is present. The handle attaches asymmetrically to the tympanic membrane. A small muscular arm deviates from the neck. The incus is not big; its short arm is directed toward the long arm of the malleus (Table 5). The tympanic ring has an asymmetrical shape. The stapes is differentiated into two crura, which form a large area between them. The base of the stapes has an oval shape. The tympanic bulla grows together with the cranial bones and participates in the formation of the cranium, as in most mammals. The tympanum is spherical in shape and is formed by periotic and tympanic bones accreted with each other.

Table 5. Weight (in g) and sizes (in mm) of the auditory ossicles

Species	Malleus				Incus					Stapes			
	I	II	III	IV	I	V	VI	VII	VIII	I	IX	X	XI
Vulpes vulpes	14.0	7.5	1.5	6.0	6.0	1.5	2.0	2.0	1.5	2.0	2.0	1.3	1.2
Canis familiaris	10.0	8.0	1.5	5.0	14.0	1.5	2.0	2.0	1.5	1.5	2.0	1.4	1.3
Mustela vison	3.5	4.5	1.5	3.0	2.0	1.2	1.3	2.0	1.0	0.5	1.5	2.0	1.0
Myocastor coypus	6.5	3.2	1.5	2.1	3.0	1.5	1.0	1.2	1.0	1.0	1.5	2.0	1.0
Enhydra lutris	5.0	6.5	2.1	2.5	5.0	1.3	2.0	1.5	1.0	2.0	2.0	1.5	1.0
Callorchinus ursinus	8.0	6.0	1.5	4.0	8.0	2.0	2.5	2.0	2.0	4.0	2.0	2.0	1.0
Eumetopias jubatus	16.0	8.0	1.5	5.0	16.0	2.0	2.5	3.0	2.0	3.0	2.5	2.0	1.0
Pagophilus groenlandicus	50.0	12.0	2.0	8.0	90.0	4.0	4.5	4.0	3.0	10.0	3.5	3.0	1.5
Pusa caspica	40.0	10.0	2.0	6.0	86.0	4.0	4.0	4.0	2.0	7.0	2.0	3.0	1.0
Phoca vitulina	51.0	13.5	2.0	8.5	139.0	5.0	5.0	4.5	2.2	13.0	2.3	3.1	1.2
Phoca insularis	54.0	13.0	2.0	8.0	143.0	6.0	4.5	5.5	2.0	15.0	4.0	3.5	1.0
Erignathus barbatus	58.0	14.0	2.5	9.0	101.0	4.0	5.0	4.0	3.0	15.0	4.0	3.5	1.5
Delphinus delphis	80.0	6.0	4.0	2.0	21.0	1.0	2.5	2.0	1.5	8.0	2.0	2.0	1.0
Tursiops truncatus	85.0	7.0	5.0	2.5	26.0	1.5	2.5	3.0	1.5	9.0	2.0	2.0	1.0

Note: I – weight; II – length of malleus; III - length of long arm of malleus; IV – length of handle of malleus; V – length of incus; VI – breadth of incus; VII – length of long arm of incus; VIII – length of short arm of incus; IX – length of stapes; X – length of footplate of stapes; XI – breadth of footplate of stapes.

In martens (*Martes*), the long arm of the malleus is dorsally and ventrally thickened. The malleus possesses a spacious articular surface for the incus. The muscular arm is significantly extended. The base of the manubrium is attached to the tympanic membrane asymmetrically. The incus is large. The stapes is differentiated into two hollow crura. Its base is small and not prominent. The head is delimited from the crura; the muscular arm of the stapes is well defined. The osseous beams are radiated from the tympanic ring.

In minks (*Mustela*), the malleus has a large head. The long arm of the malleus has a triangular form and is diminished in size. It is adjacent to, but not fastened to, the tympanic bone. The articular surface of the malleus is enlarged. The muscular arm is small and is located in the middle part of the manubrium, which widens at the end. The incus is enlarged. Its short arm is located perpendicularly to the manubrium, and the long arm stretches almost parallel to it. The long arm of the incus ends with a trough, which continues into the lenticular arm. The stapes is greatly changed; its crus are far apart from each other. The stapes is hollow; the base has a prominent surface faced toward the oval window. The foramen between the crura is sharply narrowed. The head is not detached from the body of the stapes, and the muscular arm is poorly defined (Table 5). The tympanic membrane is round, which sets it apart from that of minks. The tympanic bulla is flat and lies on the skull base, with which it grows together and forms the cranium. The tympanum is filled with osseous beams.

In a sea otter (*Enhydra lutris*), a small head of the malleus grows into a neck. The long arm of the malleus does not grow together with the tympanic bone, but is connected with it elastically. The incus is short and wide. Its long arm is located perpendicularly to the body of the incus and is sharply thickened. It ends with a little area with deepening, which is the point of the *pr. lenticularis* attachment. The stapes is differentiated into two round crura, forming a small foramen between them (Table 5). The base of the stapes is oval. Its prominent surface is faced toward the oval window. The tympanic bulla grows together with the cranial bones. The tympanum is not large due to the overgrowth of its walls. The tympanic ring is oval.

Thus, the middle ear of carnivores is characterized by various types of structure in species with different ecologies.

2.19. Pinnipeds (*Pinnipedia*)

In the order Pinnipedia, the structure of the middle ear has been studied in the suborders Otariidae and Phocidae [King, 1964; Odend'hal, Poulter, 1966; Mohl, 1967, 1968; Gracham Sylvia, 1967; Solntseva, 1973b, 1975a].

In a species of otariids, the Northern fur seal (*Callorchinus ursinus*), the tympanic bulla is small and grows together with the cranial bones completely, which allows for the possibility of osseous sound transmission. This peculiarity draws the Otariidae together with terrestrial carnivores. The malleus has a small spherical head with greatly prominent articular surfaces that are semicircular in shape. The manubrium is strongly thickened and has a triangular base. It attaches to the round tympanic membrane asymmetrically, which is stretched upon a small tympanic ring. The long arm of the malleus has a triangular shape in the form of a thin osseous formation, elastically connected with the wall of the tympanum. This process is located directly at the base of the head and leads to the *pr. gracilis*. The neck of the malleus is well defined. Rostrally curving, it grows into the main mass of the bone and joins with the manubrium. The muscular arm is located on the dorsal side of the neck of the malleus. Attached to the muscular arm is the *m. tensor tympani*, which is a short and wide muscular fascicle. The handle of the malleus is long and wide; its broadened part attaches to the tympanic membrane along all its length.

Compared to the malleus, the incus is small. It is located together with the malleus. The body of the incus is short and wide. The short arm is small and triangular. The long arm is slightly bent; its distal part ends with the lenticular arm, which is bound to the head of the stapes.

The stapes is triangular and is differentiated into a head, body and footplate. The head of the stapes is not delimited from the body and fuses with it almost completely. The body is represented by two *crura stapedis* of different thicknesses. The *anterior crura stapedis* is significantly thinner and is not crooked, as it usually appears in most mammals. The *posterior crura stapedis* is markedly thicker. There is a barely noticeable foramen between the head of the stapes. In the area of the head of the stapes, the *m. stapedis* is attached to the back crura. The base of the stapes is large and has the form of an oval. However, its thickness is not uniform; at the base of the posterior *crura* it is markedly thicker than in the area of the front attachment (Table 5).

The accretion of the malleus and incus in a Northern fur seal in the area of the incudomalleal articulation does not allow these bones to carry independent vibrations. As a result, both bones represent a single whole and function as one.

The tympanum is spherical, is formed by thick osseous walls and is covered by a thick mucous membrane, which includes numerous blood vessels. The tympanic membrane is small and thin and is rigidly fixed on the tympanic ring, increasing its resilience and creating conditions for broadband sound transmission.

Thus, on the way of adaptation to the aquatic way of life a tendency for thickening and shortening of the handle of the malleus, as well as for lengthening and thickening of its thin arm is evident. In the structure of the incus, the lengthening and thickening of the long arm is noteworthy. In the stapes, the foramen between the *crura stapedis* is sharply diminished.

In Steller sea lions (*Eumetopias jubatus*), as in Northern fur seals, the tympanic bulla is small and grows together with the cranial bones completely.

The malleus has a small head, which is greatly thickened. With its broadened base, the handle of the malleus is attached to the tympanic membrane radially. The long arm of the malleus in the form of a thin osseous formation is connected elastically to the wall of the tymapnum. The incus is massive and compact. It has two large joint surfaces for connection with the malleus. Its long arm is elongated and thickened. The short arm is thin and poorly defined. The stapes is compact and differentiated into two round crura, which form a small foramen due to their excessive thickening. The base of the stapes is oval and slightly prominent. Its prominent part is faced toward the cavity of the middle ear (Table 5).

The tympanum is not large and is covered with a thick mucous membrane. The tympanic membrane is slightly stretched into a cone. The tympanic ring is oval (Fig. 9b; 10a, b).

In different species of phocids (*Pagophilus groenlandicus, Pusa caspica, Phoca vitulina, Phoca insularis, Erignathus barbatus*), the structure of the middle ear reveals many similar traits. The malleus has a small head with two articular surfaces positioned at right angles to each other. There is a little deepening on both surfaces. The head leads into the neck. The manubrium is long and strongly thickened. The long arm of the malleus leads into the *pr. gracilis*, which is connected elastically with the wall of the tympanum. The broadened part of the base of the handle of the malleus is attached to the tympanic membrane along its entire length and is slightly stretched into a cone. The incus is large with a spacious articular surface for the malleus. Its long arm has a shape of a trough. At the distal part, it is thin and leads into the lenticular arm. The proximal part of the long arm of the incus, together with the malleus, forms the secondary articulation, described by Hyrtl [1848]. The stapes is round and heavy. Compact crura form a small foramen (Fig. 12; Table 5). The head of the stapes is not delimited from the body of the stapes, which is distinctly defined. The muscular arm is well marked. The tympanic ring is not large. The periotic and tympanic bones form an entire complex, detached from the whole occipital part, which only in the rostral part grows together with the skull. The tympanum is spherical in shape and is covered by a thick mucous membrane. Auricular muscles are well developed and have a form of short and wide muscular fascicles.

Consequently, the middle ear of otariids and phocids shows different types of structure. Interspecific differences within each suborder are not expressed to a significant degree.

2.20. Morphological and ecological correlations of the middle ear in various mammals

A combination of literature and original data allows for a comparative analysis of the structure and topography of the middle ear elements in mammals belonging to different orders.

In the class Mammalia, various degrees of knitting for the periotic and tympanic bones as well as for the tympanic bone with the cranial bones are noted. Similarly, in the orders Carnivora, Rodentia and Pinnipedia (Otariidae), the tympanic bulla grows together with the cranial bones. In phocids it is detached and does not knit with them completely. In odontocetes the tympanic bulla is separated from the skull but is attached to it with a short ligament, while in mysticetes it grows together with the cranial bones to a greater or lesser extent but does not participate in cranium formation, being separated from the cranium by a system of sinuses [Yamada, 1953a, b].

In several semi-aquatic (sea otters, pinnipeds) and aquatic (cetaceans) forms, the venous sinuses are located in the walls of the tympanum and are concentrated mainly in the osseous part of the auditory meatus.

Depending on structure, several types of the tympanum are evident [Simkin, 1977]. A spherical type is inherent to the species that use ultrasound in orientation and echolocation (forest and home mice, jerboa, red field voles, bats, cetaceans, pinnipeds). A spongy type of auditory cavity occurs in the species that live in the conditions of a dense environment (subterranean and aquatic forms) or inhabit solid and rocky substrates (gray field voles, steppe lemmings, pika, weasels, stoats, ermines). A chamber type of tympanum is inherent to ground squirrels and dormice. The tympanum of this type is divided by thin osseous partitions into a row of open-ended chambers. The coarse-cellular type is similar to the chamber type and is found in squirrels, chipmunks, marmots and martens. The tympanum is partitioned off with poorly defined osseous ribs.

In most mammals, the tympanum is covered by a thin mucous membrane due to a number of permeating blood vessels. In forms living in aquatic environments, the mucous membrane of the tympanum is strongly thickened because of the abundant blood vessels that penetrate its medium layer. In the tympanum of dolphins [Yamada, 1953a, b], which are obligate hydrobionts, a cavernous plexus is located, consisting of a dense network of blood vessels.

In most mammals, the tympanic membrane is round, slightly stretched into a cone and very thin. A significantly thickened tympanic membrane occurs in semi-aquatic and aquatic species (Fig. 9a-c; 10a, b; 12). The tympanic membrane of dolphins is not connected directly to the handle of the malleus, as it usually appears in most mammals. They are connected by a triangular ligament fixed asymmetrically to the tympanic membrane, which is round and thick. In mysticetes, the tympanic membrane consists of two parts: a fiberless "glovish outgrowth," protruding into the cavity of the external auditory meatus, and its deviating fibrous ligament, which is attached to a cylindrical handle of the malleus. The fibrous ligament of mysticetes is analogous to the triangular ligament of odontocetes (Fig. 11b). The tympanic membrane of the right whales in its structure and form occuipies an intermediary position between that of the family Balaenopteridae and that of the suborder Odontoceti [Fraser, Purves, 1960].

A reduced tympanic membrane surface is evident in the forms capable of echolocation or ultrasound orientation (shrews, bats, some species of pinnipeds, cetaceans).

Although the organization of the principal auditory ossicles is similar in all mammals, characteristic structural traits are revealed in different ecological groups [Solntseva, 1972b, 1973c-g, 1975a, 1987a]. Structural variation is apparent with changes in size, element form and weight ratio of the auditory ossicles (Fig. 13a, b; 14 a-g; Table 5), as well as in the form of their junction and attachment to the tympanum. In different orders, an accretion of the malleus and incus is noted (nutria, fur seal). In most mammals, the long arm of the malleus grows together with the wall of the tympanum, and its form is significantly varied. In some insectivores and bats, the *pars transversalis* of the malleus is well developed. The incus is small, the articular surface in the area of the incudomalleal articulation is diminished and the muscular arm is located at the occipital part of the *pars transversalis*.

In different members of the orders Rodentia, Perissodactyla and Carnivora, the long arm of the malleus is diminished or completely reduced (the genera *Cricetus, Rattus, Tapirus, Vulpes*). Also in these forms, a well-defined *pars transversalis* of the malleus and a shortened articular surface between the malleus and incus are inherent. A well-developed *pars transversalis* of the malleus and an enlarged incus are characteristic of cetaceans, in which great rigidity in junction between the malleus and incus is revealed. The handle of the malleus in cetaceans is reduced to a round prominence in the odontocetes and to a conical process in the mysticetes. An enlarged handle in the suborder Odontoceti, in contrast to the suborder Mysticeti, occurs due to the muscular arm [Yamada, 1953a, b].

In some members of the orders Artiodactyla and Carnivora, the *pars transversalis* of the malleus is reduced (Fig. 14a), but the surface of the incudomalleal articulation and the size of the incus are enlarged (*Martes, Sus*). The incus reveals variation in different mammals both in structure and size of the body and in topography of its processes. In most mammals, the long and short arms are located at an angle to the incus body. In Phocidae, they are located almost parallel to each other (Fig. 14d-f). A sharply thickened and elongated long arm of the incus is evident in cetaceans. In mysticetes, the short arm is almost reduced, depriving the incus from support on the wall of the tympanic bone and indicating that, in mysticetes, the incus can function in a slightly different way than in odontocetes [Yamada, 1953a, b].

The structure of the stapes varies significantly among species. In the order Monotremata, the stapes is not differentiated into crura and appears in the form of a column. A noticeably narrowed stapes is inherent to the mole and, especially, the genus *Heliophobius* [Fleischer, 1973a, b]. Differentiation of the stapes is absent from odontocetes (dolphins) and, as a result, the stapes attains the form of a rounded cone. The stapes of mysticetes is differentiated into crura, and the space between them is covered with an osseous plate. In dolphins, the head of the stapes, lacking a noticeable neck, goes into the stalk of an oval form, which is broadened at the base and is precisely adjusted to the oval window [Yamada, 1953a, b]. Such a location of the stapes base in the oval window provided grounds for several researchers to speculate falsely about the growth of the stapes in the oval window of the dolphin's inner ear [Hyrtl, 1984; Boenninghaus, 1903]. In pinnipeds and sea otters, the compact and heavy stapes shows strongly thickened crura, which form a small area between them. In otariids (Northern fur seal, Steller sea lion), some differentiation of the stapes into the crura is evident in the complete absence of an area between the crura. In all small mammals, the stapes is fragile and light (*Suncus, Crocidura, Chiroptera, Microtus, Spalax*), and in most mammals, the stapes is hollow. The base of the stapes may be flat, concave or oval. The *Thylogale, Heliophobius* and *Loxodonta* species show a flat base. A concave base is inherent to *Nyctalus* and *Tapirus*. A prominent base is characteristic of *Cynocephalus, Microtus, Mustela* and *Procavia* (Fig. 14c). The shape and size of the base of the stapes can also vary. In the orders Monotremata, Chiroptera and Cetacea (Odontoceti), the base of the stapes is round in shape. The elliptical contour is inherent to the stapes base of some members of Insectivora, Primates, Rodentia and Perissodactyla.

In addition to anatomical transformations of the auditory ossicles, the weight proportions of the ossicles also change, especially between the malleus and the incus. For example, in the fox, the mink and the nutria, the weight of the malleus is 2 times greater than that of the incus. In the sea otter, the fur seal and the Steller sea lion, the weight proportions are equal. In phocids, the weight of the malleus is 2 to 2.5 times less than that of the incus. In odontocetes, the weight of the malleus exceeds that of the incus (Fig. 14g; Table 5).

Thus, in mammalian species that are phylogenetically remote but have similar lifestyles, parallel traits in structure and topography of middle ear elements are evident.

The middle ear shows two types of structure. The fist one is inherent to the species capable of echolocation and ultrasound orientation (shrews, rats, bats, odontocetes). For this type, a tympanic bulla detached from the cranial bones, up to complete isolation, is characteristic. The fusion of the tympanic bulla with the skull is realized by means of connective tissue. The long arm of the malleus is thickened and rigidly grown together with the wall of the tympanum. The rigidity in the area of the incudomalleal articulation is increased, and the auricular muscles are well developed.

The second type is revealed in most mammals that do not use ultrasound for orientation. For this type of structure, fusion of the tympanic bulla to the cranial bones is inherent. The long arm of the malleus is thin, sharpened and reduced down to the *pr. gracilis*, the latter being connected elastically to the wall of the tympanum. The auditory ossicles perform independent movements, the articular surface is enlarged in the area of the incudomalleal articulation and the auricular muscles are developed to a lesser degree.

2.21. Peculiarities of middle ear biomechanics

In order to reveal the acoustic potential of the middle ear in connection with underwater sound conduction, an attempt was made to carry out the analysis of the biomechanics of this link among animals from terrestrial to aquatic forms.

As considered earlier, the auditory ossicles of aquatic (pinnipeds, cetaceans) and some semi-aquatic forms (sea otters) have specific peculiarities in structure, location and form of fixation in the tympanum.

A rotation axis of the auditory ossicles passes through the points of fastening of the incus's short process and the malleus's long process to the wall of the tympanum. In terrestrial and some semi-aquatic forms (mink, nutria), the rotation axis of the auditory ossicles is parallel to the plane of the infra-peduncle plate of the stapes; whereas in sea otters, pinnepeds and cetaceans, the rotation axis is situated at an angle to the infra-peduncle plate of the stapes. The location of the rotation axis of the auditory ossicles in aquatic forms, which becomes apparent as a result of its inclination relative to the infra-peduncle plate of the stapes, requires the introduction of the coefficient of the pressure transmission of the auditory ossicles' screw effect (K 3) for calculation [Lipatov, 1972]. The calculation of the coefficient (Table 6) of the pressure transmission (K) was carried out according to the following formula [Lipatov, Solntseva, 1972]:

$$K = K1 . K2 . K3 . K4 . K5$$

K1 is the coefficient of the tympanic membrane, and K2 is the transmission coefficient of the auditory ossicles' lever.

$$K2 = L1 / L2$$

L1 is the distance from the rotation axis to the center of pressure on the tympanic membrane, L2 is the distance from the point of the connection of the incus's long process with the stapes's head to the rotation axis, K3 is the coefficient of the pressure increase due to the screw effect, and K4 is the transmission coefficient of the tympanic membrane-ligament's lever.

$$K4 = L3 / L2$$

L3 is the distance from the point of fastening of the triangular ligament of the dolphin's tympanic membrane to the rotation axis, and K5 is the ratio of the tympanic membrane's surface area to the surface area of the infra-peduncle plate of the stapes.

$$K5 = S1 / S2$$

S1 is the surface area of the tympanic membrane in millimeters, and S2 is the surface area of the infra-peduncle plate of the stapes in millimeters.

The comparison of the obtained data shows that in typical representatives of terrestrial (fox, dog) and semi-aquatic mammals (mink, nutria), the middle ear has the lowest coefficient of transmission of the acoustic pressure (K equal to 25 to 29). In a sea otter and pinnipeds, the K value increases 1.5 to 2 times (40-60). The rise of the K value is provided by the sharp increase of the K3 value. The dolphin's middle ear has the highest K, which exceeds that of sea otters and pinnipeds 2.5 to 3 times (Table 6). Such increase of the coefficient of the transmission of the acoustic pressure in dolphins is connected with the inclination of the auditory ossicles' rotation axis and the presence of the additional lever in the form of a triangular ligament, which is assymetrically connected to the arched tympanic membrane, and also due to the peculiarities of its structure (thickening and stronger rigidity), which allows all the effort applied to the tympanic membrane to pass.

Table 6. Acoustic parameters of the middle ear

Species	Number of investigated animals	K1	K2	K3	K4	S1	S2	K5	K
Vulpes vulpes	5	0.5	2.1	1.0	-	35	1.5	23	24-25
Canis familiaris	5	0.5	2.2	1.0	-	39.3	1.6	24	26-27
Mustela vison	10	0.5	2.5	1.0	-	30	1.3	23	28-29
Myocastor coypus	10	0.5	2.4	1.0	-	15	0.7	21	25-26
Enhydra lutris	7	0.5	2.0	2-3	-	32	1.5	21	42-63
Eumetopias jubatus	10	0.5	2.1	2-3	-	43	2.0	21	44-66
Callorchinus ursinus	10	0.5	2.1	2-3	-	51	2.4	21	44-66
Pagophilus groenlandicus	10	0.5	2.0	2-3	-	50	2.5	20	40-60
Pusa caspica	10	0.5	2.0	2-3	-	40	1.9	21	42-63
Phoca vitulina	3	0.5	2.0	2-3	-	43	2.4	20	40-60
Phoca insularis	3	0.5	2.0	2-3	-	60	3.0	20	44-66
Erignathus barbatus	1	0.5	2.2	2-3	-	100	5.0	20	44-66
Delphinus delphis	10	1.0	1.2	2-3	2.2	32	1.5	21	110-166
Tursiops truncatus	10	1.0	1.2	1-3	2.2	38	1.9	20	105-158

Thus, the biomechanical peculiarities of the middle ear in mammals whose habitat conditions are mainly aqueous are directed to the increasing of the coefficient of the transmission of the acoustic pressure by the middle ear. Most likely, this determines the effectiveness of the peripheral auditory system's functioning during the mammals' orientation in the aqueous medium and greatly enlarges their perceptible frequency band.

CHAPTER 3.
STRUCTURE OF THE MAMMALS' INNER EAR

The inner ear of mammals is a most important unit of the peripheral auditory system, since the conduction tract of the auditory analyzer starts from the cells of the organ of Corti, where the auditory receptors of the VIII pair of the craniocerebral nerves are located (*n. vestibulocochlearis*).

The inner ear (*auris interna*) of terrestrial species is formed by osseous and membranaceous labyrinths. The osseous labyrinth (*labyrinthus osseus*) is located in a petrous bone and includes semicircular ducts, a labyrinthine vestibule and a cochlea.

The labyrinthine vestibule (*vestibulum*) is a globe-shaped cavity; the auditory nerve passes through its medial wall. In its lateral wall, an oval window (*fenestra vestibuli*) is located, through which the base of the stapes enters from the tympanum's side. In the caudal wall, there are openings of three semicircular ducts. The cochlea's osseous canal starts in the labyrinthine vestibule's nasal wall [Turkewitsch, 1935].

The functions of the sound-perceiving apparatus are carried out by the cochlea and the organ of Corti, in which mechanical oscillations are transformed into nervous signals. In some mammals, the cochlea makes a part of the *bulla tympanica*. In different species, the cochlea forms from 0.5 to 5 turns around its axis (*modiolus*) and is formed by spongy bone tissue which includes a cluster of ganglionic cells.

The tunnel of the cochlea has two partitions, which divide it into three independent canals (canals of cochlea): tympanic, vestibular and middle, or cochlear. The first two canals are filled with a perilymph and are connected to each other through an opening on the cochlea's top (*helicotrema*). The middle canal is filled with an endolymph and ends blindly, widening in the cochlea's apical part and repeating the number of spiral turns which form the cochlea. On the axial sections, the cochlear canal has three walls: inferior, superior and external (Diagram 4).

The inferior wall is the continuation of the spiral lamina, which starts from the modiolus and consists of two layers. The spiral ganglion's dendrites pass between them in the radial canals [Trevisi et al., 1972]. The upper layer of the osseous lamina turns into a spiral limb and the lower one into a basilar membrane. The inferior wall of the cochlear canal divides the spiral canal into vestibular (*scala vestibuli*) and tympanic (*scala tympani*) scalas of the cochlea. On the cochlea's top, the osseous lamina ends with a peculiar curve in the form of a hook (*hamulus cochleae*).

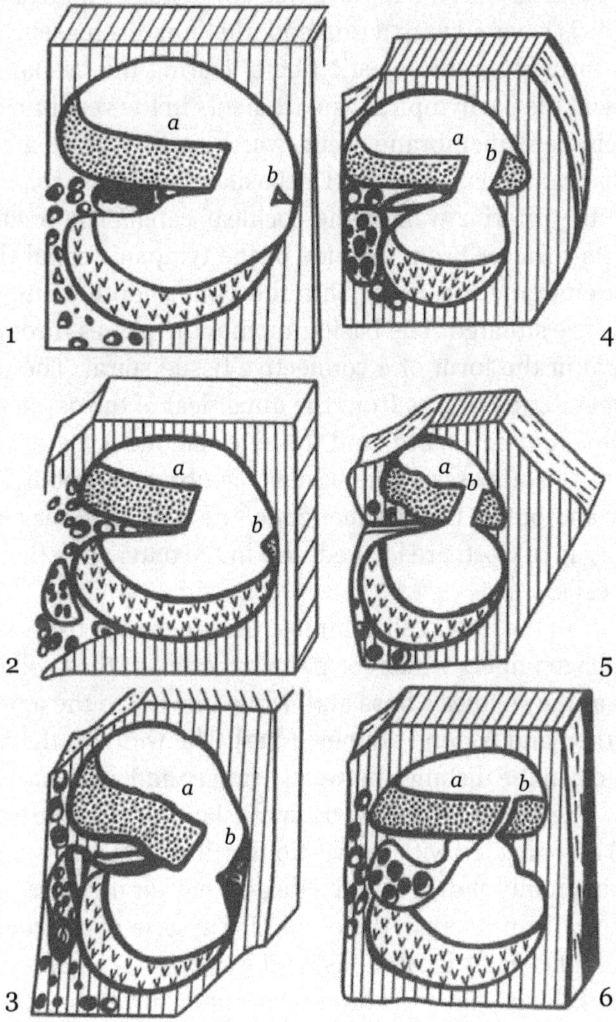

Diagram 4. Axial sections of the cochlear canal in echolocating and non-echolocating mammals:
1 — baleen whale (*Balaena mysticetus*); 2 — human (*Homo sapiens*); 3 — dog fruit bat (*Rousettus* sp.); 4 — shrew (*Crocidura russula*); 5 — bat (*Rhinolophus simulator*); 6 — harbor porpoise (*Phocaena phocaena*). a — primary osseous spiral lamina; b — secondary osseous spiral lamina. After Fleischer [1973].

The superior wall of the cochlea's canal is formed by Reissner's membrane which isolates it from the *scala vestibuli*. The external wall is the thickest and covers the upper part of the spiral ligament; this wall takes part in the production of endolymph which fills the *canalis cochlearis*.

Reissner's membrane is a thin membrane which, from the cochlear canal's side, is covered with a flat polygonal epithelium, and from the vestibular scala's side is covered with a thin endothelium of a mesenchymal origin. The layer is formed by thin elastic fibers. During the basilar membrane's oscillations and the perilymph's displacements in the vestibular scala, oscillations of Reissner's membrane occur, which, in their turn, are transmitted to the cochlear canal's endolymph. The basilar membrane (*membrana basilaris*) forms the inferior wall of the cochlear canal. The organ of Corti is situated on its surface. From the side of the tympanic canal the surface of the basilar membrane is covered with the same endothelium, under which blood vessels are situated. The basilar membrane passes through the whole cochlear canal in the form of a connective tissue spiral. The inner edge of the basilar membrane begins from the upper leaf of the osseous spiral lamina of the *habenula perforata*, and the external one is fastened in the area of the *lamina spiralis ossea*. The basilar membrane is subdivided into two zones: inner and outer. In the inner zone, the tunnel, inner and outer hair cells of the organ of Corti are located, and in the outer zone there are mainly the Hensen cells.

At the base of the outer cell columns, the inner zone turns into the outer zone. The collagen fibers which form the structure of the basilar membrane are shortest at the cochlea's base and, while moving to the apical part, they greatly lengthen and widen [Kolmer, 1927]. The width of the basilar membrane depends on the distance between primary and secondary osseous spiral laminae. The secondary osseous spiral lamina (*lamina spiralis ossea secundaria*) is connected with the vestibular fibers of the basilar membrane. It was experimentally shown that the flexible basilar membrane and the rigid osseous spiral lamina form the oscillating system of the cochlea.

The organ of Corti is located inside the *ductus cochlearis* on the vestibular surface of the basilar membrane. Its upper surface points to the *scala vestibuli*; its inner side is adjoined by the vestibular, or spiral, labium which is formed by connective tissue passing onto the inner spiral incisure (Engström, 1951, 1955).

The tectorial membrane starts from the spiral labium and is located above the elements of the organ of Corti in the form of a jelly-like lamina which stretches spirally along the organ of Corti.

According to Held's data [Held, 1926], the following structures are discernible in the tectorial membrane: the axial part, which adjoins to the top of the epithelial cells of the spiral labium; the middle part, which is freely located above the organ of Corti; and the external part, which, as it is supposed, is connected with the Hensen cells [Vinnikov, Titova, 1959, 1961].

In mammals, the organ of Corti is formed by supporting and receptor elements. The supporting elements are represented by inner frontier cells, inner cells-phalanxes, inner and outer cells-columns and Deiters, Hensen and Claudius supporting cells (Diagram 5).

The inner frontier cells border the inner side of the inner hair cells and form one to two rows of elements. With their tops they adjoin to the tops of the inner cells-phalanxes. The narrow and flat head of the inner frontier cells is tightly connected with the structural elements of the reticular membrane. The basal part of the cell is located in the area of the *habenula perforata*. The nucleus is located in the basal part of the cell [Held, 1926; Engström et al., 1963].

The inner cells-phalanxes form one row. Their upper end is connected with the reticular membrane and has the form of a phalanx, with the help of which

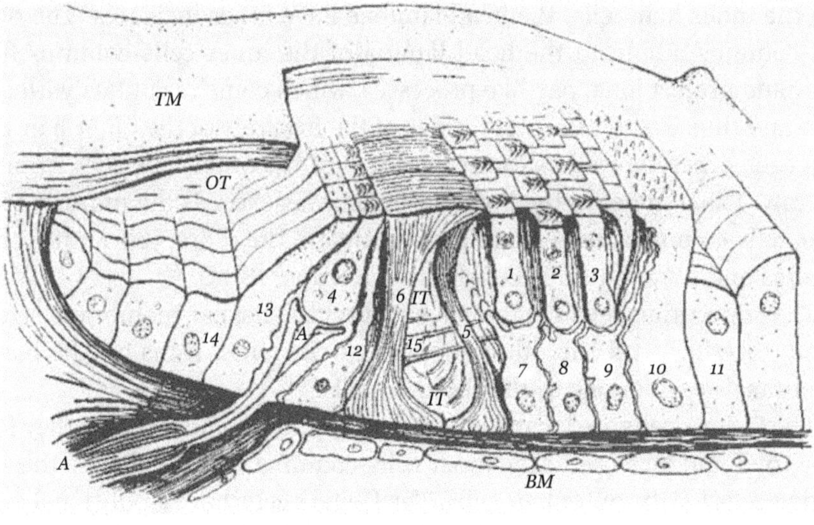

Diagram 5. The organ of Corti.
1–3 — outer hair cells (OHC) of the first, second and third rows; 4 — inner hair cells (IHC); 5, 6 — outer and inner foot cells; 7–9 — Deiters' cells; 10 – Hensen's cells; 11 — Claudius' cells; 12 — inner cells; 13 — marginal cells; 14 — cells of inner sulcus; 15 — radial tunnel fibers; IT — inner tunnel (tunnel of Corti); OT — outer tunnel; TM — tectorial membrane; BM — basilar membrane; A — afferent fiber. After Wersall et al. [1965].

the inner hair cells are separated from each other. The middle part of the inner cell-phalanx is flattened out and narrowed. The external side of the inner hair cells is adjacent to the cell's body. The basal part of the cell is located in the area of the *habenula perforata*, as well as the inner frontier cells. Between these cells the dendrites of the neurons of the spiral ganglion are located. The nucleus lies in the basal part of the cell [Held, 1926]. Between the inner cells-phalanxes and the frontier cells, there is an isolated intercellular cavity, in the apical part of which the inner hair cells are located [Vinnikov, Titova, 1961].

The outer and the inner cells-columns adjoin to each other by their tops and form a three-cornered tunnel, which is filled with an endolymph. This tunnel spirals through the whole organ of Corti. With the help of their base, the cells-columns come in contact with the basilar membrane. The cells-columns are located at a sharp angle facing each other; as a result, their tops adjoin to each other. The outer cells-columns form a larger angle of inclination than do the inner cells-columns [Held, 1926].

In place of their contact, the cells-columns form a peculiar arch (*arcus spiralis*). The apical part of the inner cells-columns forms a head lamina. In the aggregate, these laminae enable the formation of space between the inner hair cells and the first row of the outer hair cells. In the area of contact with the inner hair cells, the head laminae have inner incisures. The outer cells-columns adjoin to the head lamina of the inner cells-columns from below and project long, oar-like processes, which come in contact with each other and thus isolate the lateral sides of the first row of the outer hair cells from each other, as well as the second row of the outer hair cells from the first row. The number of the inner cells-columns exceeds the number of the outer cells-columns. For example, in a human, there are 5600 inner cells-columns and 3850 outer cells-columns [Retzius, 1884].

The supporting cells-columns located on the basilar membrane stretch it and, together with the other supporting elements, transmit the basilar membrane's oscillations to the receptor cells.

The Deiters cells are located on the basilar membrane; usually, they form three rows and lie close to the inner cells-columns. These cells are the supporting cells for the outer hair cells. The Deiters cells have a polygonal cylindrical form with a large rounded nucleus located in the basal part of the cell. In its upper part, the Deiters cell forms a phalanx-shaped incisure – the upper head, which, together with other laminae, phalanxes and outgrowths of the cells-columns takes part in the formation of the reticular membrane. Also, from the Deiters cell begins the lower head, the bowl-shaped bottom of which gives support to the base of the outer hair cells [Held, 1926].

The Hensen cells, in contrast to the Deiters cells, do not connect with the reticular membrane's structures. They adjoin to the Deiters cells and form a wide row that fastens to the basal part of the basilar membrane. The nucleus is large and rounded [Kolmer, 1927]. It has been supposed that the transportation of nutrients to the hair cells and to the Deiters cells from the vascular stria is provided with the help of the Hensen cells. [Vinnikov, Titova, 1957, 1961].

The Claudius cells are located behind the Hensen cells and are cubic in form, with distinct intercellular borders. The functional roles of the Claudius cells, as well as those of Hensen cells, are suggested to be trophic.

The outer (OHC) and inner (IHC) hair cells are the receptor elements of The organ of Corti (Diagram 6). According to Retzius [1884], there are about 12,000 outer and 3500 inner hair cells in a human's cochlea (i.e., slightly fewer than in a cat or guinea pig). The inner hair cells (IHC) lie in one row with an inclination from the modiolus, and the outer hair cells (OHC), which are isolated from the IHC by the tunnel, lie in three rows with an inclination to the opposite side. In different turns of the cochlea, the OHC differ slightly in shape, while the IHC's shape is invariable in all cochlea's turns.

The height of the cylindrical OHC's increases in the direction from the basal to the apical turn (20 and 50 μm accordingly) and their diameter is practically invariable (5 μm). The OHC's stability is provided by the system of the Deiters supporting cells and the space between them, which is filled with a liquid (Nuel's space). The receptor pole of the hair cells is formed by a cuticular lamina and sensitive stereocilia (Diagram 5).

The flagella, or stereocilia, look like peculiar rigid rods which, in their narrowest part (a neck) near the cuticular membrane, show a diameter of 0.05 μm, elsewhere about 0.2 μm. Fine structure of stereocilia is known: there is a bundle of fibrils inside, which is 30–40 A thick (collateral line in fishes) (Flock, 1965). The stereocilial membrane is part of the receptor cell's membrane, each stereocilium being fixed on the cuticular lamina with the help of a peculiar rootlet.

Each OHC has 100–120 stereocilia. The length of stereocilia in the cochlea's basal part reaches about 2 μm, increasing up to 6 μm distally. In the guinea pig, the cat, the rat or the human, the stereocilia of the middle and modiolus rows are shorter than those of the distant row. Electron microscopy shows (Jurato, 1961) that the stereocilia remain only in touch with the tectorial membrane, but the longest can be pressed inside it (Spoendlin, 1966).

The OHC contains a lot of organelles: mitochondria, ribosomes and some specific inclusions. The mitochondria are concentrated in the subapical and the basal part of the cell, as well as along the cell walls. Cytoplasmic membranes form flat cisterns which are parallel to the cell walls; the nucleus is located close to the OHC's base.

Directly behind the nucleus, the central part of the cell's cylinder also contains numerous microtubules and vesicles, features which are typical of the presynaptic endings.

The inner hair cells (IHC) are less specialized elements than the OHC. The IHC differs from the OHC in form (the IHC has a pear-shaped form), location of the filaments, distribution of the cytoplasmic organelles and the type of contact with nerve endings. In addition, the IHC's body is totally surrounded by the supporting elements. As in the OHC, sensory filaments are situated in three rows in the form of a flattened W, the number of the filaments in average amounts to 60 and in the apical turn they are longer than in the basal one. The distribution of the mitochondria in the IHC's body is not as irregular as it is in the OHC; the nucleus is located in the center of the cell. In the upper part of the cytoplasm, relative to the nucleus, numerous Golgi vacuoles are located. Under the nucleus, the membranes of the granular endoplasmic reticulum can be found, as well as mitochondria. The smooth endoplasmic reticulum is concentrated mainly in the subapical part, where it has the form of small tubules and cisterns. The receptor-neuronal contacts are not limited by the basal part of the cell only, but are irregularly located in the area of the lower two-thirds of the cell. The basal part of the IHC is often irregularly shaped with numerous and noticeable outgrowths.

The description of the cochlea's structure in most mammalian orders is cited according to G. Fleischer [Fleischer, 1973 a, b], along with the data from other investigations and our own materials.

3.1. Monotremats *(Monotremata)*

The inner ear of monotremats was studied in the typical representatives of the order – the *echidna* and the *platypus*. However, the existent data are not numerous and are cited in Denker's works [1901] and in Alexander's works [1904].

In representatives of the *Ornithorychus* genus, the cochlea is slightly bent, similar to the *lagenae* of birds. The primary spiral lamina is compact and short. The secondary lamina is entirely absent.

In representatives of the *Tachiglossus* genus, the cochlea is similar to that of the platypus. The secondary spiral lamina is also absent. In the cochlea's apical part, there is a curvature that looks like a bud.

The absence of the cochlear canal is typical of all monotremats [Denker, 1901].

3.2. Marsupials *(Marsupialia)*

The description of the marsupial's inner ear is cited in Hyrtl's works [1844] and in Denker's works [1899]. The investigation was carried out on the representatives of the *Metachirops* and *Thylogale* genera.

In the representatives of the *Metachirops* genus, the primary lamina is thickened and the secondary lamina has a peculiar structure. Between both laminae is a small space. The spiral ganglion is large, and the vestibular leaf of the primary lamina is thicker than the tympanal one.

In representatives of the *Thylogale* genus, both spiral laminae are well developed. After the first turn of the cochlea, the secondary spiral lamina disappears. Consequently, in contrast to representatives of the *Metachirops* genus, the cochlea's structure acquires features specific to this genus, whereas in the structure of the middle ear these genera do not differ.

3.3. Insectivores *(Insectivora)*

The inner ear of insectivores was studied in different representatives of this order. In the present work, we give its description using the genera *Crocidura* and *Talpa* [Denker, 1899; Alexander, 1905; Kolmer, 1913; Turkewitsch, 1934; Platzer, 1960, 1964].

In representatives of the *Crocidura* genus, the cochlea forms 2 to 2.5 turns. Both spiral laminae are well developed. Between them is a small space (up to 0.1 mm). The secondary lamina is located opposite the primary one. The cochlea's radial wall gradually turns into the secondary lamina. The modiolus is well developed.

In representatives of the *Talpa* genus, the cochlea forms 1.75 to 2 turns (Fig. 15). The primary spiral lamina is short and thin. The secondary lamina is noticeably thinner than the primary lamina and turns into the osseous leaf. In contrast to representatives of the *Crocidura* genus, the modiolus in representatives of the *Talpa* genus is isolated from the cochlea's capsule.

The basilar membrane is stretched between two spiral laminae, and its width is equal to the space between both laminae.

3.4. Chiroptera *(Chiroptera)*

The study of the chiroptera's cochlea was carried out on representatives of the Micro- and Macrochiroptera suborders [Hyrtl, 1844; Denker, 1899; Iwata, 1924; Turkewitsch, 1934; Ikeda, Yokote, 1939; Pye, 1966a, b, 1967, 1970, 1973; Henson, 1973].

In different chiroptera species, the number of the cochlea's turns can vary. Usually it is formed by 2.5 to 3.5 turns. The cochlea is large; the greatest development is reached by the basal turn of the representatives of the suborder Microchiroptera (Fig. 16a, b).

Common features in the cochlea's structure are revealed in most representatives of the Macrochiroptera suborder, with the exception of the *Rousettus* genus: the cochlea of its representatives, according to its morphological characteristics, is more similar to that in the representatives of the Microchiroptera suborder. It has been established that representatives of the *Rousettus* genus, as well as representatives of the Microchiroptera suborder, are able to use ultrasonic orientation and echolocation.

In members of the *Pteropus* genus, which possess a low-frequency hearing, the primary osseous lamina is short and the secondary lamina disappears entirely. The basilar membrane is wide.

Species of the genus *Rousettus*, in contrast to members of the genus *Pteropus*, have both spiral laminae. The primary lamina in the vestibular layer is thickened, as in most of the members of the suborder Macrochiroptera. The secondary lamina is well developed.

A narrow basilar membrane is typical of the cochlea's basal turn. The modiolus is not isolated from the bones of the cochlear capsule. In representatives of the suborder Microchiroptera (*Rhinolophus, Nyctalus*), the presence of both spiral laminae is typical. The vestibular leaf of the primary lamina is greatly thickened; the secondary lamina is located opposite the primary lamina and is well developed. The distance between the laminae is narrow, narrower in species of the family Rhinolophidae than in the genus *Nyctalus*. The spiral canal is small, and the modiolus is not expressed, while the cochlea is surrounded with a bone pierced with small canals for nerves and blood vessels.

According to the structure of the organ of Corti, all members of the suborder Microchiroptera can be subdivided into two groups. The first group is formed by species possessing high Claudius cells and a narrow basilar membrane. Among the second group are species in which the organ of Corti is characterized by short Claudius cells and a wide basilar membrane. A representative of the first group is the genus *Natalus* (*Natalus fumidirosteus*), and a representative of the second group is the genus *Myotis* (*Myotis albescens*). A compact basilar membrane is observed in both groups, but its maximum elasticity becomes apparent in *Rhinolophus ferrumequinum*.

The peculiarities of cochlear structure observed in the family Rhinolophidae have also been revealed in a genus of the family Hypposideridae. Representatives of both groups show similarities in principal cochlear structure [Fleischer, 1973a, b].

3.5. Primates *(Primates)*

Descriptions of the cochlea in primates were made using two genera of this order: *Tupaia* and *Galago* [Hyrtl, 1844; Denker, 1899; Kolmer, 1909; Haag, 1933; Meyer, 1933; Werner, 1960].

In species of the genus *Tupaia*, the cochlea forms 2.5 turns around the modiolus. Both spiral laminae are developed. The secondary lamina is shorter, less developed and slightly shifted away from the primary lamina. The distance between both laminae is wide because the primary lamina is not elongated in a radial direction. The modiolus is well developed.

In species of the genus *Galago*, as well as in *Tupaia*, both spiral laminae are present. The primary lamina is the most developed one; the secondary lamina is short and less developed. In contrast to the cochlea in *Tupaia*, species of *Galago* show the primary spiral lamina elongated in a radial direction, and, as a result, the distance between both laminae is noticeably decreased. The modiolus is perforated by canals and is entirely isolated from the bone of the cochlear capsule.

3.6. Edentates *(Edentata)*

Studies on edentate cochlea structure are few [Hyrtl, 1848; Denker, 1899].

In the genus *Bradypus*, the cochlea shows both spiral laminae, between which a wide distance is observed. The vestibular leaf of the primary lamina is greatly thickened. The modiolus is perforated and its radial process forms the wall of the tympanic canal.

Species of the genus *Tolypeutes* are also characterized by the presence of both spiral laminae. The vestibular leaf of the primary lamina is also thickened, as noted in the order Chiroptera. The secondary lamina is short but well developed. The distance between the laminae is small. The modiolus is noticeably perforated.

Species of the genus *Chlamyphorus* show an absolutely different cochlea structure as compared with *Tolypeutes*. The primary lamina is short; the secondary lamina does not form a common lamina but is located in the form of a flat osseous leaf in the cochlea's capsule. The modiolus is well developed and isolated from the bones of the cochlear capsule.

Consequently, certain genera of the edentate order reveal differences in cochlea structure that are especially evident in the secondary osseous spiral lamina.

3.7. Pangolins *(Pholidota)*

In the genus *Manis*, both cochlear canals are strongly delimited from the bones of the cochlear capsule. Both spiral laminae are present. The sharp edge of the vestibular leaf of the primary lamina is spirally involuted. The secondary lamina is thin. The distance between both laminae is wide. The spiral canal is small, and the modiolus is filled with spongy tissue [Fleischer, 1973a, b].

3.8. Lagomorphs *(Lagomorpha)*

The study of the lagomorph cochlea was mainly carried out on rabbits, as they are typical representatives of this order [Hyrtl, 1844; Turkewitsch, 1934].

In the genus *Oryctolagus*, both spiral laminae are well developed, and the distance between them is small. The width of the basilar membrane increases in the apical part of the cochlea.

3.9. Rodents *(Rodentia)*

The rodent inner ear was under study using various species of the genera *Microtus*, *Spalax* and *Rattus*, but mostly on laboratory cultures of *Cavia porcellus* [Hyrtl, 1844; Denker, 1899; Alexander, 1901, 1905; Held, 1902;

Mygind, 1945; Vinnikov, Titova, 1961; Solntseva, 1990b, 1993; Duckert, 1974; Morrison et al., 1975; Akimov, 1976].

In the genus *Microtus*, the secondary osseous lamina is short. The distance between the laminae is wide. The modiolus is underdeveloped.

The genus *Spalax* shows a quite different type of cochlea structure. Both spiral laminae are developed. The primary lamina is located opposite the secondary lamina and attains a substantial degree of growth. The secondary lamina turns into the radial wall of the cochlea, from which it does not differ in structure. A modiolus is not pronounced.

In *Rattus*, the presence of a short primary and a thin secondary lamina is typical. The distance between both laminae is wide. The modiolus is perforated.

In guinea pigs (*Cavia porcellus*), the cochlea forms 4.5 turns. The secondary osseous lamina is thin. The distance between the laminae is wide.

The hair cells of the guinea pig number a total of 8939, of which 1899 are inner hair cells and the remaining 7040 are outer hair cells. In the lower cochlear turn, the number of IHC is 874; in the second turn, 437; in the third turn, 403; in the fourth turn, 124; and in the half of the apical turn, 61.

The number of OHC also varies in the cochlea's turns, as well as directly in the cell rows [Wüstenfeld, 1957]. The IHC are situated at a distance of 9.56 μm from each other, and the OHC at 8.38 μm from each other. The width of the osseous spiral lamina is not constant along the cochlea. It decreases from the middle of the first turn to its total disappearance in the apical turn. The part of the lamina that lies between the spiral limb and the *habenula perforata* is so thin that, supposedly, it can oscillate as a part of the basilar membrane. The average length of the whole basilar membrane of the guinea pig is about 18.8 mm, and its average length in each of the four turns is 8.5, 4.8, 3.4 and 2.1, respectively. The width of the basilar membrane quickly enlarges over a 2-mm extent from 0.07 to 0.15 mm. Farther on, its width enlarges more slowly, reaching its maximum (0.25 mm) in the middle of the cochlea's apical turn, and narrowing quickly thereafter [Fernandez, 1951]. The thickness of the basilar membrane in the basal turn is 7.4 μm, and in the apical turn 1. 34 μm.

In the nutria, the cochlear structure reveals a lot of similarities with that of the guinea pig. The cochlea forms 4.5 turns (Fig. 17). The primary and secondary osseous spiral laminae are present and are located at a distance from each other. The secondary lamina is underdeveloped.

3.10. Cetaceans *(Cetacea)*

The cetacean structure of the cochlea was studied in different representatives of odontocetes (*Delphinus delphis, Tursiops truncatus, Phocaena phocaena, Lagenorhynchus obliquidens*) and mysticetes (*Balaenopteridae*) [Hyrtl, 1844; Claudius, 1858; Denker, 1902; Boenninghaus, 1903; Kolmer, 1908; Yamada, 1953a, b; Reysenbach de Haan, 1957a, b; Wever et al., 1971a, b; 1972; Fleischer, 1976; Solntseva, 1987a, 1988a, c, 1990, 1997c].

In the suborder Odontoceti (family Delphinidae), the cochlea is large and forms 1.5 to 2.0 turns. The basal turn of the cochlea is the best developed, and it embraces the apical turn from the outside (Fig. 18a, b). In the area of the basal turn, there is an additional part of the cochlea canal which forms a half-turn and goes aside from the remaining part of the cochlea. Such a structure is also typical of the mysticetes, but it is less pronounced. Both cochlear canals are incompletely parted due to separation of the primary and secondary osseous spiral laminae. The separation in their connection is provided by a bend in the organ of Corti.

A peculiar tympanic canal is reflected in the presence of a "tympanum niche" and a "basal crest." Such a widening is a typical trait of the cetaceans, while in other mammals it can be marked only at the cochlea's basal end [Fleischer, 1976].

In the cochlea's basal turn, the secondary lamina is located on the "basal crest." It is better developed than the primary lamina. The "basal crest" is located from the direction of the roof of the "tympanum niche." In the next turn, by height, the secondary lamina is noticeably weaker than the primary lamina; it is located lower than the primary lamina and is oriented in a different way. The "basal crest" and the "tympanum niche" do not rise higher than this turn. The distance between the spiral laminae is very narrow in the lower turn and increases in the turn which is located above.

The other distinguishing peculiarity of the cetaceans' cochlea is the ratio change of the tympanic scalae size and the vestibular scalae size. In the cochlea's basal end, the size of the vestibular scalae is noticeably less than that of the tympanic scalae. Farther on, both canals equalize in their size, and in the cochlea's apical turn, the vestibular scalae becomes larger than the tympanic scalae [Fleischer, 1976].

Consequently, the absolute size of the vestibular scalae does not change, whereas the size of the tympanic scalae changes harshly, decreasing from the cochlea's basal turn to its apical turn.

The cochlea of a sperm whale is typical of the odontocetes. The secondary osseous lamina almost joins with the primary lamina near the basal end, where the "basal crest" and the "tympanum niche" are well developed. The tympanic scalae is large in the initial part of the cochlea and decreases in its terminal part. As in dolphins, the canal of the cochlea juts out into the cavity of the "tympanum niche."

The basilar membrane is located between the spiral laminae and is 100 to 200 μm wide, the length of 68% from the canal's length in a sperm whale, 79-82% in a Baird's beaked whale and 81% in a pilot whale [Yamada, 1953 a, b]. The basilar membrane widens in the direction of the cochlea's top due to the decreasing of the secondary osseous spiral lamina. For the space of the basilar membrane, the size of the cells of the organ of Corti changes. Some of the Claudius cells become very high, and the others remain low. On the way from the basal turn to the cochlea's top, the orientation of the cells also changes. In the bottlenose dolphin and the Pacific white-sided dolphin, the supporting cells of the organ of Corti are large and compacted together [Wever et al., 1971a, b, 1972].

In representatives of the Mysticeti suborder, the cochlea is involuted in a different way than it is in the odontocetes. It forms 2.5 turns, growing drastically in height. In some species of mysticetes, a strong contortion of the basal turn in the area of the oval window has been observed [Fleischer, 1973a, b].

The primary osseous lamina is short. The secondary lamina relative to the primary one drastically becomes thin and narrow. In the next turn, according to height, the secondary lamina disappears. The distance between both laminae in the basal turn is much wider than that of the odontocetes. The tympanic canal is especially widened in members of the family Balaenopteridae. The vestibular canal of odontocetes is characterized by constant size, while in rorquals the size of the canal decreases. In mysticetes, the size of the spiral canal also decreases. A trabecula, located in its niche, occupies a large space. It is supposed that this enables the fine construction of the primary osseous spiral lamina in mysticetes. The primary osseous lamina of the mysticetes becomes thinner from the base to the top of the cochlea, eventually disappearing on the top, and its radial edge becomes more uneven. The secondary osseous spiral lamina disappears after the first turn [Fleischer, 1973 a, b]. The width of the basilar membrane corresponds to the distance between both spiral laminae. Its length in the rorquals amounts to 40% of the canal's length [Yamada, 1953 a, b].

Consequently, all differences between cochlear structure in odontocetes and mysticetes consist of changes in the cochlear canal's structure. The greatest differences are revealed in the structure of the primary osseous spiral lamina. In all odontocetes, the primary lamina is small relative to the cochlea's size; in mysticetes, this size is enlarged.

3.11. Proboscides *(Proboscidea)*

Data from the study of the cochlea in proboscides are not numerous [Hyrtl, 1844; Fick, 1844; Claudius, 1865; Kolmer, Eisinger, 1923].

In representatives of the *Loxodonta* genus, the cochlea is small and has a lot of features similar to those of representatives of the *Spalax* and the *Chlamiphorus* genera. The secondary osseous spiral lamina of these mammals is formed in the same way. However, it is necessary to point out the absence of the cochlear canal in the *Loxodonta* genus.

3.12. Hyracoideans *(Hyracoidea)*

The cochlea of damans is described by Hyrtl [1844]. In representatives of this order, the presence of both spiral laminae is typical, and there is a great distance formed between the laminae. The microstructure of the secondary lamina differs from that of the cochlea's capsule.

3.13. Perissodactyls *(Perissodactyla)*

The study of the cochlea of this suborder was carried out in representatives of the genus *Tapirus* and in *Equus zebra* [Hyrtl, 1844; Denker, 1889; Kolmer, 1907; Turkewitsch, 1934; Keen, 1942].

The cochlea of the genus *Tapirus* has both spiral laminae. The primary lamina starts in the cochlea's basal part at a half-turn distance from the oval window and occupies a half-turn in the apical part. The secondary lamina is evenly thickened and is located opposite the primary lamina, the vestibular leaf of which is noticeably thicker than the tympanic leaf.

Equus zebra also has both spiral laminae, but the distance between them is larger than in the genus *Tapirus*. The secondary osseous lamina is also more developed. The cavity of the tympanic scala is noticeably larger than the vestibular scala. In the basal part of the cochlea in the zebra and tapir, similar structural features are evident, whereas in the tapir's apical part of the cochlea, the secondary lamina is thicker than that of the zebra, and the distance between both laminae is narrower. The modiolus is well developed.

3.14. Artiodactyls *(Artiodactyla)*

In representatives of the genus *Camelus*, the primary spiral lamina is very thin and the secondary spiral lamina is mostly developed in the basal turn, while in the apical turn its thickness is rapidly decreased.

In species of the genus *Capreolus*, the thickened secondary osseous spiral lamina is an extension of the cochlea's capsule. The primary lamina is located opposite its free edge. The tympanic canal is reduced in size.

The cochlea of the genus *Giraffa* also has two spiral laminae, between which a large space is formed. The tympanic canal is greatly enlarged in the initial part of the basal turn, this also being typical of other mammals.

In the genus *Sus*, the cochlea is large and is formed by 3 turns. In the cochlear canal, only the primary osseous spiral lamina is well developed. The secondary lamina is found only in the cochlea's basal turn. The basilar membrane is wide since both spiral laminae in the basal turn of the cochlea diverge at a large distance from one another.

3.15. Sirens *(Sirenia)*

The study of the cochlea in sirens was carried out using manatees [Hyrtl, 1844; Claudius, 1867], in which the primary spiral lamina has a peculiar structure: it broadens and passes through a system of columns where the spiral ganglion is located. The secondary lamina adjoins only to the external wall and has the form of a thin, greatly perforated osseous leaf. The osseous cochlear canal of the cochlea is absent, this being typical of proboscides.

3.16. Carnivores *(Carnivora)*

The study of the cochlea's structure was carried out in typical representatives of terrestrial and semi-aquatic carnivores [Hyrtl, 1844; Denker, 1899; Held, 1902; Turkewitsch, 1934; Solntseva, 1987a; Dunn, Morest, 1975].

According to the type of middle ear structure, representatives of the genera *Vulpes*, *Mustela* and *Enhydra* differ strongly from each other, but they reveal a lot of similar features in cochlea structure. In all these mammals, the cochlea forms 3 turns around the modiolus (Fig. 19). In the fox and the sea otter, the cochlea has a thick secondary lamina. The distance between both laminae is small. In the cochlea of the mink, in contrast to that of the fox and sea otter, the distance between the spiral laminae quickly increases in the line of the apical part. In these species, considerable differences in cochlea structure are observed in the absolute size of the basal turn. For example, in the mink, the diameter of the basal turn amounts to only half of that in the fox and sea otter.

3.17 Pinnipeds *(Pinnipedia)*

The cochlea of pinnipeds was studied in different representatives of the Otariidae and Phocidae [Schewill et al., 1963; Harrison, King, 1965; Poulter, Thomas, 1969; Solntseva, 1973a, 1975a, b, 1990, 1998a; Ramprashad et al., 1976].

In the Phocidae, the cochlea is formed by 2.5 turns (Fig. 20). The cochlea's basal turn is noticeably enlarged. The wall between the basal and the next cochlea's turns is greatly thickened, and, as a result, these turns diverge from each other. The primary spiral lamina is thicker than the secondary. Shortly after passing through the basal turn, the secondary lamina becomes drastically thin. According to its development, it occupies an intermediate position between the echolocating and non-echolocating species. In comparison with the turn, which is located above, the cochlea's basal turn is bent. An enlargement of the tympanic scalae is typical of this turn.

The organ of Corti reveals a lot of similarities in structure among terrestrial carnivores. The number of inner hair cells in seal calves is 3654 on average, and in a ringed seal 3233. The number of outer hair cells in seals is 14,318 on average and 18,497 in a ringed seal (Table 7). The quantities of inner and outer hair cells in seals increase from the cochlea's base to its top by 21%, and in a ringed seal these quantities increase by 29% and 17%, respectively. The decrease of the quantity of receptor cells in the apical turn occurs due to the broadening of the receptor area [Ramprashad et al., 1976]. The basilar membrane of a fur seal is narrow; it width increases from the base to the top by 12.5 times.

In a Northern fur seal (*Callorchinus ursinus*), the cochlea is formed by 2.5 turns; its basal turn is greatly widened compared to the medial turn (Fig. 21). In the cochlea's passage, the secondary osseous spiral lamina is well developed. The basilar membrane is narrow and is rigidly fixed between the primary and the secondary laminae. These peculiarities of the cochlea's structure are typical in members of the species, as their auditory system is adjusted to the perception of high-frequency signals [Solntseva, 1987b, 1990a].

In a walrus (*Odobenus rosmarus divergens*), as in the majority of pinnipeds, the cochlea is formed by 2.5 turns. All cochlear turns differ slightly from each other in size. This feature of the structure is typical in species with a low-frequency hearing. The primary osseous spiral lamina is well developed; the secondary lamina is developed in the cochlea's basal turn only.

3.18. Similar features of cochlea structure in different species of mammals

The inner ear of mammals, in contrast to that of lower vertebrates (amphibians, reptiles, birds) reveals features of progressive evolution, apparent in the spiral torsion of the cochlea and the presence of the structurally complicated organ of Corti. If in representatives of the monotremats the *lagena papilla* remains, then in marsupials and placental animals the organ of Corti develops.

The periotic bone, in which the cochlea of mammals is located, knits to the tympanic bone in the majority of mammals (rodents, carnivores, pinnipeds, etc.). In odontocetes, it only partially knits to the tympanic bone and is located outside the tympanum.

The number of turns that form the cochlea can vary from 0.5 in monotremats to 4.5 to 5 turns in some rodents. In monotremats, the cochlea is slightly bent and looks like a bird's *lagena*.

A large cochlea is characteristic of all species of dolphins. The cochlear bones are localized in the periotic bone, through which a cochlea channel extends. In the Amazon dolphin (*Inia geofrensis*), the cochlea is half-covered with the medial lobe of the tympanic bone. In marine dolphins, the cochlea is separated from the other bones and is an independent structure. The cochlear canal in the river dolphin is poorly visible, and the *crista transversalis* is almost one-half the size of that in marine dolphins. In the marine dolphin, the *crista transversalis* is open and the cochlear canal is well traced.

In the majority of mammals, the size of the cochlea's basal turn differs slightly from the turn located above it. In representatives of some genera, a drastic enlargement of the cochlea's basal turn is typical (shrews, bats, odontocetes, pinnipeds).

During the comparison of the primary and secondary osseous spiral laminae, some peculiarities are evident in the cochlea's basal and apical turns (Diagram 6). The extension of the primary lamina along the cochlea's passage varies significantly less than that of the secondary lamina [Fleischer, 1973a, b].

The structure of the primary osseous spiral lamina is dissimilar in different species. The vestibular and the tympanal osseous leaves which form the primary lamina can be thick and compact, as well as less developed. For example, in a human, both leaves of the primary spiral lamina are spongy in structure, while in *Loxodonta,* the tympanal leaf is loosened [Fleischer, 1973a, b].

In most mammals, the secondary osseous spiral lamina is more developed in the cochlea's basal part than in the apical part, where it becomes thin.

In Chiroptera and Odontoceti, the secondary lamina occupies the whole cochlea's passage stretching from its basal to the apical turn. In the majority of mammals, including humans, the secondary lamina in the cochlea's basal part is well marked, and in the apical part it quickly disappears.

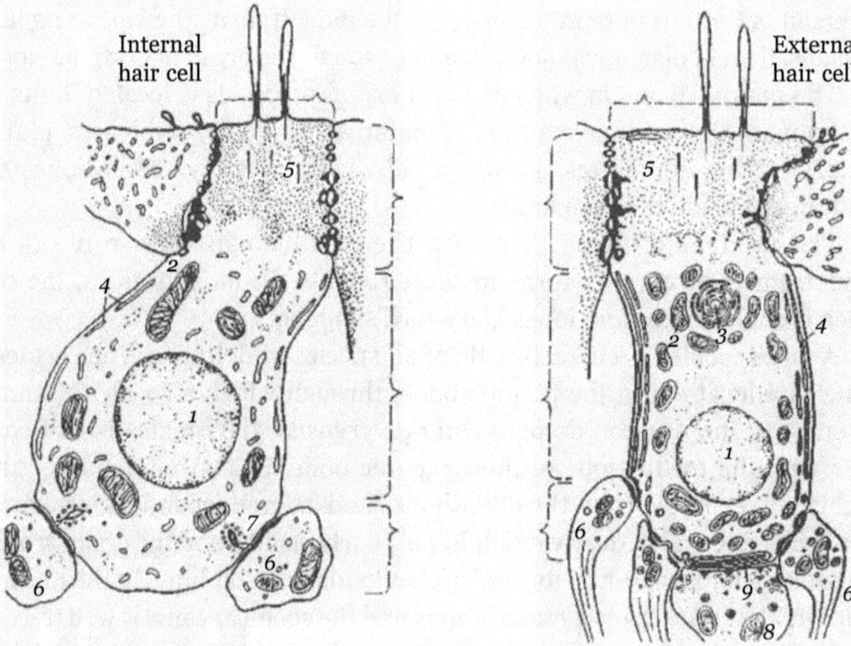

Diagram 6. Receptor cells of the organ of Corti.
1 — nucleus; 2 — mitochondria; 3 — concentric lamellar formations; 4 — near-surface cisterns; 5 — cuticular plate; 6 — efferent fiber ending; 7 – synaptic ribbon; 8 – efferent fiber ending; 9 — sub-synaptic cistern. The areas of membrane fusion in contact zones, zonulae occtudentes, joining the receptor cells to accessory cells [Culley, Reese, 1977].

The comparison of the structures of two spiral laminae shows that the species that possess a thick secondary lamina have a well-developed primary lamina. This is typical in the forms that have a narrow basilar membrane. In the forms with an underdeveloped secondary osseous lamina, the distance between the two laminae is greater (Diagram 6). A wide basilar membrane is typical in these forms (*Talpa*, *Cynocephalus*, *Spalax*, *Homo sapiens*).

In all mammals, the cochlea's basilar membrane widens heterogeneously along its length. For example, in a guinea pig, the basilar membrane is wide in the cochlea's basal part, narrow in the medial part and widens again in the apical part. In odontocetes (porpoises), the diameter of the basilar membrane changes from the basal to the apical turn 5.4 times [Kolmer, 1908], and in a Baird's beaked whale 10 times [Yamada, 1953a, b].

In the majority of mammals, the basilar membrane is up to 29 μm thick. The membrane is narrowest and thinnest in bats and odontocetes. For example, in dolphins, its thickness amounts to 6.5 μm. Among the majority of mammals, the structure of the organ of Corti and the quantity of its receptor elements reveal similar features (Fig. 22a-c; 23a-e; 24a, b; 25a-d). The outer hair cells are usually located in three rows, and the inner hair cells in one row. The total number of hair cells in a human is 14,975; in a guinea pig, 8,939; in a cat, 12,500; in a rabbit, 7,800; in a seal, 17,972; in a ringed seal, 21,792; in a bottlenose dolphin, 17,384; and in a Pacific white-sided dolphin, 12,899 [Ramprashad et al., 1976] (Table 7).

The number of hair cells usually decreases from the basal to the apical turn. Also, the volume of the nuclei of hair cells in the organ of Corti usually increases in the same direction, as well as from the inner row of the outer hair cells to the periphery [Akimov, 1976].

In some mammals, specific structural peculiarities in the supporting elements of the organ of Corti were evident. In dolphins, the supporting elements are relatively large and are compacted together [Wever et al., 1971], and in echolocating bats, the Claudius cells are enlarged [Pye, 1966a, b, 1967; Brown, Pye, 1975]. In addition, spiral ganglion cells are noticeably incrassate in echolocating bats, dolphins and Northern fur seals (Fig. 26a, b).

Consequently, a comparative and morphological analysis of the mammalian cochlea shows that there are two types of cochlear structure. The first structural type can be found in mammals whose habitat conditions are not connected with the use of ultrasonic orientation. The following features are typical of this type: a less-developed cochlear basal turn, a wide and thick basilar membrane, a large distance between the primary and secondary osseous spiral laminae, a weakly developed secondary lamina and the

Table 7. Some quantitative characteristics of the receptor and peripheral nerve elements of the mammals' auditory system

Species	Number of the cochlea's turns	Numberof the receptorsof the OHC + IHC and their sum	Neurons of the spiral ganglion		Number of the auditory nerve's fibres
			Number	Size in mcm	
Rattus norvegicus	3	1899+7040 Σ8939			3000
Cavia porcellus	4.5	1660+6150 Σ7800			
Oryctolagus cuniculus	2.5	2600+9900 Σ12 500	40 000	15 to 22 18 to 22.5	50 000
Felis domesticus	3	3232+13 497 Σ16 729			
Pusa hispida	2.5	3000+11 500 Σ14 500		16 to 25	
Pagophilus groenlandicus	2.5	3654+14 318 Σ17 972	52 000		
Lagenorhynchus obliquidens	1.75	3272+12 899 Σ16 350	65 000	22.6	
Tursiops truncatus	1.5	3451+13 933 Σ17 384	95 004 94 000	34.4 to 38.4 24 to 36	101 000 104 000
Homo sapiens	2.5	3400+13 400 Σ16 800	25 500		
Macaca mulatta		3475+11 500 Σ14 975	30 500 30 000		

1 – Bogoslovskaya, Solntseva, 1979; 2 – Bredberg, 1968; 3 – Chow, 1951; 4 – Drooglever-Fortuyn, 1914; 5 – Guild et al., 1931; 6 – Hall et al., 1974; 7 – Jacobs et al., 1964; 8 – Perkins et al., 1975; 9 – Ramprashad, 1976; 10 – Ramprashad et al., 1976; 11 – Rasmussen, 1960; 12 – Retzius, 1884; 13 – Scharf, 1950; 14 – Schuknecht, 1960; 15 – Wever, 1949; 16 – Wever et al., 1971; 17 – Wever et al., 1972; 18 – Wüstenfeld, 1957

disappearance of the secondary lamina in the cochlea's apical turns [Solntseva, 1969, 1971, 1977].

The second structural type of cochlea is found in species that use ultrasonic orientation and echolocation (shrews, rats, bats, fur seals, dolphins). For this type, the following features are typical: a noticeable enlargement of the cochlea's basal turn, a narrow and thin basilar membrane, a small distance between the spiral laminae and a well-developed secondary osseous spiral lamina.

CHAPTER 4.
ADAPTIVE PECULIARITIES OF THE MAMMALIAN PERIPHERAL PART OF THE AUDITORY SYSTEM

Basic features of structure and development of the peripheral part of the auditory system in different mammalian species are common. Nevertheless, the usage of only definite acoustic characteristics of habitat by the separate groups of animals has caused the pronounced polymorphism of all parts of the auditory system, beginning with the outer ear. The development of different habitats by mammals and the formation of various types of spatial orientation and communication have been accompanied by substantial morphological transformations of all units of the peripheral auditory system, especially of the phylogenetically young units, which are typical only in the mammalian class. During the adaptive radiation of biological forms, which are phylogenetically distant but similar in habitat conditions, similar morphological features in the structure of different organs appear [Severtsov, 1939; Schmalhauzen, 1939]. The development of the auditory receivers of the ultrasonic mammals living in different habitats has gone in the way of parallel evolution.

In this chapter we will consider the adaptive peculiarities of the outer, middle and inner ear (cochlea) structure, which have appeared as a result of the adaptation of terrestrial mammals to aerial, subterranean, semi-aquatic and aquatic habitat conditions.

The outer ear of mammals is notable for the diversity of its structure. The outer ear of the terrestrial and aerial forms includes an auricle and an external auditory meatus, which is in the form of a hollow tube with a constantly opened inlet. Such a structure of the outer ear is optimally meant for normal sound conduction in air.

Animals whose way of life is connected with aquatic living conditions acquire the tendency to close the external auditory meatus under water and open it on land. The mechanisms that open and close the outer ear are provided by different morphological transformations both of the auricle and of the auditory meatus.

In the auricle of semi-aquatic animals (mink), there are cutaneous outgrowths, which help insulate the auditory meatus from the environment during the animals' submersion. By means of contraction of the aural muscles, the cutaneous outgrowths close the inlet that leads to the auditory meatus, and the process of sound conduction is carried out when the auditory meatus is closed. The upper end of the nutria's auricle is turned toward the inlet of the auditory meatus. A hair bundle, consisting of long, rough, almost waterproof hair, is adjacent to the bottom of this inlet. At the moment of the animal's submersion, the turned upper end of the auricle bears against the upper part of the auditory inlet, by means of contraction of the aural muscles, while the lower part of the inlet closes with the help of the hair bundle. These mechanisms prevent the water from penetrating into the cavity of the auditory meatus and thus allow the outer ear to function normally under water. The auricle of some semi-aquatic insectivores (desman) is covered with thick hair, which protects it from water.

The auricle and the external auditory meatus of otariids and sea otters, which inhabit predominantly aqueous environments, are morphologically reorganized. Although these animals have turned to an aquatic way of life, they have not lost their connection with the land. As a result, their auditory system is adapted to both aquatic and aerial sound conduction. The auricle of a sea lion and a sea otter is involuted and cone-shaped. The auricle of a fur seal is bent along its long axis, which causes dense contact of its edges. In air, a partial opening of the auricle occurs, due to contraction of the aural muscles. Under water, the aural muscles relax and the edges of the auricle adjoin again, closing the auditory inlet and isolating the cavity of the auditory meatus from the environment.

The outer ear of subterranean forms (moles), pinnipeds (phocids) and typically aquatic mammals (cetaceans) consists of the external auditory meatus only. The auricle is completely reduced; therefore, the mechanisms of isolation of the auditory meatus from the environment (water) and its opening in the air are enabled by the auditory meatus itself.

The mole's auditory meatus has the form of a tube with two parts. The first part (the area of the auditory inlet) is surrounded by cartilaginous plates, which are connected with each other by sliding joints. This structure of the auditory meatus allows for the possibility of closing the area of the auditory inlet, depending on the frequency and intensity of the incoming acoustic signals. In place of the transition of the cartilaginous part of the auditory meatus to its osseous part, there is a "disk-shaped widening," which is considered essential for transferring the acoustic energy from the oscillating skull bones to the cavity of the auditory meatus and its sides [Simkin, 1977].

The external auditory meatus of phocids has the form of a double-bent tube situated diagonally directly under the skin [Solntseva, 1972c]. The surface disposition of the auditory meatus caused some researchers to suppose that this structure and disposition of the auditory meatus is enough to prevent water penetration of the external auditory meatus due to water pressure [King, 1964]. However, other researchers suppose that, because of the peculiarities of the structure and disposition of the auditory meatus, its folding occurs under water with the help of external and internal muscles in place of the double-bent turn [Rosenthal, 1825; Ramprashad et al., 1971]. We share the second hypothesis [Solntseva, 1975a]. The folding of the auditory meatus is possible because of the presence of the double-bent turn [Rosenthal, 1825], the modification of the diameter and shape of its lumen [Ramprashad et al., 1971; Solntseva, 1972c; 1975a] and the flexibility of the joints between the cartilaginous plates [Böas, 1912].

The closing of the phocid's auditory meatus by means of its folding, with the help of the aural muscles, has been experimentally confirmed [Mohl, 1968]. It has been shown that the external auditory meatus can close automatically even before the animal is submerged and last as long as the animal stays under water.

Because a fur seal inhabits water as well as land, the mechanisms that open the outer ear in air and close it under water have developed [Solntseva, 1987a]. This has been enabled by the morphological reorganization of the auricle. The auricle's bending along its long axis causes dense contact of its edges. Due to the contraction of the aural muscles, the auricle opens in air and sound conduction is carried out in the same way as it is in terrestrial mammals. The relaxation of the aural muscles causes dense contact of the auricle's edges, preventing water penetration of the cavity of the external auditory meatus. Such an assumption has been proposed based on the design data obtained from the closed auditory meatus of dolphins [Lipatov, Solntseva, 1974]. The obliteration of the dolphin's distal part of the external auditory meatus and various mechanisms that close the auditory meatus in pinnepeds [Solntseva, 1975a] are progressive adaptations for hearing under water.

Another hypothesis has recently been advanced [Lipatov, 1985], according to which the sound perception of pinnepeds depends on whether their fur is "drenched" or "not drenched." As Lipatov argues, "The reduction of the auricles at the time of turning to the aquatic way of life occurs only in those animals whose fur can be drenched and whose skin can be easily penetrated by sound waves; in this case, sound perception is enabled directly by the external auditory meatus. If fur traps an air layer inside, it serves as an acoustic shield. The sound perception enabled directly by the auditory meatus becomes ineffective. In this case, the auricles take the form of closable tubes, causing the air ducts of the auditory meatus to become longer and exceed the bounds of the fur."

There are a number of objections against this proposed hypothesis. First, this hypothesis contradicts the following known fact: other representatives of otariids (Steller sea lion, sea lion), in contrast to the fur seal, lack a dense and silky undercoat and consist mainly of the aristae [Heptner at al., 1976], but the auricle of these representatives is as well developed as the auricle of a fur seal [Solntseva, 1987 a, 1990a]. Therefore, the presence of an undercoat in the fur seal does not correlate with the development of its auricles. The author's assertions that the auricle and the external auditory meatus of the fur seal have the form of a closed tube contradict the data obtained from the histological research shown in Figure 5.

The external auditory meatus of cetaceans is constantly closed. The external auditory meatus in odontocetes and some species of mysticetes is obliterated in a certain part, but then the lumen appears again and is maintained until it reaches the tympanic membrane. The presence of earwax is typical of rorquals and humpback whales [Purves, 1955].

Concerning the question of the auditory meatus's obliteration in odontocetes, there have always been and still are very contradictory opinions. Some researchers maintain that the auditory meatus stretches in the form of a hollow, bent double tube along its whole length [Reysenbach de Haan, 1957a, b; Fraser, Purves, 1960]. According to other researchers, the external auditory meatus of odontocetes is a tube with an obliterated lumen. Such obliteration was registered in a sperm whale [Clarke, 1948; Yamada, 1953a, b], a bottlenose dolphin and a common dolphin [Belkovitch, Solntseva, 1970].

The obliteration of the auditory meatus in odontocetes allows us to suppose not only the presence of other methods of sound conduction through the outer and middle ears, but also the existence of different systems of sound conduction and sound perception that are not connected with hearing organs at all [Norris, 1964; Reznikov, 1970; Agarkov et al., 1971]. At first, there was a hypothesis that acoustic signals can penetrate the cochlea through the closed auditory meatus [Reysenbach de Haan, 1957a, b]. Norris [1964] supposed that sound waves could reach the cochlea through a fat-filled cavity in the lower jawbone. Norris's hypothesis has been experimentally confirmed [Bullock et al., 1968].

In 1971, Agarkov G.B., Solukha B.V. and Khomenko B.G. advanced the hypothesis that a reflected echo signal is accepted by a multicanal system of mechanoreceptors that are dispersed all along the head. Each of the mechanoreceptors is connected by an individual neural path with the part of the brain where image creation occurs.

According to the hypothesis of "ultrasonoscopy," a dolphin's melon represents an analogue of an optical system and epicranial air sacs are analogous to the eye's retina [Reznikov, 1970].

With the aim to study the dolphin's conduction of sound, peculiarities of the underwater audiograms were measured [Zaslavski, Babushina, 1986]; on their basis, the conclusion was made that the dolphin's body tissues should not be regarded as an accessory canal of sound conduction. However, the author's opinion was that the role of the body tissues in spatial hearing could turn out to be significant.

Contrary to these hypotheses, a number of researchers made an assumption about dolphin's conservation, the main principle of sound conduction which is typical to all mammals. However, taking into account the obliteration of the auditory meatus in dolphins, the other authors considered that sound oscillation should pass through soft tissues of the head and farther in the regular way (i.e., through the external auditory meatus) the tympanic membrane and the system of the auditory ossicles to the inner ear [Reysenbach de Haan, 1957a, b; Purves, 1967; Belkovitch, Solntseva, 1970; Solntseva, 1975a, 1990a].

In favor of this assumption based on the design data, a hypothesis has been worked out which explains the mechanism of underwater sound conduction when the dolphin's auditory meatus is closed. In order for the hearing organ to function under water, the closed auditory meatus must be filled with air [Lipatov, 1972; Lipatov, Solntseva, 1974]. As a result of the determination of the harbor porpoise's directional diagram of reception, this hypothesis was experimentally confirmed [Stosman, 1978].

The results of the biomechanical calculation of the dolphin's external auditory meatus revealed that the thickness of the auditory meatus in the line of wave transmission was much less than the wavelength (for the frequency of the best sensitivity amounts to 60 kHz); therefore, it could be considered that an alternating acoustic pressure affecting the side of the proximal part of the auditory meatus was equal along its full length. This pressure caused oscillations of the auditory meatus's sides in diametrically opposite directions; as a result, the scope change of the proximal part of the external auditory meatus occured. This produced the alternating pressure at the point of entry to the osseous part of the acoustical meatus which, being enhanced due to the resonance of the osseous part, affected the tympanic membrane. It turned out that the threshold acoustic pressure on the dolphin's tympanic membrane on the frequency of their best sensitivity was only one-half the same on the human tympanic membrane at a frequency of 1 kHz [Lipatov, Solntseva, 1974].

As it was presumed before, the sound waves travel from the front of the head through a hypodermis and a muscular tissue to the proximal part of the auditory meatus, by-passing the inlet of the auditory meatus and its distal part [Dudok van Heel, 1962; Belkovitch, Solntseva, 1970; McCormick et al., 1970; Lipatov, Solntseva, 1974].

However, according to the analysis of the researcher's basic statements concerning the ways of acoustic sound conduction into the dolphin's inner ear, and also from our own morphological data obtained later, it could be concluded that the sound waves can pass from the front of the head not through the hypodermis and the muscular tissue to the proximal part of the external auditory meatus, but by the traditional pathway: directly through the auditory inlet, the distal part of the auditory meatus and further through the area of its obliteration, which represents the "acoustically transparent" membrane formed by the epithelial cells with a thickness up to 50 to 60 μm through the proximal part of the auditory meatus to the tympanic membrane and the line of the auditory ossicles [Solntseva, 1997c].

Based on these results, it is possible to suppose that in other aquatic and semi-aquatic mammals (pinnipeds) for whom the mechanisms of closing the auditory meatus under water are typical, the normal sound conduction in an aqueous medium without loss of sensation is realized only when the auditory meatus is closed, as it is in dolphins.

This assumption was experimentally confirmed by using the method of measuring and comparing the frequency threshold curves in the caspian seal under the water and in the air; the results proved that the sound could pass from the front through the tissues of the head, but in that case the signal became abated 1.5 to 2 times compared with direct leakage [Kupin, 1986].

The tendency to shut the outer ear and its obliteration in dolphins is the progressive adaptation of the outer ear which provides for underwater hearing.

The tympanic bulla of mammals represents a periotic-tympanicum complex. The tympanic bone is formed out of the thin osseous walls which form the tympanum where the elements of the middle ear are located. In periotic bone, a cochlear bone is situated, in the thickness of which a cochlea channel is extended. This thickness of tympanic bone walls is changes sharply between species. In this way, in terrestrial and semi-aquatic species, the walls of tympanic bone are thickened as compared with those of aquatic mammals, as odontocetes. In mysticetes, the walls of tympanic bone are sharply hypertrophied.

In representatives of the family Delphinidae, the tympanic bulla has a unique structure, distinguishing it from the other species of mammals (Fig. 27a–f). The tympanic and periotic bones are partially knitted with each other in the area of *pr. posterior, pr. sigmoideus* and *pr. tubarius* (additional ossicle) [Kasuya, 1973; Pilleri, 1987; Solntseva, 2002a]. In the Amazon River dolphin *(Inia geofrensis)*, a part of marine species of dolphins (*Tursiops truncatus, Delphinus delphis, Phocaena phocaena, Stenella attenuata*), the tympanic bulla is formed by thick osseous walls. The tympanic and periotic bones in the river dolphin relatively bear against each other. In marine species of dolphins, the tympanic bulla is formed by more thin and rather fragile osseous walls. The positioning of the tympanic and periotic bones has such a pattern that it creates an impression of these bones' detachment relative to each other.

The comparative analysis of the structure of the tympanic bulla in the river dolphin and marine species of dolphin showed that in the river dolphin the tympanic bulla has species-specific features in structure that, most likely, are connected with the ecological peculiarities of the species and auditory organ functioning in the conditions of a more noisy and polluted environment.

The inner ear is located in periotic bone. The dolphin's tympanic bulla does not knit with the skull bones and is connected with them by means of the tendinous ligament [Oelschläger, 1990; Solntseva, 2001a]. In odontocetes and mysticetes, in the elements which connect the periotic bone and the skull, the numerous sesamoid ossicles which provide a certain mobility of the auditory bone in regard to the skull are located [Yamada, 1953 a, b].

The middle ear of different species of mammals is characterized by a different degree of knitting of the tympanic bulla and skull bones [Fraser, Purves, 1960]. The tympanic bulla in most species knits with the skull bones and takes part in the formation of the skull wall. It is known that such knitting can provide osseous sound conduction [Künze, Kietz, 1949].

Under the influence of the acoustic oscillations, the movements of the skull bones are transmitted to the osseous labyrinth of the inner ear and as a result to the perilymph. This does not occur in some species of bats, rodents [Pye, 1968; Fleischer, 1973a, b] and odontocetes, due to the fact that the tympanic bulla is totally isolated from the skull bones and is flexibly connected to it by means of connective tissue. As a result of such connection, both ears become independent receivers and are able to provide directional reception of acoustic signals.

The isolation of the tympanic bulla from the skull bones is provided by the sinuses, which surround the tympanic bulla from every quarter and are filled with fat emulsion foam. The foam consists of the smallest air bells which are good sound insulators, and, as a result, all acoustic oscillations which come from the skull bones do not reach the inner ear. To the cochlea, the only way is left – through the outer ear and the system of the auditory ossicles [Fraser, Purves, 1960; Solntseva, 1995]. Thanks to this separate sound reception (i.e., by each of the auditory receivers), dolphins are characterized by the so-called binaural effect which allows them to determine the direction to a sounding object [Kellogg, 1958]. In addition, the tympanic bulla in odontocetes can perform slight movements relative to the skull with the help of muscles; as a result, the stereophonic (volumetric) reception of the reflected echo signals is provided.

Only the *pars mastoidea* takes part in the rigid connection to the skull in most mammals, including the family Soricidae, while most of the tympanic bulla is connected to the skull by means of connective tissue.

On the way of adaptation to the aquatic way of life, the weight ratio of the auditory ossicles changes, which becomes apparent in the weight increase of the incus in regard to the malleus. The relative number and the form of particular elements of the auditory ossicles change. At the same time, thickening and widening of the malleus's handle occurs (pinnepeds) and its reduction in cetaceans, lengthening of the malleus's long arm and the incus's short arm occur, as well as decreasing of the inter-ear area of the stapes down to its complete reduction in odontocetes. The rigidity in the area of the incudomalleal articulation is increasing.

In non-echolocationing mammals, the structure of the middle ear reveals patterns of similarity, although the structure of the tympanum and auditory ossicles varies drastically. For this type of middle ear structure, the following is typical: knitting of the tympanic bulla with the skull bones, thinning and sharpening of the long arm of the malleus down to its reduction and changing to *pr. gracilis,* which is flexibly connected with the tympanum's wall, as well as an extension of the articular surface in the area of the incudomalleal connection. Thanks to the anatomical reorganization of the external auditory meatus and peculiarities of their location, the main principle of the sound transmission in the middle ear in different habitats remains.

As it is known, the system of the auditory ossicles is constructed according to the classical lever scheme. A rotation axis of the auditory ossicles passes through the points of the incus's short process and the malleus's long process fastening to tympanic bone. It has turned out that in most mammals (terrestrial, aerial, subterranean, semi-aquatic), the rotation axis of the auditory ossicles is parallel to the plane of the infra-peduncle plate of the stapes. In the inhabitants of an aqueous medium (sea otters, pinnepeds, cetaceans), the rotation axis of the auditory ossicles is located at an angle to the infra-peduncle plate of the stapes [Lipatov, Solntseva, 1972]. Under such location of the rotation axis, due to the screw effect, which becomes apparent as a result of the inclination of the rotation axis relative to the infra-peduncle plate of the stapes, the coefficient of pressure increase is added to the calculation of the coefficient of acoustic pressure transmission by the middle ear [Lipatov, 1972].

By means of the design data, it was shown that a middle ear in most terrestrial and some semi-aquatic forms has the lowest coefficient of the transmission of the acoustic pressure (K equal to to 25 to 29). The value of K of a sea otter and pinnepeds increases 1.5 to 2 times, compared with that in terrestrial forms, which is about 40 to 60. The increase of the K value in these animals is caused by inclination of the rotation axis of the malleus and incus relative to the infra-peduncle plate of stapes (K 3). The dolphin's middle ear has the highest K, exceeding that of sea otters and pinnepeds 2.5 to 3 times. The value of K in dolphins amounts to 110 to 166. The increase of the K value in dolphins occurs because of K3 and the additional lever in the form of a triangular ligament, which is asymmetrically connected to the tympanic membrane, and also due to the rigidity of the tympanic membrane rigidity, which allows all effort applied to it to pass.

One of the factors leading to an increase of acoustic pressure is that the amplitude of the stapes's movements is 30 times larger than that of the tympanic membrane.

Thus, the biomechanical peculiarities of the mammals' middle ear, whose way of life is connected with the inhabitating of an aqueous medium, are mainly directed to the increase of the coefficient of transmission of acoustic pressure by the middle ear. Most likely, that determines the effectiveness of the peripheral auditory system's functioning during mammals' orientation in aqueous medium and greatly enlarges their perceptible frequency band.

For maximum transmission of the energy of an incoming signal, it is necessary for specific acoustic resistance of an environment to be coordinated with the auditory receiver. Such coordination is achieved with the help of variations in the structure of the transmissive apparatus of the middle ear depending on the environment where transmission of acoustic information occurs. For the optimal reception of sound signals in aqueous medium, the sound receiver should possess a high modulus of elasticity, which is provided by the rigidity of the auditory ossicles's conjunction with each other and their fastening in the tympanum. For example, the Northern fur seal's malleus and incus form a united incudomalleal complex as both ossicles are fixedly joined with each other and function as a single unit. The broadening in ultrasonic range causes the increase of resonance frequency of natural oscillations of the auditory ossicles as well as the increase of the tympanic membrane's elasticity [Solntseva, 1990a; 1997c].

In addition to sound conduction, the middle ear of mammals carries out a precautionary function by decreasing the energy of the incoming signal to the inner ear. This function is enabled by the contraction of the middle ear muscles – *m. tensor tympani* and *m. stapedius* – which are well developed in echolocating species (dolphins, bats) [Solntseva, 1973b, 1978, 1988a, 1990c]. By means of the tympanic membrane's and the auditory ossicles's tension, the muscles of the middle ear create conditions for ultrasound conduction; relaxation of these muscles preserves the cochlea from super-intensive signals. It is assumed that the muscles of the middle ear are able to provide tuning of the auditory system to certain frequencies [Blair, 1964].

In the walls of the tympanum of the animals whose way of life is connected mainly with inhabiting an aqueous medium (sea otters, aquatic mammals), there are venous sinuses which are concentrated near the tympanic membrane. Venous sinuses are considered to be the additional formations whose function consists in leveling the tympanum's pressure while submerging under water [Tandler, 1899]. The same function is carried out by the tympanum's mucous membrane, which, in contrast to the similar ones in terrestrial and some semi-aquatic forms, is greatly thickened due to the numerous blood vessels which are situated in its middle layer.

In the tympanum of some typically aquatic species (dolphins), a cavernous plexus consisting of a thick network of blood vessels is located. A cavernous plexus, as well as venous sinuses, is considered to be the additional formation which is able to modify the volume due to its blood filling, thus leveling the pressure in the tympanic cavity when the animal submerges under water at a great depth [Solntseva, 1990a, b, c, 1995].

The structure of the tympanic membrane in different mammals is particularly worthy of notice. In terrestrial, aerial and subterranean forms, the tympanic membrane is very thin, soft, rounded and slightly elongated to a cone shape. This structure of the tympanic membrane is revealed in some semi-aquatic species (mink, nutria). In aquatic mammals, the tympanic membrane is thickened, and its shape can remain round or oval. However, in one representative of semi-aquatic species, a Northern fur seal, the tympanic membrane is thin but its size is greatly decreased and it is rigidly fixed on the tympanic annulus. This increases its elasticity and provides conditions for the transmission of broadband frequencies, including ultrasounds [Solntseva, 1998c].

The tympanic membrane of cetaceans is greatly modified. The tympanic membrane of mysticetes represents the "membranous outgrowth" with the fibrous ligament outgoing, from which it fastens to the reduced handle of the malleus.

The tympanic membrane of odontocetes is greatly thickened, rigid and does not have a direct connection with the malleus's handle. This connection is instead carried out by means of the triangular ligament, which asymmetrically fastens to the tympanic membrane. The tympanic membrane and the ligament form the additional lever that is very important for increasing the sound pressure transmission by the middle ear under water [Solntseva, 1989; 1990 b].

On the way of adaptation to the aquatic way of life, the rigidity of the incudomalleal articulation increases (sea otters, pinnipeds). Parallelism in the structure and connection of the auditory ossicles can be noticed in phylogenetically distant species (nutria, fur seal), in which the knitting of two ossicles in the area of the incudomalleal articulation occurs. Such rigidity in the articulation of the auditory ossicles in aquatic and echolocating species provides the possibility of an unhampered sound signal transmission to the inner ear as it decreases the sound energy loss in articulations and joints, thus creating optimal conditions for sound conduction. In contrast to the pendulous system of the auditory ossicles of terrestrial forms, in aquatic and echolocating species, an elastic vibrating system appears, which is able to return to the previous position after the blow of the sound wave [Simkin, 1977].

There are various structural variations in the structure of auditory ossicles in different species of mammals. However, in the row reflecting the way of adaptation to the aquatic way of life, a clear tendency is found, which includes the thickening and shortening of the malleus's handle (sea otters, pinnipeds) down to its complete reduction in cetaceans, as well as the lengthening and thickening of the malleus's thin process. In the structure of the incus, attention is paid to the lengthening and thickening of the long process. In the structure of the stapes, the inter-crura opening in dolphins decreases down to its complete disappearance.

The auditory ossicles show interspecific variability, which becomes apparent in the modifications of size and forms of the auditory ossicles themselves and their processes as well, in the weight ratio of the auditory ossicles, as well as the way of their fastening in the tympanum. The similar type of structure of the middle ear elements can be found in mammals with a high-frequency hearing (shrews, bats, rats, cetaceans, fur seal). In spite of external differences of the auditory ossicles of these animals, they possess a morphological similarity, which becomes apparent in lengthening, thickening and modification of the malleus's long process, its rigid connection with the tympanum's wall, and in the increasing of the rigidity in the area of the incudomalleal articulation.

The rigidity in the connection of the malleus's long process and tympanic bone allows for the possibility [Künze, Kietz, 1949] that the auditory ossicles's line is not functioning in cetaceans and the acoustic signal conduction is provided by means of the osseous conduction. At the same time, an opinion was advanced that the osseous conduction is impossible because of the isolation of tympanic bulla from the skull bones [Reysenbach de Haan, 1957a, b]. Seemingly immobile, the cetacean's auditory ossicles are able to pass sound oscillations to the inner ear, which was proved experimentally [Fraser, Purves, 1960; Purves, 1966] and by use of design data [Lipatov, Solntseva, 1972].

The morphological and biomechanical peculiarities of the middle ear are undoubtedly adaptive and serve for the transmission of broadband frequencies.

Therefore, turning to the aquatic way of life, the following peculiarities of the middle ear's structure are revealed: (1) modification of the relative size and forms of the particular elements of the auditory ossicles: thickening of the malleus's handle down to its complete reduction in cetaceans; lengthening of the incus's long process and decreasing of the inter-crura opening down to its complete reduction in odontocetes; (2) significant increase of the rigidity in the incudomalleal articulation; (3) modification of the weight ratio in the system of the auditory ossicles; (4) drastic thickening of the mucous membrane covering the tympanum due to the increase in amount of blood vessels in its middle layer; (5) relative thickening of the tympanic membrane especially in cetaceans due to the development of the connective tissue's elements; (6) development of the venous sinuses which are concentrated near the tympanic membrane and in the walls which form the tympanum; (7) development of the cavernous plexus which is typical to cetaceans only [Solntseva, 1990a, b].

In the inner ear, the principle of the cochlea's structure reveals similarities in almost all mammals. The number of cochlea's turns varies from 0.5 to

5. The increase of the number of cochlea's turns is explained by its morphological progress [Fleischer, 1973a,b]. However, in echolocating animals such as dolphins, the cochlea is flat and forms 1.5 to 2.0 turns only, while in bats the number of its turns amounts to 3.5. Such variability in the number of cochlea's turns in echolocating forms proves that the cochlea's height caused by the turns's increase does not influence the perceptible frequency band.

The comparative analysis shows that in echolocating forms one of the important cochlea's adaptations, which provide high-frequency hearing, is the increasing of the cochlea's basal turn. For example, the surface of the basal turn of dolphins is increased due to the "untwisting" of the cochlea up to 1.5 to 2.0 turns. The bat's cochlea reveals just the same, in spite of the fact that its promptness is increased up to 3.5.

Another important peculiarity of the cochlea's structure in echolocating mammals is a well-developed secondary spiral bone lamella. The smaller the distance between the primary and secondary spiral bone lamellae, the narrower the width of the basilar membrane becomes, and the secondary spiral bone lamella turns out to be more developed, the rigidity of which continuously and evenly decreases from the cochlea's basal turn to the apical turn (Diagram 4).

In most mammals, the structure of the organ of Corti reveals patterns of similarity. The number of receptor cells in echolocating and non-echolocating forms does not change. However, some researchers observe certain peculiarities in the structure of the organ of Corti bearing elements in dolphins and bats, in which these cells are enlarged in size and compactly located [Wever et al., 1971a, b, 1972]. The increasing number of the spiral ganglion's cells (3 times) and the enlargement of their size compared to humans testify in favor of the data concerning dolphins' and bats' high abilities to process acoustical information starting from the peripheral part of the auditory analyzer [Firbas, Welleschik, 1973].

The most interesting facts are revealed during the study of the inner ear's structure in mammals. First, this is the existence of two types of receptors belonging to a different evolutionary age (outer and inner hair cells), which are spatially separated from each other. What is more, it has an extremely small number of the receptor cells and a sufficient stability in animals of the very variable hearing specializations. The auditory system of echolocating species and animals with low-frequency hearing possesses approximately equal quantities of the hair cells and the identical character of their distribution upon the basilar membrane.

According to our point of view, the fact that most mammals, even the species with extraordinarily broad hearing abilities, have relatively few number of receptors as well as auditory nerve's fibers, is connected with a temporary specificity of the acoustic signal's perception and processing. This is a consequent receipt of information that allows simultaneous use of comparatively little quantity of parallel canals in the periphery of the auditory system [Bogoslovskaya, Solntseva, 1979].

Thus, during the adaptive specialization of representatives of different ecological groups, the outer ear is mainly subjected to the greatest morphological modifications, connected with adaptation to terrestrial, aerial, subterranean, semi-aquatic and aquatic habitat conditions due to its direct contact with the environment. As for the middle ear, in most mammals, it preserves the general basic principle of the construction; in phylogenetically distant but close by ecological specialization species, the parallel features in the development of the particular auditory ossicles' structure as well as in the way of their connection with each other and their fastening in the tympanum appear. The structure of the inner ear (cochlea) acquires some interspecific peculiarities; however, in most species with the exception of echolocating mammals (dolphins, bats), the structure of the organ of Corti is similar.

In echolocating species belonging to different ecological and taxonomic groups, the development of the peripheral auditory system has acquired general traits due to parallel evolution, in which process the conditions for their intraspecific acoustic communication in visually unfavorable conditions and due to specific medium characteristics as an acoustic communication channel have been developed.

Despite the fact that a general plan of structural organization of the auditory system's peripheral part is apparent in most mammalian species, each ecological group exhibits its own characteristic direction in the evolution of this system, which has been developing independently for functioning in habitats with specific physical properties .

Not only the structure and the development of the mammal's peripheral auditory system, which functions specifically for each group's frequency spectrum, are connected with the ecology of the species, but they are also directed to the optimization of intraspecific animal's communication in definite habitat conditions as well [Airapetiantz, Konstantinov, 1974].

Chapter 5. Innervation of the organ of Corti

The investigation of nerve components of the acoustic system's peripheral part is so difficult methodologically that a whole series of questions which were solved long ago during the investigation of the other sensory system have not yet been clarified. The basic difficulty for the morphologists lies in the fact that the organ of Corti, together with its nerve elements, is located within the osseous tissue. In addition, it is in the shape of a spirally involuted geometrical figure. These structural peculiarities create considerable difficulties during the determination of the connections between the types of peripheral and central neuron's processes and bodies of the spiral ganglion of the cochlea. Therefore, most of the work on the cochlea's innervation and the computation of the different element's quantity demands an application of special methods, including graphical reconstruction of the serial sections.

The Golgi method was and still remains the basic histological method of the organ of Corti's innervation study, which has been supplemented by the cochlea's electron-microscope investigations in the normal conditions and during the experimentally induced degenerations.

5.1. Structure of the spiral ganglion

The neurons which innervate the auditory receptor cells form a spiral ganglion: a nerve-knot of the VIII pair's acoustic part of the craniocerebral nerves. The ganglion fills the Rosental's canal in the cochlea's axis and repeats the number of its spiral turns. The ganglionic neuron has, as a rule, a widened body with two processes: peripheral and central (Diagram 7). The body is covered with a complicated myelinated membrane or capsule.

The peripheral process (dendrite) penetrates to the organ of Corti through an aperture in *habenula perforata*; a cat has 2500 of such apertures [Spoendlin, 1966]. The dendrite is covered with a myelin (medullated) membrane, which disappears only at the entrance of the *habenula*'s aperture. In terms of electrophysiology, the peripheral process as well as the central one is an axon, since the place of the myelin membrane's termination is morphologically similar to the Ranvie's interception [Engström, Wersall, 1958]. According to existing conceptions, this is exactly the area that is analogous to the initial neuron's segment, where the action potential is generated. The central processes (axons) form the auditory nerve, the fibers of which, combined with vestibular ones, enter the CNS in the area of the medulla's and pons's boundary (Diagram 7).

In the perikaryon of the cat's and the rat's bipolar ganglionic neurons, the same organelles as in other nerve cells are described [Spoendlin, 1966]. The cytoplasm contains cisterns with slit-like outgrowths. They are surrounded by a 10-nm thick membrane, on the surface of which granules of a 10-nm diameter are situated. In addition to these membranes of the granular endoplasmic reticulum, there is a system of non-granular membranes, mainly

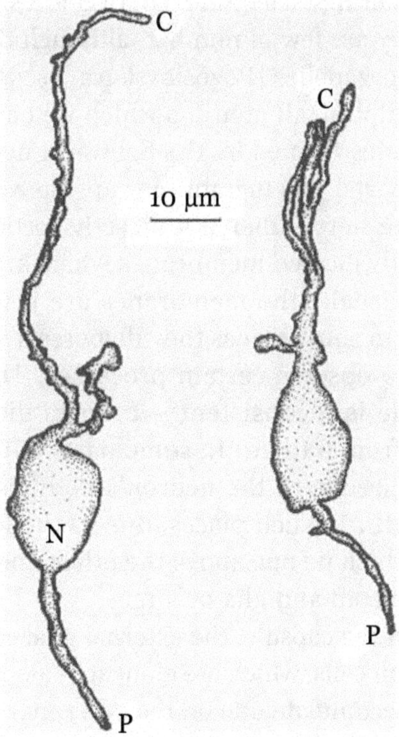

Diagram 7. Two neurons of the spiral ganglion form the middle turn of the cochlea of a six-day-old rat, showing atypical ramification of the central processes (C) near the cellular body.
P – peripheral processes (dendrites), N – cellular nucleus. Golgi method, trans-modiolar (horizontal) shear plane [Perkins, Morest, 1975].

corresponding to the Golgi complex, as well as a great amount of mitochondria with irregular internal membranes. The mitochondria are especially numerous in the area of the axon's origin, and the Golgi complex is located in the area of the dendrite's origin [Kolmer, 1927]. Between the membrane's complexes and mitochondria are numerous uncombined granules.

The bottlenose dolphin's ganglionic cells electron-microscopically represent oval perikarya with a similar electronic nucleus and cytoplasm density. Their size without capsule is 24 to 36 μm. The nucleus, with a diameter of 10 to 15 μm, contains fine-dyspersated condensed chromatin only, without large lumps. The nucleolus is often eccentrically located almost right up to the nuclear membrane; its size is 2.5–3.5 to 3.5–3.9 μm. The neuron's cytoplasm is quite uniformly filled with a large nunmber of ribosomes, which are mainly assembled in the polyribosomes-rosettes. The cisterns of the endoplasmic reticulum are ill-defined and sometimes hardly distinguishable in the ribosome's gatherings. The mitochondria are small and have dense, dark membranes. They are few in number, although there are areas with a high density of these organelles [Bogoslovskaya, 1975].

In a bottlenose dolphin, all neurons which we have investigated were enclosed in the capsules formed by the Schwann neuroglia. Such a capsule consists of two parts. The membrane capsule which is similar to the myelin capsule of the nerve fiber fits directly to the neuron's body. It contains concentrically located membranes which are appurtenant to the flat glial processes. Usually the membranes are just 10 to 20 nm away from each other, but in some places they disperse a substantial distance, and then it is easy to observe certain processes. The number of membranes in the capsule is inconsistent – even in different areas of one perikaryon it varies from 6 to 20. In some parts of the capsule, the glial process, which fits directly to the neuron's body, is distinctly widened along the major length. In such places, due to the forming of long concentric outgrowths which lie one above the other, the reciprocal covering of the neuron's cytoplasm and glia occurs.

The second part of the capsule, the external glial capsule, is formed out of the tile-like Schwann cells, which are elongated along the neuron's circle. Often two glial cells accumulate one on the other, not only with the help of their processes but their bodies as well, so that their nuclei are located very close to each other and are separated by thin tension bars of cytoplasm only. The Schwann glia nuclei's size is 4-5 to 5-12.5 μm. The glial cytoplasm is rich with ribosomes and in some cases has a greater electronic density than the neuron's perikaryon.

The neuron's capsules are comparatively thin in the places of the compact adjacency of one nerve cell to another and are greatly thickened where cells adjoin the myelin fiber bundles. The dendrite's initial parts which we investigated on the electronic diagrams are covered with the same complicated capsules.

One could think that such capsules not only isolate the intensively functioning neurons but also take an active part in the nerve cell's metabolism as it was shown for the glial satellites in other parts of the central and peripheral nervous system.

Acoustic loading can cause modifications in the ultrastructure of the ganglionic cells, affecting mitochondria, the smooth and granular endoplasmic reticulum and the nucleus. These shifts possibly represent the morphological expression of a modified cell metabolism in response to specific stimuli [Koichev, 1975]. As it was found earlier, the sound action leads to a series of essential deviations in RNA quantitative levels, on the basis of which it turns out well to predetermine the neuron's functional activity [Hyden, 1943].

There is no consensus of opinion in the literature about how many types of cells can be distinguished among the spiral ganglion neurons. Depending on methods and purposes of investigation, so many different criteria were used that the obtained results, in spite of their fundamental character, are very hard to coordinate among each other. We believe that it is very important to present here the basic points of view on this question.

The oldest and the most generally accepted discerning of the ganglionic cells is their division into two types according to their peripheral processes' distribution. The neurons with the radial distribution of dendrites innervate the inner hair cells (IHC) only; every neuron comes in contact with one or two IHC only, but at the same time, each of these cells interacts with many neurons. Such "radial" cells form the overwhelming majority of the spiral ganglion's neurons and carry out the projection of the cochlea to the cerebral centers according to the principle "point to point." The cells of the second type have dendrites which go spirally for the major space of the cochlea turn's passage and come in contact with a great amount of the outer hair cells (OHC), which, in turn, are connected with the spiral dendrites of many neurons [Smith, Sjostrand, 1961a, b]. Much less numerous "spiral" cells carry out the innervation of the organ of Corti based on a very wide convergence of their processes. The existence of the ganglionic elements of different geometry of the peripheral processes was recently brilliantly demonstrated using the Golgi method on the cochlea of young rats and cats [Perkins, Morest, 1975]. As the authors of this work noticed, in the limits of the spiral ganglion itself, both types of neurons are practically indistinguishable by the shape and the size of the cell bodies.

In the investigations of Spoendlin [1957, 1969, 1972, 1974, 1975], who researched the structure of the organ of Corti in the cat, two types of ganglionic neurons were also distinguished, but this time on the basis of dendrite myelinization. The basic (95%) is the type of cell with a myelinated dendrite, a well-discernible nucleus and a nucleolus. The second type, composing 5% of all cells, shows only a smaller neuron with an eccentric nucleus and an unmyelinated dendrite.

Finally, two types of cells are described in the spiral and vestibular ganglia of the rat according to ultrastructural peculiarities of the perikarya [Rosenbluth, 1962]. The cytoplasm of the cells of the first type contains a lot of granules of Nissle's substance and few neurofilaments; the second type is characterized by the inverse correlation of these organella. The above-mentioned bipolar neurons of the bottlenose dolphin seem to belong to the first type, based on the fine structure of their perikarya.

Rather recently, small nerve cells 12.1 to 15 µm in diameter were found in the rat spiral ganglion [Ross, Burkel, 1973]. They are round or oval in shape and can be very faintly colored with toluidine blue. Their capsule is lacking myelin. The cells possess from one to four processes which differ in length. The nucleus and the nucleolus occupy an eccentric location, the nuclear chromatin is more condensed than in the ganglion's bipolar cells and Nissle's substance is more friable. In addition, the granular endoplasmic reticulum is absent. In the area of the Golgi complex, vesicles with a dense core and a diameter of 120 to 140 nm are located. Synapses in these cells are not found. It is supposed that such small multipolar neurons composing 7 to 8% from the total amount of the nerve cells of spiral ganglion represent the third type of its elements and belong to the parasympathetic cells.

The discerning of the three types of ganglion cells was also carried out according to the other morphological criteria: the neurons with myelinated processes (92%), the neurons with unmyelinated processes (4%) and the other 4% are referred to the possible auxiliary cells [Tran-Ba-Huy, Ohresser, 1976].

Therefore, in the spiral ganglion of mammals, two or three (and maybe more) types of neurons exist. However, during the analysis of the different authors' data, it remained unclear how the cell types marked out on the basis of different structural criteria correlate with each other, since none of them coincide completely with another by the quantitative indices. The correspondence between the morphological types of the spiral ganglion's cells and the characteristics of their reactions and the responses of the auditory nerve's fibers, which were electrophysiologically registered, has not been specified until now [Kiang et al., 1965; Kohlloffel, 1974].

Such characteristics as the number of neurons, the size of the neurons, the distribution density along the cochlea's turns and the correlation with the hair cell's number are important indices of the general structural organization of the spiral ganglion and the auditory function's intensity. Unfortunately, because of the difficulties with the material's processing, the data have been so far obtained only for several mammalian species (Table 7).

In mammals with different hearing specializations, the difference in the number of ganglionic neurons is much more significant than the difference in the number of receptors. Humans and dolphins, for example, possess approximately equal quantities of hair cells, while the number of spiral neurons increases 2 to 3 times in echolocating forms. On average, the ratio of neurons to receptor cells amounts to 4-5.5 to 1 in dolphins, 3 to 1 in the seal and the cat, and 2 to 1 in humans. In the bat family Vespertilionidae, it is found to be 6 to 1 [Firbas, Welleschick, 1973]. However, this correlation is inconsistent along the extent of the cochlea, and usually it is gradually reduced from 6 to 1 in the basal turn to 3 to 1 at the apical end of the organ of Corti.

The authors who studied quantitative characteristics of the spiral ganglion used different ways to express the specific density of neuron distribution along the cochlea, calculating the average number of cells accounting for half of the cochlea's turn [Guild et al., 1931], per unit length of the cochlea [Wever, 1949; Wever et al., 1971, 1972], per unit volume of Rosenthal's canal [Schuknecht, 1953] or per unit surface [Ramprashad, 1976]. With the help of these methods, the following general rule was discovered: the largest density of nerve cells falls on the second part of the basal and the first part of the middle turns, the least on the apical turn which is well correlated with the neuron to receptor ratio. Therefore, most of the ganglionic cells are restricted to the area of the cochlea, which perceives high and middle frequencies [Kellerhals et al., 1967; Moskowitz, Liu, 1972; Trevisi, 1975]. Interestingly, in the Greenland seal, which shows different frequency characteristics of hearing in the air and under water, two maxima of neuron density were discovered, one at a distance of 1 to 1.5 mm (2620 cells/mm), the other at 20 mm (2250 cells/mm) from the basal end of the cochlea [Ramprashad, 1976].

In addition to the neurons and the satellite glia, the ganglion contains a great amount of myelinated and unmyelinated nerve fibers. Most of the fibers of the first type represent the initial part of dendrites and axons of the ganglion cells itself, as well as the efferents from the CNS bounding for the auditory receptors. At least one-half of the unmyelinated fibers come from the superior cervical sympathetic ganglion and carry out the cochlea's adrenergic innervation. The general quantity of the fibers passing inside the spiral ganglion is at a maximum in the superior basal turn and gradually decreases toward the base and toward the top of the cochlea. In all its turns, the number of the unmyelinated fibers surpasses that of the myelinated fibers 2 to 6 times. In general, the number of the first ones is constant for the space of the entire organ of Corti, but their percentage to the number of general nerve fibers, reaching a minimum in the cochlea's middle part and a maximum on its top [Paradiesgarten et al., 1976].

However, no matter how unassuming the results of the quantitative study of the spiral ganglion are, they, nevertheless, demonstrate some principles of its organization which are general to all mammals, as well as regular changes of a number of characteristics in species with intensification of the auditory function.

5.2. Afferent cochlea's innervation

The principal pattern of the innervation of the organ of Corti originated in the works of the neuro-anatomy classics [Lorente de No, 1927, 1937, 1976; Lichtensteiger, Spoendlin, 1967]. The studies of the above-mentioned researchers, with the application of different, newer methods, including transmission and scanning electron microscopy, defined more exactly many details of the peripheral processes' spatial distribution and the interrelations of the nerve endings and the hair cells. However, they did not change the traditional conceptions about the separate innervation of the OHC and the IHC by the radial and spiral fibers. The description, mentioned below, was mainly based on Lorente de No's and Spoendlin's works [1957, 1966, 1969, 1972, 1975] and on Perkins's and Morest's works [1975].

Within the limits of the osseous spiral lamina, the dendrites are located in an ordered arrangement stretching in the radial direction from the spiral ganglion toward the organ of Corti. Approaching the cochlea's top in the apical turn, the radial orientation progressively gives way to a spiral one. Just after entering *habenula perforata,* the peripheral processes lose the myelin capsules and appear in the organ of Corti in the form of the common dendrites or the unmyelinated axons. Inside the *habenula,* the fiber's diameter varies from 0.25 to 0.7 μm. In the cat, an average of 30 dendrites pass through each of the *habenula*'s 2500 apertures (i.e., about 75,000 afferent fibers get into the organ of Corti). Since the number of the ganglionic cells amounts from 40,000 to 50,000, it can be concluded that in the osseous spiral lamina itself, the dendrites slightly ramify.

It is necessary to note some peculiarities of the interaction of the afferent fibers and the auxiliary elements. In *habenula perforata,* a particular satellite was discovered whose processes accompany the radial and the spiral fibers right up to the organ of Corti's tunnel. In addition, at the output of *habenula,* the dendrites for the most of their space are completely surrounded with the cells: up to the tunnel – with the inner supporting cells, then, after the tunnel – with the Deiters cells. The spiral fibers pass into the tunnel's liquid and in the Nuel's space openly on very small parts only.

Inside the cochlea, the fiber consists of specific and nonspecific radial fibers. The first ones innervate one or more lying alongside IHC, which contact the IHC located at a considerable distance from each other. The second, smaller group is presented with the spiral fibers which anatomically combine into the external spiral bundle and, widely overlapping, innervate numerous OHC. All researchers agree with the fact that the majority of the spiral ganglion's cells have the radial dendrites and only about one-tenth have the spiral ones. However, this ratio has been demonstrated in the terrestrial mammals with only ordinary hearing characteristics, and it is unknown what it is in the echolocating species. The peculiarities of the radial and spiral fibers' passage have been most extensively studied in the rat and cat.

The radial fibers have a diameter of 2 to 3 μm, but in the apical turn thicken up to 5 μm. In the organ of Corti, each fiber subdivides into a small quantity of short branches, which, usually with the help of the varicose widenings, come in contact with two IHC only; in rare instances, the fiber is connected with one or three IHC. The nerve endings are mainly concentrated around the base of the receptor cell, but sometimes they can be situated on its sidewalls approximately at the nucleus level; even thin processes that get up to the top of the hair cell can be met. It is important to emphasize that each of the neurons with the specific radial processes interacts with 1 to 3 IHC, but at the same time, each of the IHC interacts with a lot of neurons – from 20 to 26 terminal buds can be counted on its surface. In the cat, with the help of the electron microscope, the "giant" radial fibers were found, which, according to the light-optical data, ramified within the internal spiral bundle around the bases of 10 IHC [Perkins, Morest, 1975]. Possibly, the "giant" fibers partly correspond to the unspecific radial fibers of Lorente de No [1976].

The spiral fibers are slightly thinner than the radial ones, but they have a much longer length. They pass under the bases of the IHC and cross the bottom of the organ of Corti's tunnel, whereupon they form the external

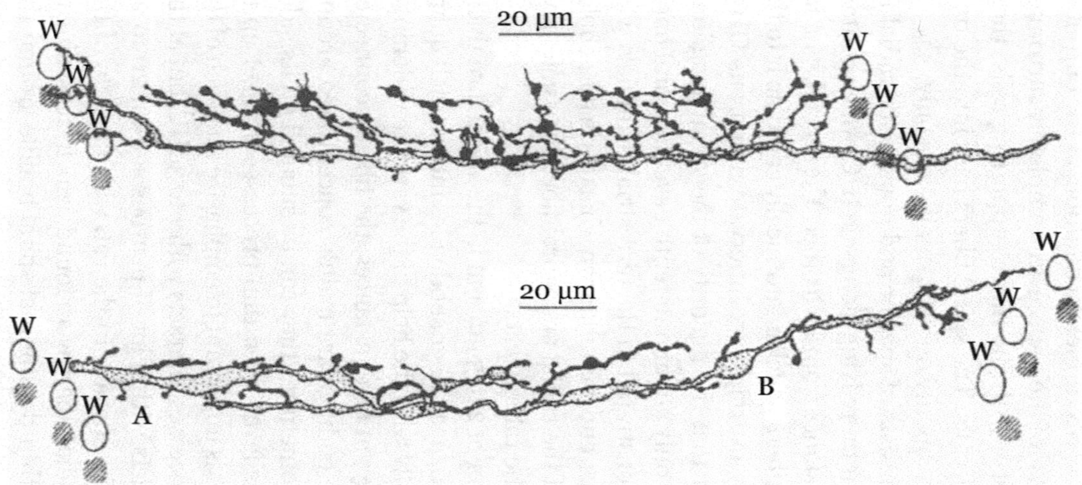

Diagram 8. Peripheral processes (dendrites) of the spiral ganglion of a neonatal cat. Golgi method in plane preparations of the apical turn of the cochlea.

Above: ending of a solitary process branching around inner hair cells. On the right and left, perikarya of hair cells marked by light circles, their bases by hatched circles, the hair drawn like a W-shaped line. Angular varicosities and short side offshoots of such a fiber end at about 60 hair cells in a rectangular area of the organ of Corti of 215 μm in length. The helicotrema is located on the right at a distance of 3.15 mm.

Below: endings of the spiral fibers from two ganglionary neurons. Both fibers branch in the inner row of the outer hair cells, contacting about 36 cells by means of varicosities and short side offshoots. Terminal branching parts of these fibers are partly non-colliding, fiber A ending below the scaled line in the place where the terminal portion of fiber B starts. However, this does not mean that sometimes only one spiral fiber innervates some individual hair cell. The terminal part of fiber A attains a length of 135 μm, that of fiber B is up to 150 μm long. The helicotrema is located on the left at a distance of 1.2 mm [Perkins, Morest, 1975].

spiral bundle. The bundle's fibers turn basally and pass along the supporting cells and the OHC's rows below their bases, sometimes stretching a distance up to 500 μm. Each fiber gradually rises to the OHC's bases and on this level gives terminal branches with the length of 85 to 365 μm. The branches possess large angular varicosities and short processes, which terminate in the OHC's basal parts (Diagram 8). Using the Golgi method, a spiral fiber was discovered in a rat, passing a distance of 670 μm from the inlet place in the middle turn to the nerve endings on the OHC. A similar strongly impregnated fiber in the apical turn of the cochlea of the cat amounted to 452 μm [Perkins, Morest, 1975].

One spiral fiber can innervate the OHC of one, two or even three rows. If the fiber innervates only one row, then it is more often the first OHC's row; if the fiber innervates two rows, then it is usually the last, second and third OHC's rows. Perkins and Morest found many nonramified external spiral fibers in the rat cochlea, whereas in the cat cochlea, all of them are ramified.

The number of the OHC innervated by one spiral fiber varies from 10 to 20; in exceptional cases, it can amount to 60. More than one spiral fiber's nerve ending is located on each OHC. The terminal ramifications of the radial and the spiral fibers end in the small widenings, which look like a bud and have some quantity of vesicles and mitochondria. As a rule, a particular nerve ending possesses more than one synaptic contact with the hair cell. Quite often, synapses plunge into the receptor's body or the outgrowths of the hair cells introduce themselves in the afferent terminals. Certain differences between the afferent terminals on the OHC and on the IHC were found: usually the first ones are smaller and contain fewer organelles [Harrison, Irving, 1964].

5.3. Efferent cochlea's innervation

The efferent innervation of the organ of Corti was described by several researchers [Maw, 1973; Rasmussen, 1942, 1946, 1953; Gacek, 1961; Jurato, 1962; Rossi, Cortesina, 1965; Smith, Rasmussen, 1968; Warr, 1975]. Later, it was comprehensively studied with the use of different methods that allowed reseaerchers to give detailed characteristics of the normal structures and of pathological changes of the efferent fibers during experimentally induced degenerations.

Summarizing all the data obtained from the literature, it can be concluded that the efferents originate in the lateral oliva, trans- and retro-olivary nuclei of the superior olivary complexes of both sides of the cerebrum, as well as in the structures adjoining this complex – in particular, in the reticular formation. Beginning from the above-mentioned areas, the axons form direct and contralateral olivocochlear tracts, which are combined into one general bundle and pass through the cerebral formations as a part of the vestibular portion of the VIII nerves, and at the periphery, only they are introduced in the auditory nerve through Oort's anastomosis [Spoendlin, 1966].

They travel into the cochlea at the level of the upper part of the basal turn and are located mainly in the intraganglionic spiral bundles, one of which rises to the cochlea's top and the other of which descends to the round window. Two distinctly different types of nerve endings were discovered [Engström, 1958]. Type I is characterized by small nerve endings with a slight quantity of vesicular structures. These nerve endings have a small contact area with the receptor elements. Type II is characterized by noticeably larger nerve endings with a large contact area. These nerve endings contain a lot of vesicles with diameters of 30 to 40 nm, and a small number of bigger vesicles with diameters of 60 to 80 nm with a dense core and numerous mitochondria. After the cutting of the olivocochlear tract, the intraganglionic bundle degenerates as well as the nerve endings of the second type, but apparently the degeneration of these terminals does not have an effect on the receptor cells at all.

The quantitative assessment of the efferent cochlea's innervation in cats revealed the following. In the basal turn on each OHC of the first and the second rows from 6 to 10 efferent and from 5 to 8 afferent terminals end. At the top line, the ratio starts to change in favor of the afferent terminals, and as early as at the end of the upper turn efferent nerve endings disappear on the OHC of the third row and in the first two rows their quantity decreases: in the middle row, 3 efferents to 8 afferents remain. In the cochlea's top, the efferents can be met on the cells of the first type only. Another picture is observed in the same cochlea's area during the comparison of the efferent and afferent terminal's contact areas. One efferent nerve ending can have a contact area 50 times larger than that of an afferent nerve ending; this is why the total area of the efferent's contacts with the hair cells exceeds that in the afferent synapses several times. Even in the cochlea's top, where there are few efferent nerve endings, these areas are at least equal.

If the average number of the efferent contacts for one OHC were equal to three, their general number in the cat's cochlea would amount to 40,000 at a primary number of 500 efferent axons of the olivo-cochlear bundle. The basic areas of ramification are situated in the osseous spiral lamina; there the efferent's number increases from 500 to 3000, and after *habenula per-*

forata, in front of the tunnel, from 3000 to 8000. In the tunnel, the ramifications are not found, and a further ramification from 8000 to 40,000 occurs after its intersection somewhere on the OHC's level.

In addition to the efferents of the external spiral bundle, which terminate on the OHC, the fibers of the internal spiral bundle, most of which are efferent, terminate on the IHC.

The axons of the efferent fibers, as well as the afferent fibers, lose their myelin capsules in the *habenula perforata*. In the organ of Corti, the typical difference of these fibers from the ganglionic cell's dendrites is the content of microtubules and multitude of neurofilaments.

With the help of the Golgi method, it was shown that, in the cochlea of young rats and cats, the individually impregnated efferent axons possess a great number of varicosities and radially project on the OHC's and the IHC's endings. In the simplest case, the thin beaded axon ends with a very small amount of branches below the IHC in the main. In a more complicated case, the thicker axon ends with numerous branches below the IHC and the OHC. The numerous efferent fibers send branches to the IHC as well as to the OHC [Perkins et al, 1975].

Interesting data on the cochlea's efferent innervation were obtained on the *Saimiri sciureus* [Nakai et al., 1974]. In adult forms, two weeks after the cutting of the intersecting olivocochlear tract on the level of the bottom of the IV ventricle of the brain, a strong degeneration of the organ of Corti was found. With the help of the electron microscope, the following was revealed within the nerve endings close to the OHC: the changing of the number of vesicles, intensive vacuolization, swelling of mitochondria, ruptures of membranes and appearance of the osmosephilic substances. In addition, in the lower part of the OHC's area, the presence of the nerve endings which degenerated after the tract's cutting was revealed. The relative quantity of the terminals located in the organ of Corti and connected to this tract varies from 50 to 90% in different parts of the cochlea, and their maximum number is usually concentrated in the cochlea's basal parts (Diagram 9).

It is of utmost importance to consider the interactions of the efferent and afferent fibers. On coming out of the *habenula perforata,* both types of fibers are in mixed, closely packed bundles, whereas in the osseous spiral lamina, they are still separated by satellite cells even after the loss of the myelin capsule. The area where both types of fibers adjoin closely to each other for a relatively long distance is the internal spiral plexus. At the entrance to the tunnel, the fibers begin to divide; the efferents go on as the superior tunnel radial fibers, which are suspended in the tunnel's liquid,

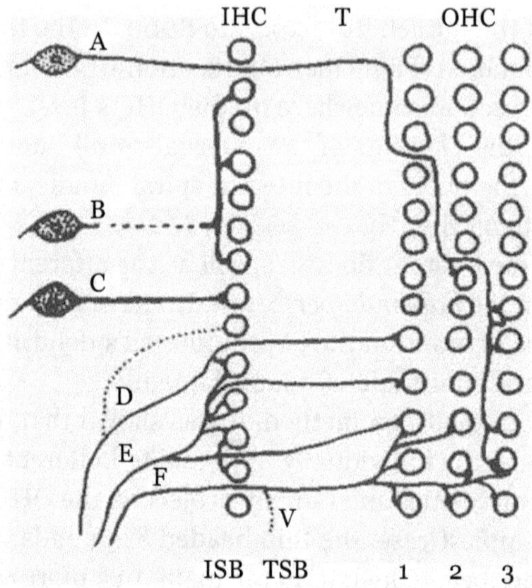

Diagram 9. General presentation of the spiral ganglion cell types and fibers innervating the organ of Corti of a young rat or a cat (basal turn of the cochlea, horizontal plane).
A — a neuron of the spiral ganglion, the dendrite of which crosses the tunnel (T) and forms collateral endings on the outer hair cells (OHC) of rows 2 and 3; B — a gigantic fiber going to a group of inner hair cells (IHC) and forming collaterals in the inner spiral bundle (ISB). For this fiber, there is no traced connection to any definite neuron of the spiral ganglion; C — a neuron of the spiral ganglion sending a process (radial fiber) to the base of the inner hair cell; D — an efferent fiber, one or more branches of which are directed to the intraganglionary spiral bundle (IGSB). Such a fiber forms a few endings with the inner hair cells and a few endings with the outer hair cells of the first row; E — an efferent fiber undergoing bifurcation in a spiral lamina and forming multiple collateral endings around the bases of the inner hair cells and a few separate endings around the outer hair cells of the first row; F — an efferent fiber which forms numerous branching endings around the outer hair cells of all three rows and sometimes sends branches (Y) to the tunnel spiral bundle (TSB) or to the inner spiral bundle [Perkins, Morest, 1975].

and the afferents are located under them and noticeably plunge into the bases of the supporting cells. In the area of the cat's OHC, the efferents and the afferents plunge separately into the processes of the Deiters cells, whereas in humans, both types of the fibers in this cochlea's area may connect with each other. Both types of the fibers come in contact with each other during the formation of the synapses on the OHC as well. In these areas, the places with a denser membrane and a vesicle's aggregation, which is typical of the synaptic zones, can be found. Quite often, one efferent ending or the fiber's widening forms such synapses with several dendrites of the ganglionic neuron. However, on the OHC's level, the efferent's synapses with the receptor cells prevail, and in the IHC's area there is an inverse situation – the efferent axon interacts there with a great amount of dendrites near the base of the receptor cell. However, the most important part of the interaction of the dendrites of the ganglionic cells and the olivocochlear axons is the internal spiral bundle. The most essential peculiarities of afferent and efferent innervation of the cochlea are presented in Diagram 9.

Anatomically, both types of fibers form four spiral bundles in the cochlea, but the specific radial fibers take an extremely small part in it: the internal ganglionic bundle, which rises in the apical direction among the spiral ganglion's cells; the internal spiral bundle, whose fibers are the terminals of the internal ganglionic bundle; the tunnel spiral bundle, which consists of very thin fibers; and the external spiral bundle [Held, 1897; Lorente do No, 1937].

The afferent and efferent fibers behave in many respects in the opposite way; the ganglionic cell's dendrites spread radially over the IHC and spirally over the OHC, whereas the efferent axons of the olivocochlear tract spread predominantly radially over the OHC and spirally over the IHC. The overwhelming majority of the afferent fibers innervates the IHC, and the minority of them innervates the OHC; the inverse ratios are shown in the efferent fibers.

Each of the OHC is innervated by several different neurons of the spiral ganglion, and one such neuron sends collaterals to many hair cells. There is a greater number of endings of different neurons on the IHC, but most of them do not ramify and innervate from one to three IHC.

The synaptic contacts exist between all hair cells and the afferent and the efferent fibers as well as between the fibers of both categories which closely adjoin to each other mainly in the internal spiral plexus, the area below the IHC.

Within the organ of Corti, the afferent and efferent fibers differ in their ultrastructure and in the number of endings.

5.4. Adrenergic cochlea's innervation

In 1962, a foolproof method for detecting the nerve system's structures that contain monoamines was suggested. The method was based on the ability of catecholamine and serotonin to form fluorescent condensation products during their reaction with formaldehyde. With the help of this method, the adrenergic fibers in the mammal's cochlea were revealed [Spoendlin, 1966; Densert, Flock, 1974]. The intensive fluorescence was observed around the basilar artery and the artery of labyrinth with its main branches, as well as in the area of the *habenula perforata,* which is not connected with vessels. The adrenergic perivascular network disappears as early as the level of the cochlea's axis and in the more peripherally located vessels. In different species, noticeable differences in the intensity of the perivascular network's fluorescence were revealed. According to this criterion, the density of the innervation of the inner ear's vessels in rats and guinea pigs is higher than in cats; however, in the peripheral vessel's ramifications, this difference disappears. The exceptionally rich adrenergic innervation was found around the rabbit's spiral vessel of the tympanic lip [Densert, 1974]. The most significant number of the fluorescent fibers not connected with vessels was revealed in the peripheral area of the osseous spiral lamina, directly in front of the *habenula perforata.* In a rabbit, a lot of adrenergic terminals around the radial nerve fibers were found in the places where they lose the myelin capsules and where, presumably, in these fibers the action potential appears. A large number of the adrenergic fibers is dispersed among the other cochlea's fibers (Diagram 10).

It was shown that the adrenergic fibers descend from the superior cervical sympathetic ganglion and, apparently, are not myelinated. They cross

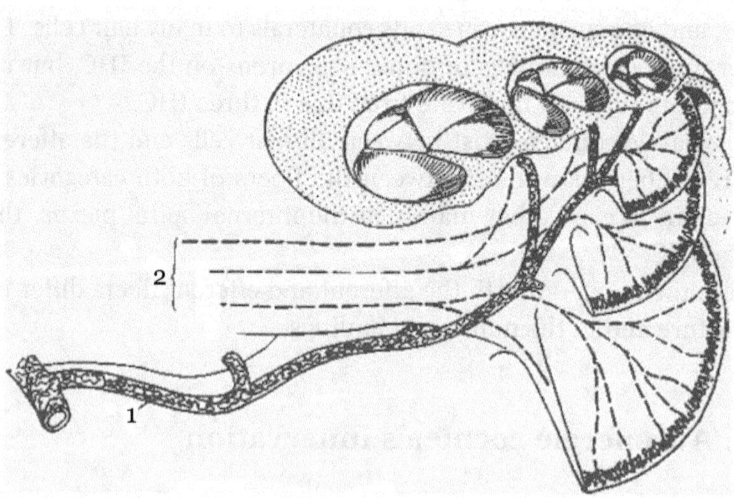

Diagram 10. Presentation of two types of adrenergic innervation of the cochlea. 1 — periarterial nervous plexus; 2 — adrenergic innervation revealed at the periphery of the auditory nerve [Spoendlin, 1966].

the spiral ganglion mainly without making contacts with its neurons [Densert, 1974; Paradiesgarten et al., 1976]. Currently, two systems of the adrenergic innervation of the inner ear have been observed: the perivascular, which disappears as early as the periphery of the cochlea's axis, and a more significant one which is formed by the adrenergic fibers independently from the vessels. Such fibers are located on the periphery of the auditory bundles and form the plexus in the area of the *habenula perforata.* The functional role of the cochlea's adrenergic innervation is still not clarified; however, it can be supposed that its influence essentially affects the processes of perception and conduction of sound stimuli.

5.5. Structure of the auditory nerve in some mammals

The auditory nerve is a complex fibrous system which consists of several components. Its main part is formed out of the axons or the central processes of the spiral ganglion's cells. They transmit the specific information to the primary acoustic centers, which are located in the CNS [Echandia, 1967]. After leaving the cell's body, the axon passes inside Rosenthal's canal, where it is usually impossible to track it at a long distance among other processes. Sometimes in young animals the axon is subdivided into several branches, which is not observed in the adult individuals [Perkins et al., 1975].

It was shown that the fibers from the apical turn pass in the center of the nerve, but beginning from the basal part, they are located on the periphery of the nerve trunk [Sando, 1965]. This fact is co-coordinated with the electrophysiological data on the tonotopical organization of the auditory nerve. Recently, new data on the fine anatomical structure of the auditory nerve were published [Gagloeva, 1976]. In the cat, monkey and human, the nerve, according to the character of the fiber's location, is divided into the dorsomedial part, which goes from the second and the third cochlea's turns, and the ventrolateral part, which goes from the basal turn [Gacek, Rasmussen, 1961]. In the comparative anatomy's row, the dorsolateral portion appears only in animals with a true cochlea, and in higher representatives its bundles are more pronounced and more strongly torsioned [Moskowitz, Liu, 1969].

The thickness of the auditory fibers in terrestrial mammals is 2 to 4 μm on average, while in a bottlenose dolphin, according to our measurements, the average is 5 to 7 μm. Out of 50 cross sections of the bottlenose dolphin's auditory fibers, 6 showed a diameter of 0.9 to 4 μm, 8 showed 4 to 5 μm, 11 showed 5 to 6 μm, 11 showed 6 to 7 μm, 12 showed 7 to 12 μm and 1 showed

13 to 16 μm. These values were obtained during the measurements of the auditory fiber's electronic diagrams and correspond only to the axoplasma's diameter. The measurements of the thickness of the myelin capsule revealed that it does not directly depend on the axoplasma's size and varies within 0.7 to 2.5 μm [Bogoslovskaya, 1975].

The study of the bottlenose dolphin's auditory nerve using longitudinal shears showed that the diamters of the fibers can change noticeably. Thus, the local narrowings from 9 to 16 times are quite usual. For example, the same fiber changes its diameter in the following way: for the space of 10 μm, the diameter widens from 0.42 μm to 6.7 μm; after 4 μm, it decreases to 0.83 μm; and in 4 μm more, it increases to 7.3. It is important to note that during all variations of the diameter, the thickness of the myelin capsule stays constant. In the longitudinally sheared fibers, small outgrowths of axoplasma, which are accompanied by a myelin capsule, are common. They can be the beginning of the internal collaterals of the auditory nerve and the collaterals outgoing to the formation of the anastomoses with the vestibular and the intermediate nerves. As reported by Gagloeva [1973], such connections between the three nerves are completely formed in humans, and in the postnatal life they increase.

It is possible to bring the results of some model studies of the nerve elements [Bogoslovskaya et al., 1973], which showed that the fiber with heterogeneous geometry should necessarily change the frequency of the impulses passing in it and thereby actively influence the character of the transmitted information, to the explanation of the role of the local changes of the auditory nerve's diameter in a bottlenose dolphin.

There is one more fact about the inner structure of the mammalian auditory nerve which is interesting. With the help of the impregnational method in a rat, the large neurons located along the whole length of the auditory nerve were found. Such interstitial cells form synapses with the collaterals of the auditory fibers, and their axons ascend to the cochlear nuclei. Similar neurons are located in the mouse's auditory nerve, but they are absent from the bat and cat [Harrison et al., 1962].

In addition to the ganglionic cell's axons, the auditory nerve contains efferent olivocochlear fibers descending to the cochlea, as well as special ascending fibers which are similar to the auditory ones. These fibers travel to the brain together with the auditory ones, but they turned out to be vestibular-posterior ampullar and saccular ones [Cajal, 1909; Kappers, 1921]. In the brain, they pass through the primary auditory centers located directly on the surface of the rope-like body and terminate in the lateral process of the superior vestibular nucleus, in the Deiters nucleus and in the reticular formation. In the auditory nerve, the adrenergic fibers, which get to the cochlea from the superior cervical ganglion, are also present. The number of the auditory nerve's fibers and the ratio of its components in different mammals have been insufficiently explored. It is only known that in the cat, approximately 1/100th of all fibers (about 500) represent the olivocochlear bundle. It can be supposed that the variations in the general number of fibers are very substantial, if compared, for example, in the rat, dolphin and human. In general, however, the brain of all mammals, even of the echolocating ones, receives much fewer auditory fibers than visual and tactile fibers. Taking into account this fact, together with the relatively fewer number of the auditory receptors, it can be concluded that the important characteristics of the signals are perceived by the auditory system mainly in series (time scanning of the image), and by the visual or the tactile systems – simultaneously (spatial scanning of the image). Therefore, a small number of the peripheral auditory channels, which in the receptor part are stable enough in most animals, is able to carry out the processing of the same body of information as it happens in the other systems, where the similar processes are performed by hundreds of thousands and even millions of receptors that work simultaneously. Probably the colossal significance of the temporal factor determined the development of a special efferent path which leads from the secondary acoustic centers of the brain directly to the receptors and to the ganglionic neuron's dendrites which innervate these receptors. The olivocochlear path is very closely connected with the evolutionarily young part of the organ of Corti, the system "outer hair cells – spiral dendrites of the ganglionic neurons." There are reasons to suppose that the elements of this system should be considered with the analysis of the acoustic signals of long duration [Shmelev, 1973].

PART II

PERIPHERAL PART OF THE AUDITORY SYSTEM OF MAMMALS IN PRENATAL ONTOGENY

Chapter 1.
Development of the auditory organ in terrestrial, semi-aquatic and aquatic mammals

Having studied the features of structural organization of the outer, middle and inner ears in representatives of various ecological groups in postnatal ontogenesis, we found it necessary to carry out a comparative and embryological research of all parts of the peripheral auditory system in prenatal ontogenesis in representatives of terrestrial, semi-aquatic and aquatic forms possessing low-, middle- and high-frequency hearing [Solntseva, 1982, 1983, 1985a, b, 1986a, b, c, 1988a, b, 1989, 1990a, 1992, 1993].

The research allowed us not only to study structural features of the auditory organ of investigated species in more detail, but also to find out the peculiarities of its formation linked to acoustic properties of their habitats, as well as to find an explanation of the appearance of some structural adaptations in semi-aquatic and aquatic forms, having defined a stage of formation [Solntseva, 1999a, b, c].

As a result of long-term researches of pre- and postnatal development of acoustical and vestibular structures in terrestrial mammals and in humans, a great number of fundamental works carried out on morphological, electron-microscopical and physiological levels were accumulated [Köllicker, 1861; Boettcher, 1869; Alexander, 1900; Held, 1909; Brunner, 1934; Cooper, 1948].

The study of the auditory organ in marine mammals (cetaceans, pinnipeds), representing an absolutely specific direction of placental animals' evolution, was started more than three centuries ago. However, these works were mainly carried out at the anatomical level and were fragmentary in nature [Eschricht, 1849; Denker, 1902; Huber, 1934; Clarke, 1948; Reysenbach de Haan, 1957a, b; Fraser, Purves, 1960; Fleischer, 1973a, b; Ramprashad, 1975]. In addition, there were no data concerning the development of the cetacean and pinniped auditory organ during an early prefetal period, according to Schmidt's [1968] periodization (i.e., beginning with the formation of an acoustic vesicle, or the stage of a bud, up to the end of the formation of basic anatomic structures of the auditory organ).

Becaused acquiring embryonic material of marine mammals was often very difficult, the organs of hearing and equilibrium in a great group of mammals remained undescribed for a long time and dropped out of the general study profile of the development of these organs in mammals as a whole. All this prevented the solution of many questions concerning structural organization of the peripheral auditory system in various groups of mammals and prevented the determination of the general developmental patterns of the peripheral auditory system in mammals as a whole.

The most important data concerning adaptive and evolutionary changes of the auditory system could be obtained only by comparative studies of embryogenesis of the system in a wide set of species that are phylogenetically close but ecologically different, as well as in species that are phylogenetically distant but have similar ways of life [Severtsov, 1939].

For the first time, we carried out a comparative and embryological research of the peripheral part of the auditory system using unique embryonic collections on pinnipeds and cetaceans, which allowed us to study the structural organization of this part in more detail, to reveal the features of similarity and distinction in formation of hearing and equilibrium organs for different stages of development, and also to determine the stages of formation of structural adaptations revealed by us earlier.

Based on the results obtained, the developmental patterns of the structures of the outer, middle and inner ears in mammals belonging to various ecological groups could be established.

While studying the development of various organs, including hearing and equilibrium organs, many researchers compared the developing structures with the length of embryos. Such an approach is misleading in the comparative analysis of prenatal development of the auditory organ in a wide set of species as their terms of pregnancy and the length of embryos for similar stages of embryogenesis are sharply different. However, in experimental embryology, conceptions about equivalent stages of development [Otis, Brent, 1954] were widely known.

To have an opportunity to compare rudiments of different species, a special research for normal development was undertaken on some laboratory animals by a group of scientists; as a result, the stages of development with common features for different species were marked out [Dyban et al., 1975].

With the intent to carry out an adequate comparison on the formation of the peripheral part of the auditory system in various mammalian species, we applied a principle to compare the developing structures of the outer, middle and inner ears at similar stages of development using known tables on normal development of laboratory animals [Dyban, et al., 1975]. In addition to this, the stages of structural formation of the outer, middle and inner ears were compared by us with the stages of mesenchymal tissue replacement by an embryonic cartilage.

As in the majority of mammals, a similarity in a sequence of development of the structures of the outer, middle and inner ears was revealed; for convenience of description, we used the known developmental stages of some terrestrial species [Dyban, et al., 1975].

The results of the comparative and embryological research demonstrated that the development of the peripheral part of the auditory system in all investigated species occured in an early prefetal period (i.e. during the formation of a cartilaginous skeleton), beginning with the formation of an acoustic vesicle (stage 13) up to the end of anatomic formation of the basic structures for the auditory system's periphery (stage 20) [Solntseva, 1983; 1985à, b; 1986 a,b,c; 1988 à, b; 1990 a,c].

Here we will examine step by step the development of the auditory organ in terrestrial, semi-aquatic and aquatic mammals.

1.1. Terrestrial mammals

1.1.1. Gray rat (*Rattus norvegicus*), guinea pig (*Cavia porcellus*), domestic pig (*Sus scrofa domestica*)

The study was carried out on the embryos of the rat, guinea pig and domestic pig, beginning with the stage of development of an acoustic vesicle (stage 13) to the complete formation of the organ (stage 21).

A pair anlage of a membranaceous labyrinth is marked at the stage of 2 to 3 pairs of somites [Wilson, 1914]. At the stage of 6 to 9 pairs of somites, the membranaceous labyrinth's bud represents an acoustic placode [Kappers, 1941]. Further, at the stage of 14 to 15 pairs of somites, an acoustic pit is formed, from which an acoustic vesicle develops at the stage of 20 pairs of somites (13th stage of development), which passes into an endolymphatic duct without any special borders. At this stage of development, an auditory ossicles' bud in the form of the mesenchyme's thickening appears.

In all investigated species, a subdivision of the acoustic vesicle into superior and anterior parts occurs at stages 14 to 15. Both parts are surrounded by mesenchyme. Further, a vestibular apparatus is formed from the superior part, and a formation of the cochlear canal begins from the anterior part (Fig. 28a, b, c), in which the base, composed by a columnar epithelium, and the roof, consisting of a cuboidal epithelium, are distinct (stage 16). In the superior part of the acoustic vesicle, a differentiation of the vestibular part into semicircular ducts and sacks is observed. In the cochlear part, only a lengthened canal is marked out; it is an endolymphatic duct, from which a cochlea will be formed later. At this stage, a subdivision of an acoustical nerve into the vestibular and cochlear branches and their ganglia is observed [Solntseva, 1990a].

At stage 16, in all mammals possessing an auricle (pinna), the first branchial cleft, widening gradually, forms a deepening, on the edge of which small protuberances start to develop, which, at stage 17, merge and form a uniform mesenchymal bud of an auricle (Fig. 29 a, b, c). At stage 16, a cartilaginous part of an external auditory meatus starts to develop from the lateral walls of the first branchial cleft. The osseous part is formed later; in immature-born species (rat, guinea pig, domestic pig), it occurs during early postnatal development. At the 16th stage, contours of the separate auditory ossicles appear, which are still continuously connected with each other. At the 16th stage, the tympanum in the form of a narrow channel located below the auditory ossicles' bud starts to develop. At this stage, there is a bud of the tympanic membrane, which is formed on the place of the contact between the endoderm of the pharyngeal recess and the ectoderm of the first branchial cleft.

At the 17th stage, each of the auditory ossicles' buds represents an independent formation; their base is formed by the immature precartilaginous tissue (Fig. 29 a, b). At this stage, the auditory ossicles' bud is isolated from the bud of the pyramid of a temporal bone. Further, these buds draw together and the stapes falls within a deepening of the pyramid. At the 17th stage, the tympanic membrane is thick and friable in all investigated species.

At the 16th stage, the cochlear canal starts to twist spirally in the inner ear, forming a lower, or basal, cochlear turn surrounded by an aural capsule consisting of a condensed mesenchyme. The development of the basal cochlear turn outstrips considerably the development of the medial and apical turns. In the cochlea, the basilar membrane is the first to form and a vascular stria is the last. The process of the formation of the cochlea's turns is accompanied by the formation of the cochlear nerve, the dorso-ventral part of which goes to medial and apical turns after their formation. The ventrolateral part of the nerve heads to the basal cochlea's turn.

At the 17th stage, the medial turn is formed and the cochlea formation is completed with an appearance of the apical turn (18th stage). At this stage, 3.0 turns of the cochlea are formed in a rat, 4.5 turns in a guinea pig and 3.0 turns in a domestic pig (Fig. 30a, b, c).

At stages 18 to 19, the pinna acquires more distinct contours in the outer ear, and its small curl appears. The auditory meatus lengthens and looks like a short, direct channel opened throughout its whole extent. By the end of the 19th stage, the lateral part of the auditory meatus is filled with epithelial cells. And only at the 21st stage of development, the external auditory meatus is completely filled with the cells of epithelium. At the 18th stage, a tympanicum is formed in the middle ear, and at the 19th stage, a turning of the tympanum around the sagittal and frontal axes of the body of the animal occurs. At this stage, formation of the elements of the auditory ossicles, which are considerably increased in size and are plunged deep in the tympanum, occurs. In the malleus, a head, neck, and handle are well expressed. In the incus, a body and both processes are formed. The stapes is subdivided into crura and a footplate. At this stage of development, the auditory ossicles are so enlarged that their size is only 4.5 to 5 times smaller than the size they reach during the definitive period (Fig. 30a, b, c). The process of cartilaginification starts in the center of each bud of the auditory ossicles and gradually extends to their periphery. The auditory ossicles are surrounded by a perichondrium, which consists of small, flat cells – chondroblasts – which do not have distinct borders. Due to the perichondrium, the locations of junctures of the auditory ossicles are well expressed.

The formation of the features connected with interposition of the auditory ossicles in the tympanum is marked at stages 18 to 19 of development (Fig. 30b). At these stages, the auditory ossicles make a turning around the sagittal and frontal axes of the body of the animal. The stapes is located more caudally relative to the malleus and incus. It is gradually unwrapping, after which its base turns out to have been revolved dorsally.

At the 18th stage, a cartilaginification of the aural capsule is marked in the inner ear. Up to the 18th stage, the cellular elements of the organ of Corti are at the same stage of differentiation in all of the cochlea's turns. At the 18th stage exactly, the columnar epithelium of the cochlear canal starts to move apart; therefore, the axial and lateral thickenings, from which the structures of the cochlear canal develop, are formed. In the place of the future organ of Corti, a certain number of oblong epithelial cells are noticeable. Of all the membranes of the cochlear canal, Reissner's membrane is the first to develop, and the vascular stria is still absent. The size of the cochlea increases at the subsequent stages of development.

At the 19th stage, a horseshoe-shaped cartilage surrounding a cavity of the auditory meatus is formed around an external auditory meatus. In the middle ear, a process of the tympanum's formation and the auditory ossicles cartilaginification continues. Replacement of a mature precartilaginous tissue by a primary cartilaginous tissue occurs non-simultaneously. So, at a given stage in the rat, the base of the auditory ossicles is formed by mature precartilaginous tissue. By this time, cartilaginification of the malleus has already been completed in the guinea pig, while in the incus and stapes, this process has only begun. In the domestic pig, the replacement of mature precartilaginous tissue by primary cartilaginous tissue has affected the malleus only, while the incus and stapes are still formed by mature precartilaginous tissue. However, in embryos of all investigated species, the replacement of mature precartilaginous tissue by embryonic cartilage finishes completely by stage 20.

At the 19th stage, the formation of the modiolus is marked in the inner ear, which is formed by dense connective tissue, with numerous blood vessels located among its bundles. At this stage, differentiation of the cochlear canal's elements and the cells of the organ of Corti has begun. Flattening of the cuboidal epithelium cells and loosening of the connective tissue adjacent to this epithelium occurs. In this place, a formation of the vestibular scala begins. The cells of the organ of Corti are differentiated from the cells of the columnar epithelium, and they separate. Differentiation of the epithelium of a vascular stria, which consists of 6 to 7 layers of cells, is marked. The spiral lamina is cartilaginous.

At the 20th stage, a replacement of cartilaginous tissue by an osseous one in the form of the small centers of ossification in tympanic bone and integumentary bones of a cranium occurs. The ossification of the auditory ossicles occurs later, beginning at the 21st stage of development.

In the cochlear canal, a differentiation of a spiral limb, a tectorial membrane, occurs. A vascular stria is formed by the undifferentiated epithelium. The future spiral incisure consists of a multi-row, high-columnar epithelium. The tunnel has not yet been formed. The size of the spiral ganglion's neurons, as well as the cochlear and vestibular branches of the auditory nerve, is enlarged. At the subsequent stages of development, the differentiation of the elements of the cochlear canal and the cells of the organ of Corti continues.

1.1.2. Greater horseshoe bat (*Chiroptera: Microchiroptera-Rhinolophus ferrumequinum*)

The Chiroptera order totals up to 1000 species which form two suborders: Megachiroptera and Microchiroptera.

A greater horseshoe bat is the representative of the Microchiroptera suborder. Inhabiting dark caves, these mammals have kept their connection with the terrestrial world. Living in constant darkness has led to a reduction of the visual apparatus. However, the other organs of sense (the sense of smell, skin sensitivity, etc.) do not carry any information about the barriers which can be met on the way of bats. Only hearing has allowed these mammals to cope with this problem. The sound reflected from the subjects gives an animal a signal in advance about all the obstacles appearing on their way. Therefore, through evolutionary processes, the ability for echolocation in bats has been developed.

The auditory system of a greater horseshoe bat is adjusted for a perception of high-frequency signals up to 80 kHz [Airapetiantz, Konstantinov, 1974]. In connection with echolocation in this species of bats, the features in the structural organization of the outer, middle and inner ears, which distinguish the bats from other mammals, appeared.

The outer ear of a greater horseshoe bat includes an auricle (pinna) and an external auditory meatus. The pinna is noted for its huge size and, in contrast to other species of bats, lacks a tragus. The auditory meatus has a flared shape.

As in other terrestrial species, the development of an auricle begins at the 16th stage in the form of small prominences located on the edge of a deepening formed due to a widening of the first branchial cleft. At the 17th stage, a fusion of these prominences occurs, and a uniform mesenchymal bud of the auricle is formed. At the beginning of the 18th stage, the auricle acquires clearer contours. By the end of the 19th stage, an external auditory meatus starts to fill up with epithelial cells. This process finishes by the 21st stage of development [Solntseva, 1990a; 1993].

In the middle ear of a greater horseshoe bat, the following features are typical: the presence of a small tympanic membrane and its vertical disposition; the reduction of the auditory ossicles' size, their thinning and the presence of deep hollows in them; the joining of the malleus and incus with each other at sharp angles in the area of an incudomalleal articulation; a compact stapes with crura that are thickened and form a small inter-crura aperture; the long process of the malleus knitted with a wall of the tympanic bone; the muscles of the middle ear considerably increased in size. These features in the structure of the middle ear provide for the transmitting of ultrasound signals.

As in other mammals, the bud of the auditory ossicles appears at the 13th stage of development in the form of the mesenchyme's thickening. At the 16th stage, the contours of the auditory ossicles become apparent, and at the 17th stage, each bud of the auditory ossicles represents an independent formation. Their base is formed by immature precartilaginous tissue.

At the 18th stage, the elements of the auditory ossicles are formed. The base of the auditory ossicles is formed by mature precartilaginous tissue. By the end of this stage, the mature precartilaginous tissue has already been replaced by embryonic cartilage.

The tympanum is formed at the 16th stage in the form of a narrow channel located below the auditory ossicles' bud. At the 18th stage, tympanic and periotic bones are formed, and at the 19th stage, a turning of the tympanum around the sagittal and frontal axes of the prefetal body occurs.

Formation of the features connected with the interposition of the auditory ossicles in a tympanum occurs at stages 18 to 19.

The bud of the tympanic membrane is formed at the 16th stage, and by the 17th stage, it is thick and friable. Its significant thinning is marked at the 18th stage; the tympanic membrane acquires a three-layer structure and lays almost vertically on the lateral surface of the middle ear's cavity.

The cochlea of a greater horseshoe bat is huge; it is formed by 3.5 turns (Fig. 31a). In the cochlea's structure, certain features aimed at the perception of high-frequency signals are marked. A significant development is reached by a lower, or basal, turn of the cochlea, as it is directly connected with the perception of high frequencies.

For the first time, Griffin [1958; 1967] noticed that in bats, the round window comes in contact with the liquid of the inner ear in a quite different place than it does in other mammals (i.e. not at the end of the cochlea, but almost at a millimeter further than its first turn). Later on, he revealed that the basilar membrane in bats is supplied with two additional thickenings in that part of the cochlea which is connected with the perception of the signals, having important biological significance for these animals. The basilar membrane is narrow and thin and is very rigidly fixed between the primary and secondary osseous spiral laminae.

At the 13th stage of development, the acoustic vesicle is formed, as well as in other mammals, which at stages 14 to 15 is subdivided into two parts. The vestibular apparatus is formed from the superior part, and the cochlear canal is formed from the inferior part (Fig. 28c).

At the 16th stage, the cochlear canal starts to twist spirally and the basal cochlea's turn is formed. At the 17th stage, the medial turn is formed, and at the 18th stage, the apical turn and a half turn. By the end of the 18th stage, the cochlea is anatomically formed with 3.5 turns.

Further along at the 19th stage, formation of the elements of the cochlea's canal and differentiation of the cells of the organ of Corti occur, beginning from the basal turn of the cochlea and gradually extending to the turns located above (Fig. 31a-g). Therefore, in all turns of the cochlea, a different degree of anatomic formation of the cochlear canal and the cytodifferentiation of the organ of Corti is noticed. Among the structures of the cochlear canal, a Reissner's membrane is the earliest to form, and a vascular stria is the latest.

Thus, the development of the auditory organ of a greater horseshoe bat reveals a great similarity with terrestrial mammals. All structures of the outer, middle and inner ears are formed from homologous rudiments, in a certain sequence and at similar stages of development.

1.2. Semi-aquatic mammals

In semi-aquatic marine mammals belonging to various orders (pinnipeds, carnivores – sea otter), there is a strong connection with two habitats: water and air. This has caused an appearance of the adaptive features in the structural organization of the auditory organ for functioning in habitats with different physical properties.

1.2.1. Bearded seal (*Pinnipedia, Phocidae-Erignathus barbatus*)

A bearded seal is the largest representative of true seals, with a body length of up to 240 to 250 cm. These seal species specialize in benthos; therefore, their way of life is connected mainly with small depths up to 100 m. Usually, a bearded seal does not overstep the limits of a continental shoal. As it has been shown by different authors, it gets food from a bottom, more often at depths of 50 to 60 m. The bearded seal from the White Sea eats basically crustaceans, gastropods, bivalves, worms and fish. Usually, in the morning, all animals are in the water and are occupied with seeking for food; in the afternoon, their activity is considerably reduced, and they get out on the ice. From the end of May until June, these animals stay on the ice longer than during the other months [Potelov, 1966; 1967].

A bearded seal is usually considered to be a settled animal that does not undertake mass and long migrations [Ognev, 1935]. According to the opinion of other researchers, it is hardly possible to consider the bearded seal to be a settled animal, as allocation of its populations is subject to appreciable changes depending on the seasons. It is connected to a great extent with changes in the ice conditions, with distribution of forages, with a choice of places for reproduction, moult and coastal rookerys. According to V.A. Potelov's data, in the second half of summer and in autumn, the basic livestock of the bearded seal is allocated in a near-shore zone and in shallow waters. With the approach of winter and the appearance of an ice cover near the coasts, the animals move away from the coasts. During all of winter, until spring, they stay outside of the fast ice (coastal ice) among drifting ice. From the second half of summer, those animals who moved away from the coasts with the appearance of ice return to them again.

Puberty comes at the age of 6 years in females, and in males at 7 years. In the White Sea, the newborns have been met in the last decade of March, in the second and third decades of April, and in the first decade of May. Whelped females are dispersed over large spaces of the drifting ice. As ice floe fragments can serve for a newborn as an ice substratum, there is an assumption that the birth of pups can occur not only on the ice floe, but also in water.

The outer ear of the adult individuals of the bearded seal is represented by the external auditory meatus only. The auricle is reduced. In the distal part, the auditory meatus is double bent and like a bent tube goes down to the tympanic bulla. The length of the meatus amounts is about 7.0 to 7.5 cm. In the area of a bend, the lumen becomes narrower; its size increases at the proximal part, and the form of its microscopic section acquires the shape of an oval.

Periotic and tympanic bones form a uniform complex separated from the whole occipital part, which only in the rostral part knits with the cranium. The tympanum is of a spherical form and is covered by a thick mucous membrane, including numerous blood vessels in the middle layer.

For the auditory ossicles of pinnipeds, an intraspecific variability is typical, distinguishing the bearded seal from the other representatives of true seals. The malleus has a small head with two articular surfaces located at almost right angles. On each of the surfaces, a small deepening is present. The handle is short and thickened. Its widened part throughout its whole length is attached to the tympanic membrane of a small diameter, a bit elongated in the form of a cone. The incus is large and heavy. Its weight exceeds that of the malleus 2 to 2.5 times. The long arm of the incus has the form of a gutter. In the distal part, it gets thin and turns into a lenticular arm. The stapes is round and heavy. Compact crura form a small inter-crura aperture. The head is not separated from the body, though it is distinctly expressed. Ear muscles are well-developed and look like short and wide muscular bundles. In the inner ear, the cochlea forms 2.5 turns. The basal turn is considerably widened in comparison with the turn located above. A wall between two turns is sharply thickened; as a result, both turns seem to be receding from each other. In a bearded seal, as well as in the majority of terrestrial mammals, at the 13th stage, the beginning of development of an acoustic vesicle in the form of a small pit located on each side of mesencephalon on the skin of a head is marked. In the middle part of the vesicle, there are three layers of epithelial cells [Solntseva, 1985 b].

At stages 14 to 15, subdivision of the inner ear into superior and inferior parts (i.e., into vestibular and cochlear apparatuses) (Fig. 32a) [Solntseva, 1985 b] is noticed. At the 16th stage, a cartilaginous part of the external auditory meatus starts to develop in the outer ear. It represents a short and direct tube (Fig. 33). In the middle ear the tympanum is not expressed. At the 16th to 17th stages, each bud of the auditory ossicles represents an independent formation; their base is formed by immature precartilaginous tissue. The malleus is small, and in comparison, the incus is large. The stapes is differentiated into crura; the bottom lamina is of an oval form. At the 16th stage, the cochlear apparatus looks like a canal. Its base is formed by a high columnar epithelium, and the top part by a low cuboidal epithelium consisting of three rows of cells. The differentiation of the cellular elements of the organ of Corti is completely absent. The nuclei of the cells are sharply basophilic, and cellular borders are not marked. At this stage, the cochlear canal starts to twist spirally, forming the basal turn of the cochlea (Fig. 34a).

At the 17th stage, the auditory meatus slightly lengthens and, at the same time, the diameter and the form of its lumen partly change. The auditory meatus is surrounded by buds of four cartilaginous laminae formed by the mesenchymal tissue.

In the middle ear, the process of formation of the tympanum, which represents a large and very narrow canal, starts. The base of the auditory ossicles is formed by mature precartilaginous tissue, the cells of which show distinct outlines. The auditory ossicles are separated by a perichondrium consisting of small, flat cells, or chondroblasts, devoid of distinct borders. The tympanic membrane is well expressed.

In the inner ear, differentiation of the cochlea is observed in the basal and medial turns. On the axial cuts, three cross sections of the cochlear canal are visible. At the given stage, the cochlea is formed by 1.5 turns only. It is surrounded by an aural capsule, the base of which is formed by mature precartilaginous tissue. The formation of the modiolus, which is represented by connective tissue elements, begins; numerous blood vessels are located between them.

Cellular elements of the organ of Corti are at about the same stage of differentiation. Cells of the spiral ganglion are compactly assembled and are sharply basophilic. Nervous fibers are not marked. Exactly at the stage of the cellular differentiation, the columnar epithelium of the cochlear canal base seems to move apart and two thickenings are formed – the larger one is an axial thickening, and the smaller one is lateral. Nuclei in the cells of thickenings are located closer to the base. They are large, with numerous nucleoli. Cellular borders are hardly marked, and the cytoplasm of the cells is light.

At the 18th stage of development, the auditory meatus is lengthened, and its cavity is filled with epithelial cells. In the middle ear, the process of formation of the tympanum continues. On the cross-section cuts, it has the form of an irregular oval. The auditory ossicles are increased in size and differentiated into the elements constituting them. The incus is large and exceeds the size of the malleus almost twice. In the area of the incudomalleal joint, the auditory ossicles are connected with each other almost at right angles, which is also marked in definitive forms. Around chondrocytes, the pericellular substance increases, a process of isogenous groups' formation occurs and the basic substance is formed. The chondroblasts of the perichondrium have distinct contours and are located more rarefiedly. In maturing auditory ossicles, a heterochrony is found. As the malleus and incus become more mature, their base is formed by primary cartilaginous tissue, while the base of the stapes is formed by mature precartilaginous tissue (Fig. 34b, c, d).

At the 19th stage, the pericellular substance grows around the chondrocytes of the aural capsule in the inner ear and isogenous groups are formed. The cochlea is differentiated into 2.0 turns (Fig. 34b-e), the inferior of which is considerably increased. Cellular elements of the organ of Corti are located more rarefiedly and acquire orderliness in disposition. A sharply expressed nucleous basophilia, found in earlier embryos, is absent. This is the initial stage of differentiation of the cells of the organ of Corti. The cells of a spiral ganglion are at the former stage of differentiation. Cells are gathered compactly and have no distinct borders.

At the 19th stage, the auditory meatus is formed basically and filled incompletely with epithelial cells. In the middle ear, the process of tympanum formation and auditory ossicles cartilaginification continues. In the tympanum, an interposition of the auditory ossicles is the same as in definitive forms. Subsequent differentiation of the auditory ossicles into forming elements continues. The malleus has a small head and a thickened handle. The incus is large relative to the malleus. Short and long processes of the incus are well visible. The stapes is differentiated into two thick crura, forming a small inter-crura aperture. Around the bottom lamina, the ring-shaped ligament is formed.

In the cochlear apparatus, the process of the cartilaginification of an aural capsule has ended. Chondrocytes have a rounded form with an eccentrically located nucleus. In the intercellular substance, differentiation of the collagen fibers occurs. The cells of the organ of Corti are located rarefiedly, and their nuclei are less basophilic. The process of cellular differentiation starts. Reissner's and basilar membranes are formed, the last of which is located under the cells of axial and lateral thickenings.

In an embryo at the 20th stage of development, the auditory meatus is considerably lengthened, widened and bent. Throughout its whole length, a diameter and a form of its lumen change. The basic process of its formation has ended. In the middle ear, the size of the auditory ossicles increases; their base is formed with a primary cartilaginous tissue. The tympanum is completely formed. In cartilage, an intercellular substance grows. The number of ossification sites in tympanic and periotic bones increases. The widened part of the malleus's handle is connected with tympanic membrane throughout its length as in adult forms. The formed ear muscles are represented by thickened muscular bundles of a triangular form.

In the inner ear, an aural capsule is formed by a slightly modified cartilage. The size of the cochlea is increased. Cellular differentiation of the organ of Corti starts in the cochlea's basal turn and gradually spreads ahead to the apical turn, which has also been noticed in terrestrial mammals. A vascular stria formation begins. It is possible to see on a cross section of the cochlear canal that in various turns of the cochlea, the elements of the cochlear canal and the cells of the organ of Corti are at different stages of differentiation. On the section through the basal cochlea's turn, one can observe the beginning of tunnel formation in a form of appearance of a small groove between the axial and lateral thickenings. In the epithelial cells of the axial thickening, the nuclei are located rarely, and their karyoplasm is light. Above these cells, a dark stria, which gives rise to tectorial membrane formation, is located. The future inner hair cells are formed from the cells of an axial thickening; similarly, the future outer hair cells are formed from the epithelial cells of the lateral thickening. The spiral ganglion's neurons have distinct borders and are located at a certain distance from each other. Their nuclei are light, with numerous nucleoli. Nervous fibers are marked. The process of differentiation of the basic elements of the organ of Corti has begun.

In an embryo at the 20th to 21st stages of development, a formation of the tympanic and vestibular scalae starts in the inner ear. The process of formation of the tectorial membrane has ended. The formation of a spiral incisure, spiral limb and a vascular stria continues. The tunnel is completely formed. The basic stage of a cellular differentiation of the receptor elements of the organ of Corti has ended. The formation of the supporting cells of the organ of Corti continues.

Thus, separate components of the peripheral auditory system of a bearded seal develop in the same sequence as they do in terrestrial forms. However, in a bearded seal, unlike terrestrial species, the auricle is reduced. In the middle ear, there is a completely new additional structure which is not found in terrestrial forms – the venous sinuses – the functional importance of which is linked to a pressure equalization in the tympanum during an animal's diving.

1.2.2. Ringed seal (*Pinnipedia, Phocidae-Pusa hispida)*

A ringed seal is a representative of true seals. The animals are of a middle and small size, and they are the smallest in the family. Distribution of these species of true seals is observed in arctic and temperate zones of the Northern Hemisphere: the seas of the Atlantic, Pacific and Arctic oceans, in which the animals adhere mostly to continental areas. Sexual dimorphism is practically not expressed. Usually, adult males are slightly larger than females and do not always differ in color.

These seals vary in terms of ecology. Some of them are ascribed to a coastal zone and during reproduction and moltings are connected with the land. Others prefer a pelagian way of life and are connected with the ice, on which they reproduce and moult. They are essentially gregarious animals, though during the fattening period and in whelping season they hold aloof. Whelping mainly occurs from the end of January until July; during the mating season, formation of harems does not occur. In development of an embryo, a latent period lasts for 1.5 to 3 months. A molting follows the mating, after which an intensive fattening period starts. The food of a ringed seal consists of fish, crustaceans and cephalopods.

A ringed seal is characterized by special, narrowly located areas of winter gatherings, where the seals reproduce and moult. In summertime, these animals form concentrated gatherings on coasts in rookerys [Fedoseyev, 1964; 1965].

The outer ear of a ringed seal, as inherent to other representatives of true seals, includes an external auditory meatus only. The external auditory meatus looks like a lengthened S-shaped tube, which diameter and form vary throughout its whole length. The length of the auditory meatus amounts to about 6.5 to 7.0 cm, and its lumen is narrow in the area of a bend double. In the proximal part, the size of the auditory meatus's lumen increases considerably and looks like a right oval. In the osseous part of the auditory meatus, there are numerous venous sinuses.

The development of the outer ear in a ringed seal, as well as in other mammals, starts in an early prefetal period at the 16th stage of development [Solntseva, 1988 b]. In a ringed seal, in contrast to other representatives of true seals, a rudimentary pinna, which does not develop in the fetal period and is absent in definitive forms, is formed at early stages of embryogenesis (Fig. 35). The cartilaginous part of the auditory meatus, which is represented by a small deepening, is formed out of the lateral walls of the first branchial cleft.

By the beginning of the 18th stage, the auditory meatus is lengthened and looks like a straight, short canal opened throughout its whole extent. By the end of the 19th stage, the proximal part of the auditory meatus is filled with epithelial cells; the distal part is completely filled with such cells only by the end of the 21st stage of development. By the moment of birth, a resorption of the epithelial cells occurs and the cavity of the auditory meatus, opened from the lumen up to the tympanic membrane, appears. Beginning at the 18th stage, species-specific features in the structure of the auditory meatus are formed in a ringed seal, similar to other representatives of pinnipeds. The auditory meatus is considerably lengthened and narrowed in the distal part, acquiring the S-shaped form. At the 20th stage, a widening of the proximal part of the auditory meatus occurs. The formation of four cartilaginous laminae surrounding the auditory meatus is carried out at the 19th stage of development.

The middle ear of a ringed seal includes the same components as those found in the majority of mammals. Unlike terrestrial species, the auditory ossicles of a ringed seal are massive. The malleus has a small head, and its handle is considerably thickened and lengthened. The widened part of the handle throughout its whole length is attached to the tympanic membrane, which is greatly thickened and elongated in a cone shape. The incus is heavy and has two big articular surfaces for the malleus. The stapes is massive and rounded, with two compact crura forming a small inter-crura aperture. In the area of the incudomalleal joint, the auditory ossicles are connected with each other at right angles, as occurs in other representatives of true seals. The tympanum is spherical and is covered with a thick mucous membrane, including numerous blood vessels in the middle layer.

The middle ear is laid by a protrusion of the first pharyngeal recess, the entoderm of which is transformed into a common tube-tympanic protrusion. As in the majority of mammals, the bud of the auditory ossicles is noticed at the 13th stage in the form of a mesenchymal condensation in a ringed seal. At the 17th stage of development, each of the buds of the auditory ossicles is represented by an independent formation; their base is formed by immature precartilaginous tissue.

At the 18th stage, a formation of structural features of the auditory ossicles occurs. At the 18th to 19th stages, structural features connected with interposition of the auditory ossicles in the cavity of the middle ear are marked. Their turning around the sagittal and frontal axes of the body of the animal. At these stages of development, the tympanum looks like a long, narrow canal. The bud of the tympanic membrane is seen at the 16th stage. At the 17th stage, the tympanic membrane is thick and friable. It thinning starts at the 18th stage. At the 19th stage, the tympanic membrane acquires a three-layer structure and is located almost horizontally on the lateral surface of the cavity of the middle ear.

The cochlea of a ringed seal is formed by 2.5 turns. The basal turn differs insignificantly from the turn located above. The secondary osseous spiral lamina is not developed. The anatomic structure of the cochlea, as well as the structure of the organ of Corti, is similar to that typical to the majority of mammals possessing low- and medium-frequency hearing.

In the majority of mammals, a subdivision of an acoustic vesicle into the superior and inferior saccules occurs at the 14-15th stages of development. From the inferior saccule, a cochlear canal and sacculus are formed, and from the superior saccule, the utriculus and semicircular ducts. At the 16th stage of development, a base and a roof are well expressed in the cochlear canal. As in other mammals, the base is formed by a columnar epithelium, and the roof by a cuboidal epithelium. The cochlear canal twists spirally, forming a lower, or basal, turn of the cochlea [Solntseva, 1999b].

At the 17th stage of development, a medial turn is formed in the cochlea. Anatomic formation of the structures of the cochlear canal is marked. The cellular differentiation of the organ of Corti is not revealed. At the 18th stage, the apical turn of the cochlea is formed; as a result, the cochlea is formed by 2.5 turns, as it occurs in definitive forms. Cells of the organ of Corti are at the same stage of differentiation in all turns of the cochlea. In the basal turn of the cochlea, the formation of the tympanic and vestibular scalae starts.

In the cochlea, at the 19th stage of development, cellular differentiation of the organ of Corti is marked. The cells of the organ of Corti are differentiated from the cells of columnar epithelium. The process of differentiation starts from the basal turn of the cochlea and gradually spreads over the medial and, further, over the apical turns. At the 20th stage of development, the anatomic formation of the cochlea has ended. In the cochlear canal, the formation of a spiral limb, a vascular stria, tectorial and Reissner's membranes is marked. The size of the cochlear and vestibular branches of the acoustical nerve is increased. The cellular differentiation of the organ of Corti continues.

1.2.3. Steller sea lion (*Pinnipedia, Otariidae-Eumetopias jubatus*)

The Steller sea lion is one of the largest representatives of order Pinnipedia, the largest in the family Otariidae. The average body length of males is up to 330 cm, of females about 230 cm. In the waters of Russia, the greatest numbers of Steller sea lions occur in the Kuril Islands [Perlov, 1970].

Steller sea lions are considered to be settled animals that do not migrate for long distances; however, constant migrations in the areas of all rookerys have been observed.

Coastal rookerys of Steller sea lions are subdivided into harems, where reproducing females, harem males and pups concentrate, and into single-animal rookerys. Usually, rookerys are located in remote places, such as uninhabited islands or rocky capes. The areas occupied by rookerys can look like separate boulders or large stony surfaces of rocks.

Adult Steller sea lions ususally eat fish and cephalopods. Food fields are located at the distance of approximately 50 to 70 miles from coastal rookerys. The depth in these areas is about 200 m, but more often the Steller sea lions feed in a coastal zone.

In coastal rookerys, about 5 to 20 females fall on one bull during the harem's formation. Each harem occupies a certain site on the comparatively flat places at a height up to 10 to 15 m above sea level. In spring, the sexually mature large males (bulls) appear first at coastal rookerys. The sexually mature females come later. On a rookery, a female gives birth to a pup some time later after the arrival; she does not leave the pup for several days. Then females search for food in the sea and return back to feed the pups.

The latent period of the embryonic development of Steller sea lions lasts for 2.5 to 3 months; the embryo develops during 9 to 10 months. Before the birth, embryos reach the length of up to 100 to 120 cm (i.e., about half of the mother's length). For the first hours of its life, a newborn pup is completely helpless, but soon it starts to move on a rookery independently. Within 25-30 days after its birth, a pup goes freely into the water. Puberty begins at 9 years. Females of Steller sea lions feed pups with milk until the age of one year. The population of Steller sea lions is very small; their reproduction rate is slow, which is why an account of the population including division into age and sexual groups is rather important.

The structure of the outer ear of the Steller sea lion includes an auricle and an external auditory meatus. In semi-aquatic species, as opposed to terrestrial mammals, the auricle adopts new structural features. In a Steller sea lion, the auricle looks like a thick dermatoid fold, twirled in a cone, the top of which is caudally inverted (Fig. 1b). The same structure of the auricle is marked in the representative of marine mammals from the Carnivora order – the sea otter (*Enhydra lutris*).

The external auditory meatus of a Steller sea lion has the form of a bent tube which is about 9 cm long and is surrounded with horseshoe-shaped cartilage. The diameter of the lumen of the auditory meatus changes insignificantly from the lumen up to the tympanic membrane, but the form of the lumen remains oval (i.e., keeps the features typical of terrestrial species). In air, due to the contraction of the auditory meatus's muscles, the auricle edges are partly open; under water, the muscles relax and the edges of auricle adjoin densely, thus preventing water penetration into the cavity of the auditory meatus [Solntseva, 1998a].

At the 16th stage of development in a Steller sea lion, as well as in other mammals possessing an auricle, the first branchial cleft, widening gradually, forms a deepening, at the edge of which small protuberances start to develop, which, at the 17th stage of a normal development, merge and form a uniform mesenchymal bud of an auricle. The cartilaginous part of an external auditory meatus starts to form from lateral walls of the first branchial cleft at the 16th stage of development. The osseous part of the auditory meatus in a Steller sea lion, as in other immature-born species (pinnipeds, cetaceans, ungulates), finishes its formation by the moment of birth.

By the beginning of the 19th stage of development, the auricle of a Steller sea lion acquires more distinct contours. The auditory meatus is lengthened. By the end of the 19th stage, the lateral part of the auditory meatus is filled with epithelial cells, which close the lumen of the auditory meatus completely by the 20th to 21st stages of development. In a Steller sea lion, the resorption of epithelial cells of the auditory meatus occurs by the moment of birth, as it does in other mature-born mammals. The formation of the species-specific features of the outer ear is marked at the 19th stage, as well as in the majority of species.

In order to adapt to an aquatic way of life, a tympanic membrane is formed in the middle ear of semi-aquatic forms, which in a Steller sea lion is rounded and thickened considerably. Thus, the rigidity of the connection and fastening of the auditory ossicles in the tympanum increases. Venous sinuses, concentrated near the tympanic membrane and in the walls forming the tympanicum, develop, and the relative size and form of the auditory ossicle elements change. Most likely, an elastic and thickened tympanic membrane prevents its rupture under quick pressure changes, to which marine mammals are constantly subjected.

The middle ear is an additional organ of a labyrinth and is laid by a protrusion of the first pharyngeal recess, the endoderm of which will be transformed into a common tube-tympanic protrusion. All elements of the middle ear develop from mesenchyme. The bud of the auditory ossicles in a Steller sea lion, as well as in other mammals, appears at the 13th stage of development as a mesenchymal condensation. In this bud at the 16th stage, contours of the auditory ossicles appear (Fig. 36a, b, c). At the 17th stage, all buds of the auditory ossicles are represented by independent formations; their bases are composed of immature precartilaginous tissue. The tympanicum is formed at the 16th stage as a narrow canal located below the buds of the auditory ossicles. At the 19th stage, the tympanum's turning around the sagittal and frontal axes of the prefetal body occurs. The formation of the structural elements of the auditory ossicles is marked at the 18th stage. The auditory ossicles are increased in size and are plunged deep in the tympanicum. In the majority of mammals, cartilaginification of the auditory ossicles starts in the center of each bud of the auditory ossicles and spreads gradually to the periphery. In a Steller sea lion, as well as in other pinnipeds, maturation of the auditory ossicles occurs non-simultaneously. The malleus matures first of all, then a body of the incus, while the lenticular arm of the incus, as well as the stapes, are formed by mature precartilaginous tissue. However, in the majority of species, by the 20th stage, replacement of the mature precartilaginous tissue by an embryonic cartilage has been completed. Formation of the features linked to interposition of the auditory ossicles relative to each other is marked at the end of the 16th and the beginning of the 17th stages of development. At these stages, the auditory ossicles turn around the sagittal and frontal axes of the animal's body. The tympanic membrane is formed in an area of contact of pharyngeal recess's entoderm and ectoderm of the first branchial cleft. Its bud appears at the 16th stage; at the 17th stage, a Steller sea lion's tympanic membrane, as well as in the majority of species, is thick and friable. Significant thinning of the tympanic membrane is marked by the end of the 16th and the beginning of the 17th stage of development (Fig. 36c). At the same stage, the tympanic membrane acquires a three-layer structure and is located almost horizontally on the lateral wall, forming one of the sides of the tympanic bone. The development of the venous sinuses in a Steller sea lion, as well as in other representatives of pinnipeds, is marked at the 20th stage (i.e., during the replacement of cartilaginous tissue by osseous tissue in integumentary bones of the cranium). The process of ossification of the auditory ossicles starts during the 21st to 22nd stages of development.

In a Steller sea lion, as well as in other representatives of otariids and phocids, the cochlea is formed by 2.5 turns. In species possessing high-frequency hearing (dolphins, bats), the sharp increase of the basal cochlea's turn and the significant development of a secondary osseous spiral lamina in the cochlear canal are typical, while in a Steller sea lion, similar to other representatives possessing low-frequency hearing, the size of the basal cochlea's turn practically does not differ from the turn located above, and the secondary osseous spiral lamina is not developed. The structure of the organ of Corti is the same as in most of mammals possessing low-frequency hearing. As in other mammals, the pair anlage of a membranaceous labyrinth is marked at the stage of 2 to 3 pairs of somites. At the stage of 6 to 9 pairs of somites, the bud of the membranaceous labyrinth represents an auditory placode. Further along, at the stage of 14 to 15 pairs of somites, an auditory pit is formed, from which, at the stage of 20 pairs of somites (13th stage of development, the stage of a forelimb bud), an auditory vesicle develops, which passes into the endolymphatic duct without clear borders.

In a Steller sea lion, as in most mammals, at the 15th stage, subdivision of the acoustic vesicle into superior and inferior parts occurs. At the 16th stage of development, a cochlear canal starts to twist spirally, forming a lower, or basal, turn of the cochlea. At the 17th stage, a medial turn of the cochlea is formed (Fig. 36d), and anatomic formation of the cochlea is completed with the formation of an apical turn at the 18th stage of development. The cochlea is formed by 2.5 turns (Fig. 36d). At the same stage, cartilaginification of the aural capsule begins. Structures of the cochlear canal and cellular elements of the organ of Corti are at approximately the same stage of development in all turns of the cochlea. At subsequent stages, the size of the cochlea increases.

At the 19th stage, a formation of an axis, or modiolus, of the cochlea, consisting of a connective tissue, as well as an anatomic formation of the cochlear canal's elements and a differentiation of cells of the organ of Corti, begins. The cuboidal epithelium's cells flatten; a loosening of connective tissue adjacent to this epithelium also occurs. At these locations, a formation of tympanic and vestibular scalae begins. The differentiation of the cells of the organ of Corti begins from the basal turn of the cochlea and is spread gradually onto the turns located above. As a result, in all turns of the cochlea, a different degree of anatomic and cellular differentiation of the cochlear canal's elements is marked. The cochlear and vestibular branches of the auditory nerve are well developed (Fig. 36e, f).

At subsequent stages, the size of the cochlea increases and cellular differentiation of the organ of Corti continues, which in a Steller sea lion, as well as in other mature-born species, is completed by the moment of birth. As in terrestrial species, Reissner's membrane is formed first of all structures of the cochlear canal, and a vascular stria is the last to form.

Thus, as in most mammals, in a Steller sea lion anatomic structures of the cochlear and vestibular parts are separated from each other simultaneously and reveal general basic features in structure at the early stages of development. In the first half of an early prefetal period, the outer, middle, and inner ears of a Steller sea lion are also similar in development to terrestrial species. The species-specific features in structural organization of hearing are formed in the second half of an early prefetal period, depending on ecological specialization of each species.

1.2.4. Walrus (*Pinnipedia, Odobenidae-Odobenus rosmarus divergens*)

A walrus is a unique representative of this genus. According to a habitus, these animals differ from all representatives of pinnipeds. The body is large and massive, with thick, folded skin and wide extremities . On a firm substratum, the walruses move hardly, leaning on all four extremities. In water, walruses are fast and mobile, but they get out on ice floe with great difficulties with the help of canines and four flippers.

The pacific walrus (*Odobenus rosmarus divergens*) is the largest form of the species. The body length amounts to 450 cm in males and 367 cm in females. The habitat is relatively shallow waters rich with bottom mollusks and crustaceans. During winter, walruses live only on ice. In summer, walruses form coastal rookerys on sandy or pebble shallows on coasts of the continent or islands.

Feeding objects have been investigated insufficiently. Within the first two years, pups feed on their mother's milk and start to feed independently only in the third year. The base for food is formed by bottom invertebrate animals. A walrus gets food from the bottom, digging the ground with canines. A walrus gets food at the depths from 30 to 50 m. However, the depth to which a walrus submerges during its feeding can reach 180 m.

Individual sites of separate individuals or groups are not present. In summertime, walruses form coastal rookerys. The Pacific walrus spends most of its time on ice. Male rookerys are few in number, are mixed and can reach several hundreds of animals. Rookerys are located along the edge of drifting ice. Usually, animals lie down on strong, long-term smooth or hummocky ice floes [Popov, 1959, 1960].

The Pacific walrus winters in shallow waters of the southeast part of the Bering Sea. Coupling occurs during May and at the beginning of June. During this time, in some females, embryos at the earliest stages of development were found. The whelping occurs during the same period the coupling does and lasts for a month. Pregnancy lasts for 12 months.

During the reproduction period, walruses do not form harems and stay in family groups consisting of 3 to 6 animals (male, female and pups). The pups suckle for 2 years. The females which whelp annually feed two pups simultaneously – the current year's and the last year's pups. [Krylov et al., 1964; Krylov, 1965].

The walrus's slow rate of reproduction determines a very slow recovery of its depleted population, and a number of measures for the protection of its stock directed to the preservation of the animals are necessary.

In the walrus's embryos (Fig. 37) at the 16th stage of development, an external auditory meatus looks like a short and direct tube. The tympanum is not formed. The buds of the auditory ossicles appear at the 13th stage; they are formed by a mesenchymal tissue and are connected continuously with each other [Solntseva, 1986 b; 2001 c]. At the 16th stage, the contours of the auditory ossicles' buds, the form of which differ greatly from corresponding ossicles of more mature embryos, are hardly found. The differentiation of the auditory ossicles into other elements is also absent (Fig. 38b, c). The cells of mesenchymal tissue of the buds of the middle ear's elements have large nuclei which occupy most of a cell. The subdivision of the inner ear into cochlear and vestibular apparatuses, as in other species, occurs at the 14th to 15th stages of development (Fig. 32b). On a cross section, the cochlear canal looks like an irregular oval, whose base is formed by a high columnar epithelium. The upper wall of the cochlear canal is formed by three rows of the cuboidal epithelium's cells. Further along, from the cells of a high columnar epithelium, the elements of the organ of Corti develop, and from the cells of a cuboidal epithelium, Reissner's membrane and the underlayer of the lateral internal wall are formed.

In an embryo at the 17th stage, the auditory meatus is lengthened, and the form and diameter of its lumen change in its different parts. The buds of the auditory ossicles, in which chain their juncture borders are clearly discerned, slowly plunge into the forming tympanum. The form of the auditory ossicles already corresponds to the one they will have during the definitive period. The base of the auditory ossicles is formed by a mesenchymal tissue (Fig. 38c).

At the 16th stage, in the inner ear around the cochlear canal, the process of formation of an aural capsule begins, the base of which is formed by a mesenchyme. At the end of this stage, the process of transformation of mesenchyme into the mature precartilaginous tissue starts, the cells of which acquire more distinct forms. The cochlear canal twists spirally, forming the basal turn.

At the 17th stage, the medial turn (Fig. 39a-f), and at the 18th stage, the apical turn, are formed. The basic process of the cochlea's formation has ended. Similar to other species, the development of the basal turn outstrips the development of the following turn, which is located above. In size, the basal turn differs insignificantly from the apical turn. The elements of the organ of Corti are at approximately identical stages of cellular differentiation. The formation of the basilar and Reissner's membranes begins; these membranes, opposite other details of the cochlea, start to develop first.

At the 18th stage, significant changes in the development of the auditory meatus are not marked. However, in the lateral part of the auditory meatus, the epithelium's cells, which gradually fill up this part of the auditory meatus, are found, and by the 19th stage they fill up its lumen completely.

In the middle ear, an increase of the tympanum and the auditory ossicles, which are formed by a mature precartilaginous tissue (Fig. 40a), occurs. However, unlike the previous stage, the auditory ossicles undergo the process of morphological differentiation into the elements forming them. In the ear muscles, their differentiation continues. They are represented by long and wide muscular bundles.

In the inner ear, the cochlear canal forms 2.5 turns (Fig. 40b, c). On the axial cuts of the cochlea, four cross sections of the cochlear canal are marked. On the section through the basal turn of the cochlea, two protuberances, which are formed by a multi-row epithelium, can be seen. Between the protuberances, a small deepening is located, from which the tunnel will be formed later. The aural capsule is formed by a mesenchyme, and transformation process of mesenchyme into mature precartilaginous tissue begins from the vestibular apparatus. At this stage, the cellular elements of the organ of Corti are involved in the process of morphological differentiation. Cells are located more rarefiedly. Cells' nuclei are large, have an oval form and numerous nucleoli. However, the typical patchiness in the arrangement of the outer hair cells is completely absent.

At the 19th stage, the size of the auditory meatus increases. The size of the auditory ossicles and tympanum compared to the previous stage is increased insignificantly. The tympanic muscle of the middle ear is long and wide; it lies deeply in the bone canal which separates the cochlear apparatus from the vestibular one. The stapedius muscle is short and wide. The differentiation of the middle ear's muscles continues. The auditory ossicles are differentiated into the elements forming them. The base of the auditory ossicles is formed by a mature precartilaginous tissue. The interposition of the auditory ossicles in tympanum is the same as in definitive forms (Fig. 40d, e).

According to its formation, the inner ear practically does not differ from the previous stage. The modiolus is at the initial stage of morphological differentiation. The differentiation of the cochlear canal's elements and the cells of the organ of Corti is marked. The tympanic and vestibular scalae are formed (Fig. 40d, e).

At the 20th stage of embryo development, the auditory meatus becomes longer, and the form of its lumen differs considerably in the distal and proximal parts. The cartilaginous part of the auditory meatus is filled completely with epithelial cells. The size of the tympanum and the auditory ossicles is increased. The base of the auditory ossicles is formed by a mature precartilaginous tissue with distinct cellular borders. At the given stage, the process of cartilaginification of the auditory ossicles begins, which initially covers their periphery and then spreads gradually to the center. Around the chondrocytes of the embryonic hyaline cartilage, the pericellular substance is located. The cells' nuclei are large; they are surrounded by a thin layer of cytoplasm and are located at some distance from each other. The auditory ossicles are surrounded by a perichondrium consisting of small flat cells – chondroblasts – without distinct borders. Due to the perichondrium, the borders in the junction of the auditory ossicles are marked precisely. Tympanum formation and formation of the tympanic membrane, which is thinned, continues. In the wall surrounding the tympanum, the first centers of ossification appear.

In the inner ear, the cochlear canal forms 2.5 turns like in adult forms. On the axial cuts of the cochlea, five cross sections of the cochlear canal are visible (Fig. 40c). The process of cartilaginification of the aural capsule has been completed. There is a differentiation of the collagen fibers in the intercellular substance of the embryonic hyaline cartilage. The morphological differentiation of cells of the organ of Corti continues. The location of cells is more ordered. Nuclei of the cells have rounded and oval forms. Below the forming outer hair cells, the cells with large nuclei are located. These are future Deiters cells. Below them, the cells with pyknotic nuclei and a light cytoplasm are located. Under these epithelial cells, the basilar membrane is located. The future Hensen and Claudius cells are represented in the form of a cuboidal epithelium located in one to two rows.

In the cochlear canal, differentiation of the spiral limb, vascular stria and spiral incisure goes on. The spiral limb is formed by oblong cells. The vascular stria is formed by an undifferentiated epithelium, the cells of which have transparent nuclei of an oval and rounded form. Nuclei contain one centrally located nucleolus. The outlines of cells are hardly visible. The basilar membrane, on which the cells of the organ of Corti are located, is composed of a forming connective tissue. The cuboidal epithelium of Reissner's membrane consists of light and dark cells. In a spiral labium, in the place of the future spiral incisure, a multi-row layer of the high columnar epithelium is located. Nuclei of these cells have a round-oval form and are transparent. Cellular borders are weakly marked. The tunnel is not yet formed. However, the bases of the supporting cells have already diverged. Thus, the internal cells-columns are located vertically, and the external cells-columns form a corresponding angle for the tunnel. The cochlear and vestibular branches of the auditory nerve are well expressed. The neurons of the spiral ganglion are increased in size, are located at a small distance from each other and have distinct borders. The modiolus is penetrated by the dendrites of the spiral ganglion and has a more compact base.

At the 21st to 22nd stages of development, the formation of the outer and middle ears has basically ended. The structures of the middle ear continue to increase in size. The number of the ossification's sites in tympanic and periotic bones has considerably increased. The auditory ossicles are formed by an embryonic hyaline cartilage. The head of the malleus is small; its handle is short and thick, which is not typical of terrestrial forms. The widened part of the malleus's handle is fastened throughout its length to the forming connective-tissue tympanic membrane of a rounded form. In comparison to the malleus, the incus is large and heavy. Its size exceeds that of the malleus almost twice. The stapes is differentiated into two crura and a rounded footplate, around which a ring-shaped ligament is formed. The compact crura form a small inter-crura space. The interposition of the auditory ossicles in the tympanum is the same as in adult forms. The articular surface of the malleus is connected at almost right angles with the articular surface of the incus. Ear muscles are completely formed and are represented as long muscular bundles.

In the inner ear, the aural capsule is formed by the embryonic hyaline cartilage, in which there is an increase of the intercellular substance and the capsules of the cartilaginous cells are clearly prominent. The process of formation of the isogenic groups of chondrocytes continues. The cochlear canal forms 2.5 turns and is located deep in the periotic bone. The differentiation of the rostral and the paddle-shaped processes of the cells supporting the organ of Corti begins. The supporting cells have an oblong form with a nucleus located in the basal part of a cell.

Nuclei include one to two nucleoli. The outer hair cells densely adjoin each other. With their help, a characteristic angle starts to be formed relative to the tunnel. The rounded bases of the outer hair cells are located on top of the Deiters cells, the basal parts of which lie on the forming basilar membrane. The phalanx processes of the Deiters cells surround the outer hair cells and approach their tops – a place of the phalanxes' buds. The differentiation of the Hensen and Claudius cells, which are located in several rows, continues. The cells include large oval nuclei and have a narrow band of cytoplasm. Reissner's membrane is formed by the epithelial and connective-tissue layers. The cells of the epithelial layer have a transparent cytoplasm and include rounded and oval nuclei. The basilar membrane consists of the connective tissue, the elements of which are located in three to four rows. The basic process of cellular differentiation in the organ of Corti has finished.

1.3. Aquatic mammals

1.3.1. Toothed whales (*Cetacea: Odontoceti*)

Spotted dolphin (*Stenella attenuata*), White whale (*Delphinapterus leucas*)

Cetaceans represent one of two orders of recent mammals that have completely adapted to an aquatic way of life. Along with general morphological features typical in all representatives of the order, species-specific features in the structure of organs and systems have appeared in each of the suborders. Mysticetes have kept the olfactory analyzer and have not acquired an ability to echolocate, but odontocetes, having lost the olfactory analyzer, have acquired amazing abilities for echolocation.

Well-developed hearing of odontocetes is supplemented with special organs of sound signal generation. The combination of a perfect auditory receiver with organs of sound signal production has provided odontocetes with unlimited opportunities for orientation in water environment by a reflected echo signal.

The spotted dolphin (stenella) and white whale (beluga), related to the family Delphinidae, reveal a lot of similar features in structure and development of the peripheral auditory system; therefore, the results of development of the auditory organ in these two species are given in uniform description.

The spotted dolphin (*Stenella attenuata*) belongs to the genus *Stenella*. This is a small dolphin. The length of a male body amounts to 165 to 208 cm, a female body to 180 to 201 cm. The constitution is slim, like the constitution of a common dolphin (Fig. 41). In the waters of Russia, their presence has not been established. They inhabit the southern half of the Atlantic, Indian and Pacific Oceans [Heptner et al., 1976].

The body structure of cetaceans is typical of mammals, but at the same time, many of their systems and organs have undergone essential morphological and functional reorganizations. This is connected with the fact that the aqueous medium, according to its physical and chemical properties, strongly differs from the air, which is the habitat for the majority of mammals. First, water possesses an increased thermal capacity, 20 times that of air. The density of water surpasses the density of air 800 times, and during the animal's submersion, the hydrostatic pressure increases 1 atm for each 10 m of depth. This forced changes in the body's form, as well as the functional reorganizations of respiratory, cardiovascular and analyzing systems and, first of all, the peripheral auditory system.

Dolphins do not undertake far migrations; the majority of them are settled animals and are ascribed to a certain water area. High speed of movement is reached by a well-streamlined body form, well-developed muscles of a tail stock and a specificity of the tegument's structure, its damper properties.

A spotted dolphin produces signals that sound like whistles and have an average duration of 0.5 sec. The basic energy of whistles falls at components with the frequency of 4 to 7 kHz. Some signals turn from whistles to grinding sounds. It has been shown that stenella's whistles are similar to the whistles of other species, though the frequency range of their whistles is a little bit lower than in bottlenose and common dolphins [Airapetiantz, Konstantinov, 1974].

In dolphins, coupling and birth terms last up to 2 to 3 months during the spring and summer. Pregnancy lasts up to 11 months. Birth occurs under water. Immediately after birth, a calf is quite adapted to an independent movement and follows its mother right away. All cetaceans are known to be mature-born species. The calf feeds on the mother's milk under water. In cetaceans, the sexual dimorphism becomes apparent in the various lengths of male and female bodies.

A white whale (*Delphinapterus leucas*) is a unique representative of the genus, having a completely white body that distinguishes this species from other marine mammals. Usually, it inhabits coastal zones. On the Far East, the maximum body length of white whale males reaches 600 cm, of females up to 500 cm. A white whale lives in the cold waters of the Northern Hemisphere where it is spread circumpolarly. The white whale of the White Sea is of the smallest form; the length of the adult's body amounts to 253 to 376 cm. We have investigated the beluga whale of the White Sea. The food objects of the White Sea beluga whales which live near Ñanin are lumpfish and flounder, and in the second half of June, herring and flounder [Kleinenberg et al., 1964].

A white whale is a typically gregarious animal and usually is met in groups of a different size. The population of the White Sea beluga inhabits the White Sea beginning from early spring until late autumn, when it leaves for the Barents Sea, where it winters until the freezing is over. Spring migrations of animals start when most of the sea is still filled with dense, drifting ice. In summer, white whales are distributed on all water areas of the sea and also come into numerous gulfs and bays [Kleinenberg et al., 1964].

As well as the period of coupling, the period of birth is extended and birth occurs beginning in early spring and continues during all summer months. Pregnancy for white whales lasts for 11 to 12 months, and, sometimes, even for 13 to 14 months. As the period of birth is extended, it is possible at any time to observe embryos of various sizes.

Vision and hearing in a white whale are well developed. The eyesight serves at a close distance only. By means of hearing, animals orient themselves in very different situations, perceiving sounds of various intensity. The range of the uttered sounds is very wide. The sound frequency varies from several up to 10 and even 20,000 Hz. Sounds play the role of various signals and are used for echolocation.

The stenella's and beluga's outer ear, as in other representatives of cetaceans, includes the external auditory meatus only, as the auricle is completely reduced.

At the 16th stage, from the lateral walls of the first branchial cleft, the cartilaginous part of the external auditory meatus starts to develop in the form of a short and slightly bent tube opened along the whole extent. The osseous part of the auditory meatus forms later. Filling of the cartilaginous part of the auditory meatus with epithelial cells occurs at the 19th stage, and its total closing by these cells comes to an end by the 21st stage of development [Solntseva, 1983; 1999 a].

At the 19th stage, the formation of the species-specific features of the outer ear in the form of a noticeable narrowing of the auditory meatus in its distal part is marked. Further along, the auditory meatus lengthens and acquires a double-bent shape typical of definitive forms. At the 20th stage, the auditory meatus widens in the proximal part. At the later stages, the increase in the absolute size of the auditory meatus, which is carried out proportionally to the growth of the prefetal body, is marked.

All elements of the middle ear develop from mesenchymal and mesodermal elements. In a spotted dolphin and in a white whale, as well as in other species of mammals, the bud of the auditory ossicles appears at the

13th stage in the form of a mesenchymal condensation. At the 14th to 15th stages, the contours of the auditory ossicles' buds are visible, but their form is not similar to that they will adopt at the later stages of development. Junction between the auditory ossicles is continuous. The differentiation of the auditory ossicles into the elements forming them is absent. In the mesenchyme, from which the buds of the auditory ossicles consist, large nuclei, occupying most of a cell, are found (Fig. 42a-d).

In the bud at the 16th stage, the contours of the auditory ossicles are revealed, and the process of the tympanum's formation begins. The buds of the auditory ossicles slowly plunge into the depth of the tympanum (Fig. 43 a). The base of the auditory ossicles is formed by a mature precartilaginous tissue, the cells of which acquire more distinct outlines (Fig. 42b). At this stage, the process of differentiation of the precartilaginous tissue to the embryonic hyaline cartilage begins. The process of cartilaginification begins in the center of each bud of the auditory ossicles and spreads gradually to their periphery. Around the chondrocytes, the pericellular substance is located. Cells' nuclei are large and surrounded by a narrow band of cytoplasm and are located at a distance from each other. The auditory ossicles are surrounded by a perichondrium, which consists of small, flat cells whose chondroblasts have no distinct borders. Due to the perichondrium, the areas of the auditory ossicles' junctures are clearly visible. At the given stage of development, the differentiation of the tympanic membrane-ligament, stapedius and tympanic muscles begins. The tympanum is represented by a narrow blind canal located below the buds of the auditory ossicles.

At the 17th to 18th stages of development, there is a formation of tympanic and periotic bones. At the 19th stage, the tympanum's turning around the sagittal and frontal axes of the animal's body occurs. The location of the auditory ossicles in the tympanum is similar to definitive forms (i.e., the malleus and incus are connected with each other at right angles) (Fig. 43a, b, c).

At the 18th stage, the formation of structural elements of the auditory ossicles is marked. The auditory ossicles are increased in size. In the malleus, a head, neck and handle are well expressed. In the incus, a body and both processes are formed. In the stapes, there is no differentiation into crura; therefore, the stapes acquires the form of a smoothed cone. The base of the auditory ossicles is formed by a mature precartilaginous tissue. Further along, the replacement of the mature precartilaginous tissue by the embryonic hyaline cartilage occurs. The process of cartilaginification of the auditory ossicles starts in the center of each bud and spreads gradually to their periphery (Fig. 41c, d).

The formation of the features connected with interposition of the auditory ossicles in the tympanum of a spotted dolphin and a white whale is marked at the 16th stage of development, whereas in the majority of the species which do not possess abilities for echolocation, it occurs at the 18th to 19th stages of development.

The tympanic membrane is formed in the place of the contact of the entoderm of a pharyngeal recess and the ectoderm of the first branchial cleft. The bud of the tympanic membrane appears at the 16th stage, and at the 17th stage it is thick and friable. At the 18th stage, the tympanic membrane becomes considerably thinner, acquires a three-layer structure and is located almost horizontally on the lateral surface of the middle ear's cavity. The ligament connecting the tympanic membrane with the handle of the malleus is formed (Fig. 43c, d; 44c). In odontocetes and mysticetes, the tympanic membranes reveal similarity in structure at similar stages of development, whereas during the fetal period they adopt species-specific features.

The formation of a cavernous plexus is marked at the 18th to 19th stages. The development of the venous and peribullar sinuses occurs a little bit later, beginning at the 21st stage of development. The replacement of the cartilaginous tissue by the osseous tissue is marked at the 20th stage in the form of separate centers of ossification in the integumentary bones of the cranium, tympanic and periotic bones. The initial ossification of the auditory ossicles is marked at the 21st stage. The process of formation of the ear muscles and a ring-shaped ligament of the stapes has ended. The aural capsule is formed by a slightly differentiated cartilage.

In the inner ear at the 13th stage, the acoustic vesicle develops; its subdivision into superior and inferior parts occurs at the 15th stage, as in other mammals. The vestibular apparatus is formed from the superior part, and the cochlear canal is formed from the inferior part (Fig. 32c, d).

At the 16th stage, the cochlear canal starts to twist spirally, forming a lower, or basal, turn of the cochlea, which is surrounded by the aural capsule consisting of a compact mesenchyme. Cellular elements of the organ of Corti are at approximately identical stage of differentiation.

At the 17th stage, the next turn of the cochlea is formed (Fig. 43a). The process of formation of turns is accompanied by the formation of a cochlear nerve (*n. cochlearis*). The dorso-medial part of this nerve goes to the apical turn, and its ventro-lateral one to the basal turn, the size of which considerably surpasses those of the turn located above (Fig. 43b).

At the 18th stage, the cartilaginification of an aural capsule begins. The cochlea is formed by 2.0 turns (Fig. 43c, d; 44a, b). The beginning of the cellular differentiation of the organ of Corti is marked. At this stage, the columnar epithelium of the cochlear canal moves apart; therefore, two thickenings are formed – an axial and a lateral – from which the elements of the cochlear canal and the organ of Corti are formed (Fig. 44a, d). Nuclei of the cells of the future organ of Corti are large, have an oval form and include numerous nucleoli. Cells have not formed a typical mosaic in their location yet. Their cytoplasm is light and hardly visible. The structure and location of these cells enable us to suppose that further along the outer hair cells will be formed from them. Under these cells, the cells with large nuclei including nucleoli are located. Most likely, these are Deiters cells; under them, the cells with small pyknotic nuclei are located. The cytoplasm of these cells is light and forms a narrow band. These are epithelial cells, under which the basilar membrane is located. The Hensen and Claudius cells are presented by a cuboidal epithelium located in one to two rows. Reissner's membrane is formed. The formation of a cavernous plexus is marked (Fig. 43d).

At the 19th stage, the supporting cells, as well as the receptor cells of the organ of Corti, are involved in the process of differentiation. Cells are located in order, forming three rows. The nuclei of the cells have a rounded and oval form. The outlines of the cells are clearly visible. The neurons of the spiral ganglion are large and densely adjoin each other, forming characteristic clusters. The formed modiolus is penetrated by numerous blood vessels.

At the 20th stage, the formation of the supporting elements of the organ of Corti (outer cells-columns) continues. These cells have a thin, bent body with a nucleus located in the basal part of a cell. Cells of a cylindrical form with a rounded base are differentiated simultaneously. These are the internal hair cells. Cells of the precartilaginous tissue acquire more distinct borders, and in the central sites, they increase in size. The quantity of the intercellular substance grows. The cochlea is increased. In the cochlear canal, the differentiation of a spiral limb, a vascular stria and a spiral incisure begin. The spiral limb is represented by the cells of an extended form. The vascular stria is formed by the undifferentiated epithelium. The future spiral incisure consists of a multi-row, high columnar epithelium including transparent nuclei of an oval and rounded form. The nucleus includes one nucleolus which is centrally located. The outlines of the cells are hardly visible. The basal membrane, on which the cells of the organ of Corti are located, consists of a connective tissue. The tunnel is not yet formed. The sizes of the neurons of the spiral ganglion are increased, and nerve fibers are clearly visible. Their nuclei are large, have a rounded form and are eccentrically located.

At the 21st stage, the connection of the tympanic membrane-ligament with the handle of the malleus is clearly visible. The process of formation of ear muscles and a ring-shaped ligament of the stapes has finished. The aural capsule is formed by a slightly differentiated cartilage. In comparison with the previous stage, the cochlea acquires a much larger size. The basic process of the cellular differentiation of the organ of Corti has ended. In the cartilage there is an extension of the intercellular substance, and the capsules of the cartilaginous cells are clearly visible. The process of formation of the isogenic groups of chondrocytes occurs. In the cochlear canal, the process of formation of the tunnel begins. The cells of the spiral ganglion are located more rarefiedly.

In the white whale's embryo, at a length of 250 mm, the auditory ossicles are connected at right angles in the area of the incudomalleal joint. In the malleus, the centers of ossification have appeared. The tympanic membrane is thick; it is connected with the handle of the malleus with the help of a ligament. The long process of the malleus knits with the wall of tympanic bone (Fig. 45a, b). The structures of the cochlear canal are basically formed. The tympanic and vestibular scalae, the cochlear canal, Reissner's membrane and the spiral ligament are distinctly visible; the differentiation of the cells of the organ of Corti continues (Fig. 45c).

In the process of embryogenesis of the peripheral part of the stenella's and beluga's acoustic analyzer, the morphological differentiation and maturation of the structures of the outer, middle and inner ears occurs in the same sequence as in terrestrial forms [Altmann, 1950; Titova, 1968; Solntseva, 1983, 1999 a; Zharskaya, Solntseva, 1986, 1989].

Thus, in the early embryogenesis of a spotted dolphin and a white whale, the original features of the auditory organ's formation are found, which are connected with the way of life as well as with the perception of frequencies of a wide range. Adaptive features in the structure of the auditory organ are revealed at different stages of embryogenesis, even at the earliest, in spite of the fact that the development in the mother's womb occurs without direct influence of the environmental conditions.

1.3.2. Mysticetes (*Cetacea, Mysticeti*)

Minke whale (*Balaenoptera acutorostrata*)

The minke whale is a representative of mysticetes and is widely distributed in the Southern Hemisphere; it also can be found beginning at the tropical zone in the North up to an ice border in the Antarctic waters. Migrations of these whales have not been investigated because of the absence of whaling directed toward this species. From December until March, the minke whales can be found throughout the whole water area of the Antarctic region. During this period, they stay in groups formed by up to 20 whales.

During late autumn, minke whales leave the Antarctic waters and migrate to the warmer areas. In the Antarctic waters, krill is the food basis for the minke whale. There is an assumption that in the more northern latitudes (warmer zones), various species of gregarious fishes can serve as its food.

Reproduction of the minke whale has been investigated insufficiently. Males reach a sexual maturity at the body length of about 7.1 m, females at about 7.9 m (at the age of 7 to 8 years). It is supposed that pregnancy lasts about 1 year and occurs almost annually. Birth occurs from June to July. A newborn has the length of up to 2.7 m [Sleptzov, 1965].

Mysticetes, in contrast to odontocetes, produce long, low-frequency signals which absolutely differ from the sounds produced by odontocetes. Usually, these are low-frequency cries and groans, the frequency of which is not higher than 15 kHz.

Low- and middle-frequency signals spread under water to a distance of up to 15 to 20 km, and ultrasonic signals up to 3 to 5 km. Therefore, for the more distant transfer of sound information, it is necessary to use low-frequency signals. As it is known, the speed of a sound grows with an increase in pressure. There are depths where infrasounds are transmitted with maximal speed. These are so-called sound channels in which the sound is practically not absorbed and spreads over thousands of kilometers.

In the minke whale, as in all other representatives of the order Cetacea, the auricle is reduced. Passing through the hypoderm, the external auditory meatus is restricted beginning from an opening up to a tympanic membrane and, after passing through a fatty layer, is obliterated. The extent of such obliteration can reach several centimeters, and then the lumen appears again and remains up to the tympanic membrane [Fraser, Purves, 1954; Purves, 1958].

A distinctive feature of the external auditory meatus in mysticetes is the presence of an ear cerumen, the form of which usually corresponds to the form of the auditory meatus's lumen (conic). The base of such a cone is rounded according to the form of a "glove finger" – a tympanic membrane. In the middle ear, a tympanic bone forms a heavy compact tympanic bulla, which only in two points knits with the periotic bone. The principle of the auditory ossicles' structure reveals the features of similarity in all cetaceans; however, mysticetes have peculiar structural features. The cochlea of mysticetes is twirled differently than in odontocetes. It forms 2.5 turns, increasing in height. In some species of mysticetes, in the area of an oval window, a strong curvature of the basal turn is observed.

The external auditory meatus starts to be formed at the 15th to 16th stages of development (Fig. 46a); it looks like a short and direct tube opened along the whole extent. At the later stages, an increase in the size of the auditory meatus is marked, and at the 19th stage, it is possible to observe a noticeable narrowing of the distal area of the auditory meatus's cartilaginous part. By the beginning of the 20th stage, the lateral part of the auditory meatus starts to be filled with epithelial cells, and by the end of the 21st stage, the cartilaginous part of the auditory meatus is already completely filled with these cells. Resorption of the epithelial cells occurs in the fetal period. The osseous part is formed later (i.e., beginning from the second half of an early prefetal period).

At the 14th to 15th stages of development, the buds of the auditory ossicles and tympanum are located at a distance from each other. Contours of the auditory ossicles are not visible. The mesenchyme, of which the buds of the auditory ossicles consist, includes large nuclei occupying most of the cell [Solntseva, 1985 a].

The inner ear, being the most ancient formation in vertebrates compared to the outer and the middle ears, is the first to develop at the stage of 2 to 3 pairs of somites in the form of a pair anlage. At the stage of 14 to 15 pairs of somites, the auditory pit is formed; an auditory vesicle is formed from it at the 13th stage, which at the 14th to 15th stages is subdivided into superior and inferior parts surrounded by a mesenchyme (Fig. 32e). The cochlear canal is represented by a tube, the base of which is formed by a multi-row epithelium. The superior part of the cochlear canal consists of three rows of the cuboidal epithelial cells. The cochlear ganglion is not expressed, and nerve fibers are not visible.

In embryos at the 16th stage, the buds of the auditory ossicles and tympanum, as well as in the previous embryos, remain at a distance from each other. However, at the given stage of development, the outlines of the auditory ossicles appear, though their connection is still continuous. The formation of the borders between the auditory ossicles starts. The base of the auditory ossicles' and tympanum's buds is formed by a mesenchymal tissue. By the end of this stage, the auditory ossicles are represented by independent formations.

In the inner ear, the formation of a lower, or basal, turn of the cochlea is typical. The size of the cochlear apparatus considerably exceeds that of the vestibular apparatus (Fig. 46b). The cochlear canal is increased in size. A differentiation of the cellular elements of the organ of Corti is absent.

In an embryo at the 17th stage, the formation of the external auditory meatus (Fig. 46a), which is lengthened and bent in the direction of the middle ear, continues. The differentiation of the auditory ossicles into the elements forming them has already ended (Fig. 46b, c, d), while the formation of the tympanum continues. The base of the auditory ossicles is formed by a mature precartilaginous tissue, the cells of which have distinct outlines.

In the inner ear, the aural capsule is formed. The cochlea is differentiated into the basal and medial turns (Fig. 46b). Between the aural capsule and the Eustachian tube, numerous blood vessels, as well as a single, large venous sinus, are located. At this stage, the process of formation of the modiolus starts; the modiolus is formed by numerous bundles of collagen fibers, between which separate blood vessels are located. Reissner's membrane is formed.

In an embryo at the 18th stage, the diameter and the extension of the auditory meatus increase. The areas of junctures of the auditory ossicles, which are formed by a mature precartilaginous tissue, are clearly visible (Fig. 46e, f, g). The process of maturation of chondrocytes begins in the center of the auditory ossicles and spreads gradually to their periphery. Nuclei of cells are large and eccentrically located. The basic substance of chondrocytes slowly accumulates. The first sites of ossification in the tympanic bulla have appeared. The auditory ossicles are increased in size; the areas of their junctures are clearly visible. The long process of the malleus knits with the wall of tympanic bone (Fig. 46f). The size of the cochlea is twice the size of the vestibular apparatus (Fig. 46e, f, g).

In the inner ear, the cochlea is differentiated into 2.5 turns, as it occurs in definitive forms (Fig. 46h). It is well noticeable on the axial cuts of the cochlea, where five cross sections of the cochlear canal are visible. Cellular elements of the organ of Corti continue to remain at approximately identical stages of differentiation.

In an embryo at the 19th stage, the auditory meatus is increased in size. The process of formation and ossification of the walls of the tympanic and periotic bones continues. The auditory ossicles are increased in size and completely differentiated into the elements forming them. Interposition of the auditory ossicles and their fastening in the tympanum is the same as in definitive forms. The malleus has a large head with a big articulate surface for the incus. The neck of the malleus is poorly marked. The connection of the malleus with the tympanic membrane is carried out by means of a ligament formation. The *pr. gracilis* of the malleus is lengthened and greatly thickened. The incus is massive. Together with the long's arm lengthening and thickening, the short arm is considerably reduced. The malleus and the incus are connected with each other at right angles, preventing them from the possibility of independent vibrations. The stapes is differentiated into crura, a head and a bottom lamina, around which the ring-shaped ligament of the stapes is formed. At this stage, the process of transformation of the mature precartilaginous tissue into the primary cartilaginous tissue begins. The pericellular substance is located around the chondrocytes. The nuclei of the cells are large and surrounded by a narrow band of cytoplasm. The auditory ossicles are surrounded by the perichondrium, consisting of small flat cells – chondroblasts. The inner ear is increased in size. The base of the aural capsule is formed by a primary cartilaginous tissue.

In an embryo of the 20th to 21st stages, the absolute size of the auditory meatus, a ditory ossicles, tympanum and the cochlea continues to increase. The increase in this size occurs proportionally to the growth of the embryo. The tympanum is completely formed, and the number of sites of the ossification of its walls increases (Fig. 46i).

In the inner ear, the process of the cellular differentiation of the organ of Corti starts. However, the cells have not formed a typical mosaic in their location yet. Their cytoplasm is light. Nuclei are large, have an oval form and include numerous nucleoli. The structure and arrangement of these cells enables us to assume that, further along, the outer and inner hair cells will be formed from them. Under the prospective outer hair cells, the cells with large nuclei are located. It is possible that the Deiters cells will be formed from them, as well as the Hensen and Claudius cells.

Thus, during the early embryogenesis of the peripheral auditory system of a minke whale, the morphological differentiation and maturation of the structures of the outer, middle and inner ears proceed at the same sequence as in terrestrial, semi-aquatic and aquatic mammals [Solntseva, 1993, 1999c].

The adaptive features of the structurally functional organization of the peripheral part of the auditory system of the minke whale, which are connected with the aquatic way of life and ability to a low-frequency location, appear a-t the earliest stages of embryogenesis and continue to develop in a fetal period and in the early postnatal ontogeny [Solntseva, 1985a].

1.3.3 Comparative analysis of development of the auditory organ in mammals with different ecologies

For the first time, a comparative and embryological research of the peripheral auditory system, which was carried out on representatives of terrestrial, semi-aquatic and aquatic mammals with the use of unique embryonic collections of cetaceans and pinnipeds, has allowed us to study the structural organization of the outer, middle and inner ears in representatives of various ecological groups in more detail, and also to determine stages of formation of morphological adaptations and other basic structures revealed by us earlier [Solntseva, 1992, 1993].

The pair anlage of a membranaceous labyrinth is marked at the stage of 2 to 3 pairs of somites [Wilson, 1914]. Further along, at the stage of 6 to 9 pairs of somites, the auditory placode is formed [Kappers, 1941], and at the stage of 14 to 15 pairs of somites, the auditory pit is formed, from which an acoustic vesicle develops at the stage of 20 pairs of somites (forelimb bud, stage 13) [Titova, 1968]. At the same stage, in all investigated species, the bud of the auditory ossicles in the form of a mesenchymal condensation was found.

In studied species of mammals, a subdivision of the acoustic vesicle into superior and inferior parts is marked at the stages 14 to 15. The cochlear canal is developed from the inferior part, while the vestibular apparatus is formed from the superior part. In the cochlear canal, the base formed by columnar epithelium, and the roof, which consists of cuboidal epithelium, are well discernible (st. 16). At the same stage, the cochlear canal starts to twist spirally, forming a basal turn of the cochlea. The capsule, surrounding the cochlear canal, consists of a condensed mesenchyme.

In all studied species, at the 16th stage, from the first branchial cleft, a deepening, on which edge small protuberances arise, is formed. From the lateral walls of the first branchial cleft, the cartilaginous part of the auditory meatus starts to develop.

The middle ear of mammals is laid by a protrusion of the first pharyngeal recess, the endoderm of which will be transformed into a common tube-tympanic protrusion. It has already been mentioned that in all investigated species, the bud of the auditory ossicles appears at the 13th stage in the form of a condensation of a mesenchyme and is located separately from the temporal pyramid bud. At the 16th stage, in the buds of the auditory ossicles, their contours, which are connected among themselves continuously, become visible. At this stage, the tympanum is represented by a narrow, blind canal, which is located below the bud of the auditory ossicles. The bud of a tympanic membrane is marked.

At the 17th stage, in the inner ear, the medial turn of the cochlea is formed. In the outer ear, by means of fusion of the protuberances, which are located at the edges of a deepening, a common mesenchymal bud of an auricle is formed. In the middle ear, each of the buds of the auditory ossicles is represented by an already independent formation; their base is formed by the immature precartilaginous tissue. In all species, the tympanic membrane is thick and very friable. The size of the tympanum is increased.

At the 18th stage, in the inner ear of most studied species, anatomic formation of the cochlea comes to the end, and the apical turn is formed. The cartilaginification of an aural capsule is marked. The elements of the organ of Corti are at the same stage of cellular differentiation in all turns of the cochlea. At this stage of development, the columnar epithelium of the cochlear canal moves apart; therefore, two thickenings – an axial and a lateral – are formed. In the outer ear, the auricle acquires more distinct contours, its small curl appears. At the same stage, in some representatives of phocids (ringed seal), the formation of the auricle, which in the fetal period does not develop and in adult forms is absent, has been marked.

In all investigated species, the cartilaginous part of the external auditory meatus is represented by a short, direct canal. Its osseous part is formed later: in immature-born mammals (rats, bats) in an early postnatal development, and in mature-born species (cetaceans, pinnipeds, artiodactyls) before birth. Filling of the auditory meatus with epithelial cells starts in its distal part and spreads gradually to the proximal part. In the middle ear, the formation of structural elements of the auditory ossicles, which are increased in size and are plunged in the depth of a tympanum, is marked. In the malleus, a head, neck and handle are well expressed. In the incus, the body and both processes (long and short) are formed. The stapes is subdivided into crura and a footplate. At this stage of development, the auditory ossicles are so increased that their size is only 4.5 to 5 times smaller than that they will adopt during the definitive period. The base of the auditory ossicles is formed by a mature precartilaginous tissue, but already at the end of this stage, the replacement of the mature precartilaginous tissue by an embryonic hyaline cartilage is marked. The process of cartilaginification starts in the center of each bud of the auditory ossicles and spreads gradually to their periphery. The auditory ossicles are surrounded by a perichondrium, which consists of small flat cells – chondroblasts. Due to the perichondrium, the places of juncture of the auditory ossicles are well noticeable. At the same stage, the formation of the features connected with interposition of the auditory ossicles is marked. The auditory ossicles change position due to their turning around the sagittal and frontal axes of the animal's body. The stapes is located more caudally in relation to the malleus and incus. As a result of the turning, the bottom lamina of the stapes appears to be located caudally. The tympanic membrane is considerably thinned, has a three-layer structure, and is located almost horizontally on the lateral surface of the middle ear's cavity.

In odontocetes and mysticetes, in the early prefetal period, the tympanic membranes-ligaments reveal similarity in structure at similar stages of development, whereas in the fetal period, their structure acquires species-specific features [Solntseva, 1992].

At the 19th stage, in the inner ear, the differentiation of elements of the cochlear canal and cells of the organ of Corti is marked. The flattening of cells of a cuboidal epithelium and loosening of the connective tissue adjacent to this epithelium occurs. In this place, the formation of the tympanic and vestibular scalae begins. Cells of the organ of Corti are differentiated from the cells of a columnar epithelium, which move apart relative to each other. The differentiation of elements of the organ of Corti begins from the basal cochlea's turn and spreads gradually to the turns located above. As a result, in all turns of the cochlea, a different degree of the anatomic and cellular differentiation is marked.

By the end of the 19th stage, the lateral part of the external auditory meatus is filled with epithelial cells. In cetaceans and pinnipeds, the formation of species-specific features of the outer ear is marked. In pinnipeds, the noticeable narrowing of the cartilaginous part of the auditory meatus in its distal part is observed. In odontocetes, phocids and walruses, the auditory meatus is considerably lengthened, bent and acquires an S-shaped form. Around the auditory meatus, the formation of cartilage is marked. In phocids, around the auditory meatus, four buds of the future cartilaginous laminae, which have different shapes, are formed. In the middle ear, the formation of the tympanum and cartilaginification of the auditory ossicles continues. In different species, the replacement of the mature precartilaginous tissue by a primary cartilaginous tissue occurs non-simultaneously.

For example, in a gray rat, the base of the auditory ossicles is still formed by a mature precartilaginous tissue at this stage. In a guinea pig, the cartilaginification of the malleus has finished, while in the incus and stapes, this process has just begun. In a domestic pig, only the malleus is cartilaginous. In bats, as well as in rats, the auditory ossicles are formed by a mature precartilaginous tissue. In pinnipeds, the maturation of the auditory ossicles occurs non-simultaneously. First, the malleus matures, then the body of the incus, while the lenticular processes of the incus and stapes are formed by a mature precartilaginous tissue. In cetaceans, on the contrary, the stapes and the incus mature first, whereas in the malleus, the process of cartilaginification only begins. However, by the 20th stage, the replacement of the mature precartilaginous tissue by an embryonic cartilage has been completed.

At the 20th stage, in the inner ear of dolphins and bats, a substantial growth of the cochlea's size relative to the size of the vestibular apparatus (2 times) occurs. In all investigated species, in the cochlear canal, the differentiation of a spiral limb, a vascular stria, and a tectorial membrane begins. The vascular stria is formed by an undifferentiated epithelium. The future spiral incisure consists of a multi-row, high columnar epithelium. The size of the spiral ganglion's neurons is increased. The cochlear and vestibular branches of the auditory nerve are well expressed. The differentiation of elements of the cochlear canal and the organ of Corti continues. In the outer ear of odontocetes, phocids, and walruses, the widening of the proximal part of the auditory meatus is marked. In all investigated species, an increase of the absolute size of the auditory meatus occurs proportionally to the growth of the prefetal body. At the given stage, the basic process of formation of the outer ear comes to an end.

In the middle ear of odontocetes, the formation of a cavernous plexus is marked. In prefetal pinnipeds and cetaceans, in the osseous part of the auditory meatus, the formation of the venous sinuses is marked. The development of the peribullar sinuses located between the cranial wall and tympanic bulla is typical of cetaceans only. In odontocetes (white whale), the separation of the tympanic bulla from the cranium is noticeable. At the 20th stage, the second half of an early prefetal period comes to an end. The basic process of formation of the outer, middle and inner ears has finished [Solntseva, 1992; 1993].

In studied species, by the beginning of the 20th stage, the auditory meatus is completely filled with epithelial cells. In mature-born species, these cells are resorbed by the moment of birth; in immature-born species, the process of resorption finishes only during early postnatal development.

For the first time, we have identified the source of the epithelioid obliteration in the distal part of the auditory meatus of some dolphins and the white whale [Solntseva, 1992; 1999c]. In odontocetes, the structure of the auditory meatus differs from that of all investigated species of mammals. Only in odontocetes, the auditory meatus has a strongly pronounced S-shaped form. At some distance from the lumen, the cavity of the auditory meatus disappears; as a result, its two parts are formed – distal and proximal. As we have already mentioned, at the 20 stage, the auditory meatus is completely filled with epithelial cells, which in immature-born species are resorbed completely in the early postnatal ontogenesis only, and in mature-born species by the moment of birth. In mature-born odontocetes, the complete resorption of epithelial cells occurs in the proximal part of the auditory meatus only, while in the distal part, a piece of the embryonic epithelial obliteration is left, not being exposed to a resorption, and later the epithelial tissue of adult forms, which has been found out by us earlier, is formed upon its base [Solntseva, 1971].

The results of comparative study of the peripheral part of the auditory system's development in representatives of various ecological groups have shown that formation of the structures of the outer, middle and inner ears in the early prefetal period occurs at a similar sequence and at approximately similar stages of development. The greatest similarity in the formation of the peripheral auditory system of mammals is marked in the first half of the early prefetal period. Species-specific features in the structural organization of the auditory organ are formed in the second half of the early prefetal period, depending on an ecological specialization of species. The process of cellular differentiation of the organ of Corti and resorption of epithelial cells of the auditory meatus in the mature-born animals (artiodactyls, pinnipeds, cetaceans, etc.) finishes, basically, by the moment of their birth [Solntseva, 1993].

In the immature-born species (rat, mouse), the differentiation of elements of the cochlear canal, cells of the organ of Corti, and also the resorption of the epithelium of the auditory meatus comes to an end only at the 20th day of early postnatal development, and in bats by the 25th to 30th days [Airapetiantz, Konstantinov, 1974], since a part of their fetal period is completed only after birth.

In echolocating forms (bats, dolphins) belonging to different taxonomic and ecological groups, the development of middle and inner ears acquired general properties due to parallel evolution, during which the development of traits for their intraspecific acoustic communication have been created in conditions adverse for vision, and in connection with specific properties of the environment as a channel of acoustic communication.

Based on the results obtained, the following general trends in the development of the peripheral auditory system in representatives of various ecological groups have been determined:

1. In the first half of the early prefetal period (stages 13 to 16), the peripheral auditory system shows structural features common to most mammals;

2. Species-specific features in the structural organization of the peripheral auditory system are formed in the second half of the early prefetal period (stages 18 to 20), depending on an ecological specialization of species;

3. The morphological features of the mammalian peripheral auditory system, which were formed in the early prefetal period, continue to develop in the late prefetal and fetal periods, and during an early postnatal ontogenesis.

Chapter 2. Development of the Vestibular Apparatus in Terrestrial, Semi-Aquatic and Aquatic Mammals

Study of the development of the auditory and vestibular analyzers has allowed us not only to reveal the structural features of these organs in investigated species in more detail, but also to find out the features of their formation in connection with properties of habitat, and also to find an explanation for occurrence of some structural adaptations in the peripheral auditory system in semi-aquatic and aquatic forms, having determined the stages of their formation.

As a result of long-term research of pre- and postnatal development of the vestibular structures in terrestrial mammals, a lot of fundamental works at morphological, electron-microscopical, and physiological levels led in laboratory animals and humans have been collected [Köllicker, 1861; Boettcher, 1869; Alexander, 1900; Held, 1909; Brunner, 1934; Cooper, 1948; Titova, 1968].

Having studied the structural and functional organization of the auditory organ in pre- and postnatal ontogenesis in representatives of different ecological groups of mammals, we turned to a study of another component of the inner ear, the vestibular apparatus, the diverse functions of which in the majority of mammals have a paramount importance in comparison with the auditory function of the cochlea.

In the available literature on the study of prenatal development of the vestibular apparatus in pinnipeds and cetaceans, we failed to find any data that could answer the basic questions concerning the structural organization of the organ of equilibrium and its development at the earliest stages of embryogenesis.

To solve the problem, we have carried out comparative embryological research of the vestibular apparatus, beginning with the formation of an acoustic vesicle, or the stage of a forelimb bud, to the end of the formation of basic anatomic structures, with the purpose to reveal the features of similarity and distinction in formation of the structures of the organ of equilibrium at different stages of development in prenatal ontogenesis in species with different ecological specializations [Solntseva, 2000 a, b; 2001 b].

The inner ear, in which the cochlea and the vestibular apparatus are located, is located in a petrous part (pyramid) of a temporal bone. Because of the complexity of its structure, it is also called a labyrinth. There are two labyrinths: osseous and membranaceous. The osseous labyrinth includes a cochlea, a vestibule and three semicircular canals. The vestibule looks like an oval cavity. It is located between the cochlea and the semicircular canals. The membranaceous labyrinth is located inside the osseous labyrinth, and its form usually repeats the form of a labyrinth but is significantly smaller in size. The walls of a membranaceous labyrinth are formed by a dense connective tissue.

In the osseous vestibule, two membranaceous vesicles are located – a saccule and an utricle. In the osseous cochlea, there is a membranaceous canal of the cochlea, and in the osseous semicircular canals, there are membranaceous semicircular canals. The space between the osseous and membranaceous labyrinths is filled with a liquid – a perilymph. The membranaceous labyrinth also contains a liquid – an endolymph. On the internal surface of the membranaceous vesicles of a vestibule and semicircular canals, there are special formations named as auditory spots and ampullar cristae. The sensory cells, which the fibers of the vestibule nerve approach, are located there. Vestibule and semicircular canals form the so-called vestibular apparatus, or organ of equilibrium. It is represented by a closed system of membranaceous saccules and canals, and consists of a round saccule (*sacculus*), an oval saccule (*utriculus*), and three semicircular canals, which are located in three mutually perpendicular planes. These canals open into the utriculus. At the front end, each semicircular canal forms a widening – ampulla – which is directly connected with the utriculus [Brant, 1970].

All structures of the membranaceous labyrinth include receptor organs: the auditory spots – maculae – which are located in the sacculus, utriculus and *lagena*, peculiar to monotremats; and ampullar cristae, located in the ampullae of the semicircular canals. All these structures differ among themselves in their form, structure and location; however, they have a common origin and are developed from the common anlage of all terminal organs – the *macula communis*.

In the majority of mammals, the receptor spot of the utricular macula is located almost horizontally, and the receptor spot of the saccular macula is located almost vertically, forming a right angle relative to each other. The utricular macula is located in the bottom of the utriculus, as well as on its front and medial walls. The receptor epithelium of the macula has the form of an oval spot and lies on the hardly concave basal membrane. The saccular macula lies on the medial wall of the sacculus in a vertical position, forming with the utricular macula an angle of 90 degrees. Usually, in mammals, the saccular macula is smaller than the utricular macula and has an oval form. As a rule, the base of the saccular macula is located on the vestibular osseous wall. The receptor epithelium of both maculae is single-layered and cylindrical. The receptor epithelium is surrounded by the supporting cells, which have a long and oblong body. Nuclei of these cells lie at different heights, creating the impression that it is a multi-layer epithelium. Hair, or receptor, cells can be jug-shaped or cylindrical, with a large nucleus located at the cell base.

The cristae located in the broadened parts of the semicircular canals, termed ampullae, are referred to as ampullar cristae. Usually, the following cristae are distinguished: an anterior vertical, a posterior vertical and a horizontal (or lateral) one. The receptor epithelium of each crista can be orientated differently, i.e. covering either both slopes of the crista and its apex or clothing only the medial part. As a rule, in most mammals, the receptor epithelium of cristae is not divided; it covers all of crista's crest continuously. The subdivision of the sensory epithelium has been revealed in bats [Iwata, 1924] and cats [Retzius, 1884].

The epithelium of each crista consists of a two-layer receptor epithelium surrounded by a layer of columnar epithelial cells of a transitional epithelium [Kolmer, 1927] located on the periphery of a crest. Receptor and supporting cells lie on a basal membrane and clearly differ among themselves.

Between the receptor and supporting cells, the nerve fibers, which innervate hair cells, pass. Above the crista's receptor epithelium, a cupula is located; due to its large water content in the fixed preparations, it looks like a contracted formation.

The vestibular apparatus has an important role in sensation of body location and movement in space. The receptors of the vestibular apparatus, which react to the change of body positioning in space, are located in the ampullar cristae and auditory spots, which are located in the membranaceous vesicles of a vestibule and in the membranaceous semicircular canals. In case of changing the head positioning or rate of body movement, the endolymph pressure upon sensitive cells of cristae and maculae changes, which, in turn, causes the excitation of the vestibule nerve. The excitation of the vestibule nerve is transferred to the brain. In the cerebral cortex of the brain, a sensation of body positioning in space arises. Together with it, a reflex change of a muscular tonus of different groups of muscles occurs. Due to this mechanism, in mammals, the body equilibrium in space is maintained.

2.1. Terrestrial mammals

2.1.1. Gray rat (*Rodentia-Rattus norvegicus*), Guinea pig (*Rodentia-Cavia porcellus*), Domestic pig (*Artyodactila-Sus scrofa domestica*)

In most mammals, the pair anlage of a membranaceous labyrinth is marked at the stage of 2 to 3 pairs of somites. Further along, at the stage of 6 to 9 pairs of somites, the auditory placode is formed, and at the stage of 14 to 15 pairs of somites the auditory pit is formed, from which, at the stage of 20 pairs of somites (the stage of a forelimb bud), the acoustic vesicle develops. The acoustic vesicle looks like a small pit located on each side of a mesencephalon. The wall of the acoustic vesicle consists of a single-layered epithelium. Further along, the pit increases in size and plunges deeply into an ectoderm. Thus, it is possible to distinguish the endolymphatic canal, which is outlined in the top part of the pit, and in the bottom part to distinguish the beginning of the development of a cochlear canal. In the middle part of the auditory vesicle, two to three layers of epithelial cells are distinct. This epithelial thickening of the medial wall of the auditory vesicle is a anlage of all terminal organs of a labyrinth – the *macula communis*. The macula and the auditory vesicle increase in size and are simultaneously divided into the top and bottom parts. By means of an epithelial bridge, these parts appear to be temporarily connected with each other. Further along, the epithelial bridge is replaced by an indifferent epithelium, and two neuroepithelial spots are formed; one is located in *pars superior*, and the other in *pars inferior*. The anlage of terminal organs, which is located in *pars superior*, gives rise to the development of the utricular macula and *crista ampullaris* of the anterior vertical and horizontal semicircular canals. The anlage of terminal organs in *pars inferior* forms a process inward and backward in the ampulla of the posterior vertical semicircular canal, forming the ampullar crista. The other part of this anlage grows in length and is divided into two anlages – small superior and big inferior. From the first one, the saccular macula is formed; the second one develops further, forming the anlage of the organ of Corti [Titova, 1968].

In the investigated species, the cochlear apparatus is represented by a slit-shaped canal, in which both the base formed by columnar epithelium and the roof consisting of cells of cuboidal epithelium are distinguishable.

At the 14th to 15th stages of development, the subdivision of the inner ear into two parts occurs; the superior part forms the utriculus of both semicircular canals, and the inferior part forms the sacculus and the cochlear canal. Both parts are surrounded by a condensed mesenchyme (Fig. 47a). After the separation of the acoustic vesicle from an ectoderm, it is covered by a multi-row epithelium. Further along, from the lengthened epithelial cells, the epithelium of a basilar membrane and a vascular stria is formed, and from the more rounded epithelial cells, the organ of Corti develops. In terrestrial species, the formation of the vestibular part considerably outstrips the development of the cochlear part.

At the 16th stage, in the superior part of the acoustic vesicle, the differentiation of the vestibular part into the structures forming it (sacculus, utriculus and semicircular canals) begins. In the cochlear part, the lengthened endolymphatic canal, from which the cochlea will be formed later, is distinctly visible. The sites of the top part of the superior saccule's wall become thicker, and from them the flat recesses, whose opposite walls adjoin each other, are formed. Further along, the places of adhesion resolve, and semicircular canals are formed from the external parts of the recesses [Burlet, 1934]. The anterior and posterior vertical semicircular canals develop from a common anlage, and their back ends fall into the middle part of the forming utriculus. The other ends of the semicircular canals fall directly in the utriculus; as a result, the expansions – ampullae – of the semicircular canals are formed.

At this stage, the subdivision of the auditory nerve into the cochlear and vestibular branches with their ganglia has already begun. In the vestibular part, the epithelium is grouped in two to three lines, while the cochlear part is formed by a single-layered epithelium. At the end of this stage, a complete disappearance of the acoustic vesicle occurs, and two totally independent parts are formed – vestibular and cochlear. In the formed vestibule, two groups of cells can be distinguished; from them, the sacculae and three ampullae of the semicircular canals are formed. The ampullae have a flat form and a hardly visible auditory crest, the epithelium of which consists of several layers. The sensory epithelium of the cristae is noticeably divided into receptor and supporting cells. In the apical part of the receptor cells, a vacuolization, which can be connected with the beginning of cupula formation, is marked.

The cochlear canal starts to twist spirally, forming a lower, or basal, turn of the cochlea surrounded by an aural capsule, which consists of a condensed mesenchyme. At the given stage, the formation of the cochlea follows the formation of vestibular structures.

At the 17th stage, the utriculus and sacculus are already separated from each other (Fig. 47b). A nerve approaches each of them. The multi-row epithelium has transformed into a three-row epithelium. All neuroepithelial formations – the auditory spots of saccules of a vestibule, the auditory crest of ampullae of semicircular canals, and the organ of Corti – start to develop from a great number of epithelial cells, which are disorderly and form approximately 8 to 10 rows of these cells. In the process of development, the quantity of epithelial cells considerably decreases. By the end of the first half of an early prefetal period and the beginning of the second half, the form of cells and the quantity of their rows change; the cells become more mature.

At the given stage of development, the lumen of semicircular canals increases. The utriculus acquires a more oval form, and the sacculus a rounded form (Fig. 46b). In the utricular macula, the initial cellular differentiation of the sensory epithelium into the receptor and supporting cells is marked.

The connection between the utriculus and sacculus is provided by a narrow canal – *ductus utriculosaccularis* – in which *ductus endolymphaticus* also opens.

In the cochlear part, the second (medial) turn of the cochlea is formed. The process of formation of separate structures of the cochlea begins; during it, a basic, or basilar, membrane is first formed, and only then a vascular stria. In the basilar membrane, the epithelial cells are located on its surface. These cells are elongated, and the nuclei are rounded and lie at the cell base.

The cochlea innervation is complicated. Several nerve trunks are connected to all of the cochlea's turns and give rise to a number of nerve fibers, which take spiral and radial directions. Spiral ganglion cells reach maturity, which is maintained until the birth of the fetus.

At the 18th stage of development, the ampullae of semicircular canals have an oval form (Fig. 47b). On the bottom wall of the ampulla, a crest has appeared; on its surface, a multi-row epithelium with a cupola, starting to be formed, is located. The receptor spot of the utricular macula acquires a more horizontal position in relation to the receptor spot of the saccular macula, which lies almost vertically. As a result, both spots form a right angle relative to each other. The macula represents a receptor formation covered by a columnar epithelium. Each macula is an independent organ and performs a very specific function.

In the formed ampullae of the semicircular canals, the auditory, or ampullar, cristae are located, the receptor epithelium of which is similar to the receptor epithelium of maculae. Above the surface of a sensory epithelium, an otolithic membrane is located. At the given stage, in all investigated species, apart from the anterior and posterior vertical semicircular canals, the formation of the horizontal semicircular canal continues. The differentiation of the sensory epithelium of maculae and cristae in the utricular and saccular maculae is marked.

At the given stage of development, the formation of the cochlea finishes with the formation of the last apical turn. In *Sus scrofa domestica* and *Rattus norvegicus,* the cochlea is formed by 3 turns (Fig. 47c, d). In *Cavia porcellus,* the height of the cochlea reaches 4.5 turns. The elements of the cochlear canal are in the process of formation. The cells of the organ of Corti are at the same stage of differentiation in all turns of the cochlea. From the columnar epithelium of the cochlear canal, two thickenings are formed; from them, the structures of the cochlear canal and the cells of the organ of Corti are formed.

At the 19th stage of development, the differentiation of the sensory epithelium into the receptor and supporting cells occurs simultaneously in several sites of maculae and cristae, covering most of their surface. The structure of the cells, which form a mosaic distribution pattern, as well as the cells of the organ of Corti, is well visible. The neurons of the vestibular ganglion contain large, oval nuclei with expressed nucleoli. The size of all structures of the vestibular apparatus is considerably increased. The size of the utriculus surpasses the size of the sacculus. The basic process of formation of the neuroepithelial cells in the saccules of the vestibule and otolithic membrane has ended. In the utricular and saccular maculae, and also in the ampullae of the semicircular canals, the epithelial cells transform into the differentiated epithelium. Nerve fibers connect with the cells of the maculae. The development of the utricular macula considerably outstrips the development of the saccular macula.

At this stage, the distinctions between the receptor and supporting cells are marked. Receptor cells are located in one row in the top part of the sensory epithelium; they have a cylindrical form with large, light nuclei of a rounded form. Nuclei usually lie in the basal part of the cells. Cellular borders of the supporting cells are hardly visible. The top parts of the supporting cells are sharply narrowed and are located between the bodies of the receptor cells. Nuclei of the supporting cells are dark and small, and are located in the bottom part of the cells. The structure of the receptor epithelium of cristae is very similar to that of maculae. As it has been marked, the basic difference between the receptor epithelium of maculae and the receptor epithelium of cristae is that in maculae the receptor epithelium is strongly vacuolated [Titova, 1968]. Above the surface of the receptor epithelium, a well-developed cupula with a smooth base is visible. The cupula consists of twisted bars.

In the cochlea, the formation of structures of the cochlear canal continues. The flattening of cells of a cuboidal epithelium and the loosening of a connective tissue adjacent to this epithelium occurs. In this place, the tympanic and vestibular scalae are formed. The differentiation of the cells of the organ of Corti starts in the basal turn of the cochlea from the moment of moving apart of the cells of a columnar epithelium and spreads gradually to the turns located above. As a result, in all turns of the cochlea, a different degree of the anatomic and cellular differentiation is observed.

At the given stage, a tectorial membrane, a spiral incisure, a basilar membrane and the Claudius cells are formed. The formation of the vascular stria and further differentiation of the hair, as well as of supporting Hensen and Deiters cells, continue.

At the 19th to 20th stages of development, the vestibular apparatus is twice the size of the cochlea. All elements of the vestibular apparatus are formed and well marked.

Sensitive and supporting cells are located more rarefiedly, forming free spaces between them. The size of the ampulla of the semicircular canal considerably increases; the canal widens and becomes more rounded (Fig. 47d). The cellular differentiation of the sensory epithelium of maculae, cristae, and the cells of the organ of Corti continues. The cochlear and vestibular branches of the auditory nerve are formed. At the given stage, the basic process of anatomic formation of the structures of the inner ear has ended.

Thus, the formation of the structures of the inner ear in terrestrial mammals occurs in the early prefetal period and is extended in time, which is caused by a heterochrony in the inner ear development. In early embryogenesis, the auditory and vestibular structures are simultaneously separated from each other, and investigated species of mammals reveal similar features in structure (stages 13 to 15). Specific features in the structural organization of the inner ear are formed in the second half of an early prefetal period at similar stages of development and in a certain sequence [Solntseva, 2000b; 2001b; 2002b].

In terrestrial mammals, the initial cellular differentiation of the sensory epithelium occurs in the utricular macula, which indicates the important role of organ gravitation in the vital functions of these mammals.

2.1.2. Greater horseshoe bat (*Chiroptera: Microchiroptera-Rhinolophus ferrumequinum*)

At the 13th stage of development, in the middle part of the acoustic vesicle, two to three layers of epithelial cells are well discernible. This epithelial thickening of the medial wall of the acoustic vesicle is a anlage of all terminal organs of a labyrinth – *macula communis*. The macula and the acoustic vesicle increase in size and are simultaneously divided into superior and inferior parts. By means of an epithelial bridge, these parts are temporarily connected with each other. Later, the epithelial bridge is replaced by an indifferent epithelium and two neuroepithelial spots are formed, one of which is located in *pars superior*, and the other in *pars inferior*. The anlage of terminal organs, which is located in *pars superior*, gives rise to the development of *macula utriculi, crista ampullaris*, and the anterior vertical and horizontal semicircular canals. In *pars inferior*, the anlage of terminal organs gives a process inward and backward to the ampulla of the posterior vertical semicircular canal, forming the ampullar crista. Another part of this anlage grows in length and is divided into two anlages – small superior and big inferior. The saccular macula is formed from the first anlage; the second develops further, forming the anlage of the organ of Corti [Titova, 1968].

At the 14th to 15th stages of development, the subdivision of the inner ear into two parts occurs; from the top part, a utriculus and both semicircular canals are formed, and from the bottom part, a sacculus and the cochlear canal are developed. After the separation of the acoustic vesicle from an ectoderm, it becomes covered by a multi-row epithelium. Further along, from the lengthened epithelial cells, the epithelium of a basilar membrane and a vascular stria is formed, and from the more rounded epithelial cells, the organ of Corti develops.

At the 16th stage, in the top part of the acoustic vesicle, the differentiation of the vestibular part into structures forming it (sacculus, utriculus and semicircular canals) begins. In all investigated species, the top part of the walls of the saccule start to thicken; from them, the flat recesses, the opposite walls of which adjoin each other, are formed. Later, the places of adhesion resolve, and the semicircular canals are formed from the external parts of the recesses [Burlet, 1934]. The anterior and posterior vertical semicircular canals develop from the common anlage, and their back ends fall into the middle part of the forming utriculus. The other ends of the semicircular canals fall directly into the utriculus, as a result of which the thickenings – ampullae – of the semicircular canals are formed.

At this stage, the subdivision of the auditory nerve into the cochlear and vestibular branches with their ganglia has already begun. In the vestibular part, the epithelium is grouped in two to three lines, while the cochlear part is formed by a single-layered epithelium. At the end of this stage, the full disappearance of the acoustic vesicle occurs, and two completely independent parts are formed – vestibular and cochlear. In the formed vestibule, two groups of cells can be distinguished; saccules and three ampullae of semicircular canals are formed from them. The sensory epithelium of cristae is noticeably divided into the receptor and supporting cells.

At the 17th stage, the utriculus and sacculus are already separated from each other. A nerve connects each of them. All neuroepithelial formations – the auditory spots of saccules of a vestibule, the ampullar cristae of ampullae of semicircular canals, and the organ of Corti – start to develop from a great number of the epithelial cells that are disorderly and form approximately 8 to 10 rows. In the process of development, the quantity of epithelial cells decreases considerably. By the end of the first half of the early prefetal period and the beginning of the second half, the form of cells and the quantity of their rows change, and the cells become more mature.

At the given stage of development, the lumen of the semicircular canals increases. The utriculus acquires a more oval form, while the sacculus acquires a rounded form (Fig. 31a, e, f). In the utricular macula, the initial cellular differentiation of the sensory epithelium into the receptor and supporting cells is marked.

The connection between the utriculus and sacculus is provided by a narrow canal – *ductus utriculosaccularis* – into which *ductus endolymphaticus* also opens.

The cochlea's innervation is complicated. Several nerve trunks are connected with all the cochlea's turns and give from themselves a set of nerve fibers, which adopt spiral and radial directions. Cells of the spiral ganglion reach maturity, which is maintained until the birth of the fetus.

At the 18th stage of development, the ampullae of the semicircular canals have an oval form. On the bottom wall of the ampulla, the crest has appeared; on its surface a multi-row epithelium with a cupola, which starts to be formed, is located. The receptor spot of the utricular macula acquires a more horizontal position in relation to the receptor spot of the saccular macula, which lies almost vertically. As a result, both spots form a right angle relative to each other. The macula represents a receptor formation covered by a columnar epithelium. Each macula is an independent organ and performs a very specific function.

In the formed ampullae of the semicircular canals, the auditory, or ampullar, cristae are located, the receptor epithelium of which is similar to the receptor epithelium of maculae. Above the surface of a sensory epithelium, an otolithic membrane is located. As opposed to the anterior and posterior vertical semicircular canals, the formation of the horizontal semicircular canal continues. The differentiation of the sensory epithelium of maculae and cristae in the utricular and saccular maculae is marked.

At the 19th stage of development, the differentiation of the sensory epithelium into the receptor and supporting cells occurs simultaneously in several sites of maculae and cristae, covering most of their surface. The structure of the cells, which form a mosaic distribution pattern, as well as the cells of the organ of Corti, is well visible. The size of all structures of the vestibular apparatus is considerably increased. The size of the utriculus surpasses the size of the sacculus. In the utricular and saccular maculae, and also in the ampullae of the semicircular canals, the epithelial cells transform into the differentiated epithelium. Nerve fibers are connected with the cells of maculae. The development of the utricular macula considerably outstrips the development of the saccular macula.

At this stage, the distinctions between the receptor and supporting cells are marked. Receptor cells are grouped in one row in the top part of the sensory epithelium; they have a cylindrical form with large, light nuclei of a rounded form. Nuclei usually lie in the basal part of cells. Cellular borders of the supporting cells are hardly visible. The top parts of the supporting cells are sharply narrowed and are located between the bodies of the receptor cells. Nuclei of the supporting cells are dark and small, and are located in the bottom part of the cells. The structure of the receptor epithelium of cristae is very similar to that of maculae. As it has been pointed out, the basic difference between the receptor epithelium of maculae and the receptor epithelium of cristae is that in maculae the receptor epithelium is strongly vacuolated.

Above the surface of the receptor epithelium, a well-developed cupula with a smooth base is visible. The cupula consists of twisted bars.

At the 20th stage of development, the vestibular apparatus is 2 to 3 times larger than the cochlea. All elements of the vestibular apparatus are formed and well marked. Sensitive and supporting cells are located more rarefiedly, forming free spaces between them. The size of the ampulla of the semicircular canal increases considerably; the ampulla widens and becomes more rounded. The cellular differentiation of the sensory epithelium of the maculae, cristae and the cells of the organ of Corti continues. The cochlear and vestibular branches of the auditory nerve are formed.

Thus, the formation of the structures of the vestibular apparatus in a greater horseshoe bat, as well as in terrestrial mammals, occurs in the early prefetal period. In early embryogenesis, the auditory and vestibular structures are simultaneously separated from each other and reveal similar features in structure (stages 13 to 15). Specific features in the structural organization of the inner ear are formed in the second half of an early prefetal period at similar stages of development and in a certain sequence.

The initial cellular differentiation of the sensory epithelium occurs in the utricular macula, which indicates the important role of organ gravitation in the vital functions of these mammals. developed semicircular canals are responsible for the air speed of these mammals in different directions, especially during their hunting of insects.

The large cochlea of a greater horseshoe bat and the small, but well-developed, vestibular apparatus indicate that the cochlear, as well as the vestibular, structures are vitally important for this species. The hypertrophied cochlea indicates an important role of hearing in vital functions of a greater horseshoe bat and, undoubtedly, is an adaptation to echolocation and perception of frequencies of a wide range, including ultrasounds.

2.2. Semi-aquatic mammals

2.2.1. Bearded seal (*Pinnipedia: Otariidae-Ergnathus barbatus*)

In a bearded seal, as well as in terrestrial mammals, at the stage of 20 pairs of somites (forelimb bud, stage 13), an acoustic auditory vesicle develops; at the stages 14 to 15, it is subdivided into superior and inferior saccules. The utriculus and semicircular canals are formed from the superior saccule, and from the inferior saccule, the cochlear canal and sacculus are developed. The wall of the acoustic vesicle consists of a single-layered epithelium; the epithelial thickening of the medial wall of the auditory vesicle is observed further along; it is a anlage of terminal organs of a labyrinth – *macula communis*. The macula and the acoustic vesicle increase in size and are then simultaneously divided into two parts – superior and inferior. By means of an epithelial bridge, these parts are temporarily connected with each other. Later, the epithelial bridge is replaced by an indifferent epithelium, and then two neuroepithelial spots are formed, one of which is located in *pars superior*, and the other in *pars inferior*.

The anlage of terminal organs is located in *pars superior* and gives rise to the macula utriculi, crista ampullaris, and the anterior vertical and horizontal semicircular canals. The anlage of terminal organs in *pars inferior* gives inward and backward process into the ampulla of the posterior vertical canal, forming its ampullar crista. The other part of this anlage grows in length and is divided into two anlages – small superior and big inferior. From the first one, the saccular macula is formed; the second develops further and forms the anlage of the organ of Corti [Alexander, 1900].

At the 16th stage, the semicircular canals represent hollow tubes with well-expressed ampullae. The cochlear canal has a slit-like form and twists spirally, forming the basal turn of the cochlea, which is surrounded with an aural capsule consisting of a condensed mesenchyme. In the cochlear canal, the base, which is formed by a columnar epithelium, and a roof, which consists of a cuboidal epithelium, are well discernible [Solntseva, 1997 a].

At the 17th stage of development, the lumens of the semicircular canals are considerably increased. The formation of the posterior vertical and horizontal semicircular canals continues. In the anterior vertical semicircular canal, an ampulla is formed (Fig. 48a). The cellular differentiation of the sensory epithelium into receptor and supporting cells is marked neither in sccular macula, nor in utricular macula. In the cochlea, the medial turn is formed. The differentiation of the receptor and supporting cells of the organ of Corti is absent.

In a bearded seal, at the 18th stage, as in other mammals, the receptor spot of the utricular macula acquires the horizontal position in relation to the vertically located receptor spot of the saccular macula. As a result, both receptor spots form a right angle relative to each other (Fig. 48b). Despite their structural similarities, each macula is an independent organ and performs a very specific function.

In the ampullae of semicircular canals, the ampullar cristae are located, the receptor epithelium of which, according to its structure, is similar to the receptor epithelium of the maculae. The crista lies perpendicularly to the surface of a semicircular canal. The sensory epithelium of the cristae covers its entire crest and both lateral surfaces. The cochlea finishes the anatomic formation with the formation of the apical turn. In adult forms of the bearded seal, the cochlea forms 2.5 turns, which are already anatomically formed at an early stages of embryogenesis.

At the 19th stage of development, all structures of the organ of equilibrium are considerably increased in size – the sacculus, utriculus, semicircular canals, ampullar cristae and ampullae of semicircular canals. In terrestrial mammals, the utriculus has an oval form and its size prevails over the size of sacculus, which has a rounded form. In a bearded seal, the size of the utriculus and sacculus and also their form (oval) possess some features of similarity (Fig. 48c). At the given stage, the cellular differentiation of the receptor and supporting elements is marked. In terrestrial mammals, the earlier differentiation of the sensory epithelium is marked in the utricular macula [Titova, 1968], and in aquatic mammals (dolphins) in the saccular macula [Solntseva, 1996]. In the representatives of semi-aquatic forms, the bearded seal, the cellular differentiation of the sensory epithelium is marked simultaneously in the utricular and saccular maculae.

In the cochlea, the differentiation of elements of the cochlear canal and the cells of the organ of Corti is marked. Flattening of the cells of a cuboidal epithelium and loosening of the connective tissue adjacent to this epithelium occur, and then in this location, formation of the tympanic scala starts, and a little bit later the vestibular scala is formed. The differentiation of the cells of the organ of Corti starts in the basal turn of the cochlea and spreads gradually to the turns located above. As a result, in all cochlea turns, a different degree of the anatomic and cellular differentiation is marked; the basal turn of the cochlea is the most anatomically formed one.

In a bearded seal, at the 20th stage of development, the size of the vestibular and cochlear parts reveals similarity, whereas in terrestrial mammals, the vestibular apparatus is twice the size of the cochlear apparatus, and in the echolocating dolphin (*Stenella attenuata*) the vestibular apparatus is one-half the size of the cochlear apparatus [Solntseva, 1996]. The size of structures of the vestibular and the cochlear apparatuses is increased. The cellular differentiation in the ampullar crista of the horizontal semicircular canal is marked. The size of the spiral ganglion's neurons is increased. The cochlear and vestibular branches of the auditory nerve are formed. The cellular differentiation of the organ of Corti and the sensory epithelium of maculae continues.

Thus, in the early embryogenesis in a representative of semi-aquatic forms – the bearded seal – a vestibular apparatus includes the same components that are typical of other mammals. In terrestrial mammals, the initial cellular differentiation of the sensory epithelium occurs in the utricular macula. The earlier cellular differentiation of the gravitation organ (*utriculus*) points at its vital necessity for terrestrial forms. The simultaneous cellular differentiation of the saccular and utricular maculae into receptor and supporting elements, and also the similarity in size of the vestibular and cochlear parts of the inner ear in the bearded seal, can serve as a base for an assumption that in semi-aquatic forms the organ of gravitation and organ of vibration are equally important, but each of them is adapted for the functioning in a habitat with definite physical properties. In the bearded seal, the early cellular differentiation of the saccular and utricular maculae indicates that the vibration and gravity receptor organs finish their formation by the moment of birth and are completely ready to function in an early postnatal ontogenesis.

2.2.2. Ringed seal (*Pinnipedia Phocidae-Pusa hispida*)

The vestibular apparatus of a ringed seal includes the same structures that are typical of terrestrial, semi-aquatic and aquatic mammals. In most mammals, subdivision of the acoustic vesicle into superior and inferior saccules occurs at the 14th to 15th stages of development. The cochlear canal and sacculus are formed from the inferior saccule, and the utriculus and semicircular canals are formed from the superior saccule.

In the cochlear canal, at the 16th stage of development, the base and the roof are well expressed. As in other mammals, the base is formed by a columnar epithelium, and the roof by a cuboidal one. The cochlear canal twists spirally, forming a lower, or basal, turn of the cochlea. In the vestibular apparatus of a ringed seal, as well as in the majority of species, the thickening of sites of the top part of the wall of the superior saccule occurs, from which the flat recesses are formed. Further along, opposite the walls of the recesses adjoining each other, the places of adhesion are formed; after their resorption from the external parts of the recesses the semicircular canals are formed [Kolmer, 1927; Burlet, 1934]. The anterior and posterior semicircular canals are formed from a common anlage, and their back ends fall into the middle part of utriculus. The other ends of the semicircular canals fall directly into the utriculus, forming the widenings – ampullae – of the semicircular canals.

At the 17th stage of development, in the cochlea, the medial turn is formed. The anatomic formation of structures of the cochlear canal is marked. The cellular differentiation of the organ of Corti is absent.

In the vestibular apparatus, the size of the semicircular canals is considerably increased, and the anatomic formation of utriculus continues. In the anterior vertical semicircular canal, an ampulla is formed. Together with the ampullae, the formation of the posterior vertical and horizontal semicircular canals continues. As well as in other representatives of phocids (*Erignathus barbatus*), at the end of the given stage, differentiation of the sensory epithelium of the saccular and utricular maculae into the receptor and supporting cells occurs [Solntseva, 1997 a].

As well as in definitive forms at the 18th stage, the apical turn of the cochlea is formed; as a result, the cochlea is formed by 2.5 turns. The cells of the organ of Corti are at one stage of differentiation in all turns of the cochlea. In the basal turn of the cochlea, formation of the tympanic and vestibular scalae starts. As well as in other species, in the vestibular apparatus, the receptor spot of the saccular and utricular maculae are located at right angles relative to each other. The maculae are receptor formations; they are similar in structure, but each of them performs a very specific function. In the anterior and posterior vertical semicircular canals, the ampullar cristae are marked; the structure of their receptor epithelium is very similar to the receptor epithelium of maculae. The sensory epithelium covers the whole crest and both lateral surfaces of the crista. In a ringed seal, the ampullar cristae reveal features of similarity with those of terrestrial species.

At the 19th stage of development, in the cochlea, cellular differentiation of the organ of Corti is marked. The cells of the organ of Corti are differentiated from the cells of a columnar epithelium. The process of differentiation starts in the basal turn of the cochlea and spreads gradually to the medial and, further, to the apical turn. In the vestibular apparatus, the size of the semicircular canals, sacculus, utriculus and the ampullar cristae is considerably increased. The formation of the horizontal semicircular canal and its ampulla has ended. The differentiation of the sensory epithelium of the saccular and utricular maculae continues, and also the differentiation of the sensory epithelium into receptor and supporting cells in the ampullar cristae of the anterior and posterior vertical semicircular canals has started.

At the 20th stage of development, anatomic formation of the cochlea has finished. In the cochlear canal, the formation of a spiral limb, a vascular stria, tectorial and Reissner's membranes is marked. The size of the cochlear and vestibular branches of the auditory nerve is increased. The cellular differentiation of the organ of Corti continues.

In a ringed seal, the size of the vestibular apparatus surpasses the size of the cochlea twice, as in terrestrial and other semi-aquatic species. The differentiation of the sensory epithelium of the ampullar cristae of the horizontal semicircular canal is marked. In comparison with the previous stage of development, all vestibular structures are insignificantly increased in size.

The comparative analysis of the development of the auditory and vestibular structures of the inner ear of a ringed seal shows that these structures are simultaneously separated from each other at the 14th to 15th stages of development, as it occurs in most mammals.

At the 16th stage of development, in the outer ear, a rudimentary auricle and a cartilaginous part of the auditory meatus are formed. In the middle ear, the bud of a tympanic membrane and auditory ossicles is marked. In the inner ear, the cochlear canal twists spirally, forming the basal cochlea's turn; in the vestibular apparatus, formation of the semicircular canals and utricular macula is marked.

At the 17th stage, in the outer ear, formation of the cartilaginous part of the auditory meatus continues. In the middle ear, the auditory ossicles are represented by independent formations with clear borders in the places where they join with each other. In the inner ear, the medial turn of the cochlea is formed, and cellular differentiation of the organ of Corti is absent. In the vestibular apparatus, the ampulla of the anterior vertical semicircular canal is formed, and differentiation of the sensory epithelium of the saccular and utricular maculae into the receptor and supporting cells is marked.

At the 18th stage, in the outer ear, the rudimentary pinna is well expressed; the auditory meatus is lengthened, sharply narrowed in its distal part and has an S-shaped form. In the middle ear, thinning of tympanic membrane is marked; the interposition of auditory ossicles in the cavity of the middle ear is the same as in definitive forms. In the inner ear, the apical turn of the cochlea is formed. In the basal turn of the cochlea, the tympanic and vestibular scalae are formed. In the basal turn of the cochlea, the initial cellular differentiation of the organ of Corti is marked in the form of separation of cylindrical cells. In anterior and posterior vertical semicircular canals, the ampullar cristae are marked; the structure of their receptor epithelium is very similar to the receptor epithelium of maculae. As in other mammals, the sensory epithelium covers the whole crest and both lateral surfaces of the crista. The receptor spots of the saccular and utricular maculae are located at right angles relative to each other.

At the 19th stage, in the outer ear, the cartilaginous part of the auditory meatus is filled with epithelial cells. The formation of four cartilaginous laminae surrounding the auditory meatus is marked. In the middle ear, a turn of the auditory ossicles around the sagittal and frontal axes of the animal's body occurs. In the cochlea, the cellular differentiation of the organ of Corti and the anatomic formation of structures of the cochlear canal are marked. In the vestibular apparatus, formation of the horizontal semicircular canal, as well as the formation of the ampulla, has finished. The differentiation of sensory epithelium of the saccular and utricular maculae continues, and also an initial differentiation of sensory epithelium into the receptor and supporting cells in ampullar cristae of anterior and posterior semicircular canals is marked.

At the 20th stage, the distal part of the auditory meatus is filled with epithelial cells. In the middle ear, the tympanic membrane has a three-layer structure and is located horizontally on the lateral wall of the middle ear's cavity. In all turns of the cochlea, the cellular differentiation of the organ of Corti continues. The size of the vestibular apparatus surpasses the size of cochlea twice. Differentiation of the sensory epithelium of ampullar cristae of horizontal semicircular canal is marked.

Thus, as comparative analysis shows, as in most of mammals, development of the hearing and equilibrium organs in a representative of semi-aquatic mammals – the ringed seal (*Pusa hispida*) – occurs in a series. However, in a ringed seal, the differentiation of the sensory epithelium into receptor and supporting cells in the saccular and utricular maculae, as well as in other representatives of phocids, occurs simultaneously and considerably earlier than the differentiation of the cells of the organ of Corti. In a ringed seal, the simultaneous differentiation of the sensory epithelium in the saccular and utricular maculae indicates that the vibration and gravity receptor organs carry out significant functions and are vitally important for these species. The substantial growth of the size of the vestibular apparatus relative to the size of the cochlea can allow for the assumption that functions of the gravitation organ have a paramount importance in comparison with other functions of the inner ear structures [Solntseva, 1999 b].

2.2.3. Steller sea lion *(Pinnipedia: Otariidae-Eumetopias jubatus)*

In a Steller sea lion, as in terrestrial and other semi-aquatic species of mammals, the pair anlage of a membranaceous labyrinth is marked at the stage of 2 to 3 pairs of somites. At the stage of 6 to 9 pairs of somites, the bud of the membranaceous labyrinth represents an auditory placode. Further along, at the stage of 14 to 15 pairs of somites, the auditory pit is formed, from which, at the stage of 20 pairs of somites (13th stage of development, the stage of a forelimb bud), an acoustic vesicle develops, which without distinct borders passes to an endolymphatic duct. In a Steller sea lion, as well as in other mammals, at the 15th stage, subdivision of the acoustic vesicle into the superior and inferior parts occurs. The vestibular apparatus is formed from the superior part, and, from the inferior part, the cochlear canal develops.

At the 16th stage of development, the cochlear canal starts to twist spirally, forming a lower, or basal, turn of the cochlea (Fig. 49a). At the 17th stage, the medial turn of the cochlea is formed and an anatomic formation of the cochlea finishes with the formation of an apical turn at the 18th stage of development. At the given stage, the cochlea is formed by 2.5 turns. At the same stage, cartilaginification of an aural capsule begins. Structures of the cochlear canal and cellular elements of the organ of Corti are at approximately the same stage of development in all turns of the cochlea. At subsequent stages, the size of cochlea increases. At the 19th stage, the formation of the cochlea's axis consisting of connective tissue begins; also, an anatomic formation of elements of the cochlear canal and differentiation of the cells of the organ of Corti begins. The cells of a cuboidal epithelium flatten; there is also a loosening of the connective tissue adjacent to this epithelium. In these places, formation of the tympanic and vestibular scalae starts. The differentiation of the cells of the organ of Corti begins in the basal turn of the cochlea and spreads gradually to the turns located above. As a result, in all turns of the cochlea, a different degree of anatomic and cellular differentiation of elements of the cochlear canal is marked. At subsequent stages, the size of the cochlea increases and the cellular differentiation of the organ of Corti, which in a Steller sea lion comes to an end by the moment of birth, continues.

The vestibular apparatus of a Steller sea lion includes structures that are also typical of other species. It consists of three semicircular canals (two vertical and one horizontal), located in three mutually perpendicular planes, and membranaceous saccules – a round sacculus and an oval utriculus. In the widened parts of semicircular canals – ampullae – the receptor structures, the ampullar cristae, are located. The ampullae of semicircular canals are connected with the base of the utriculus. The receptor structures of the sacculus and utriculus are represented by auditory spots, or maculae: a *macula utriculi*, located at the base of the utriculus, a *macula neglecta*, located on the medial wall of the utriculus, and a *macula sacculi* [Titova, 1968].

As a result of the auditory vesicle's subdivision into the superior and inferior saccules at the 14th to 15th stages of development, the cochlear canal and sacculus are formed from the inferior saccule, and the utriculus and semicircular canals from the superior saccule.

At the 16th stage of development, the similarity with terrestrial species in the formation of elements of the vestibular apparatus is revealed in a Steller sea lion. The thickening of sites of the upper wall of the top saccule with the subsequent formation of flat recesses, the opposite walls of which adjoin closely to each other, is marked. Further along, as has been shown, these places of adhesion resolve, and from the external parts of the recesses, the semicircular canals are formed. The posterior ends of two vertical semicircular canals fall into the middle part of the utriculus (Fig. 49a). The other ends of these semicircular canals fall into the utriculus and the widenings – ampullae – are formed.

At the 17th stage of development, the size of the semicircular canals, sacculus, utriculus and ampullar cristae (Fig. 49b) increases. The initial cellular differentiation in the utricular macula that is typical of terrestrial mammals is marked. Further along, the cellular differentiation of the sensory epithelium of the saccular macula into the supporting and receptor elements, and then into the ampullar cristae, is carried out. In different species, the size of the lumens of the semicircular canals (two vertical and one horizontal) varies. In a Steller sea lion, as well as in the species whose way of life is to a great degree connected with a stay on a firm substratum, the semicircular canals are considerably widened in diameter. In aquatic forms, the size of both vestibular apparatuses relative to the cochlear part and its separate components, including semicircular canals, is considerably reduced.

At the 18th stage of development, the receptor spot of the utricular macula acquires a more horizontal position relative to the receptor spot of the saccular macula, which is located almost vertically. As a result, both spots are located at right angles relative to each other, as in other mammals. Each macula is an independent organ and performs a very specific function. The receptor epithelium of the ampullar cristae reveals similarity in its structure with the receptor epithelium of the maculae. The sensory epithelium of the ampullar crista covers its crest and both lateral surfaces. The earlier formation of the ampulla is marked in the posterior vertical canal, while in the anterior vertical and horizontal semicircular canals, the formation of the ampullae occurs at the 19th stage of development.

At the 19th stage of development, in a Steller sea lion as well as in terrestrial mammals, the differentiation of the receptor epithelium occurs simultaneously in several sites of maculae and cristae, spreading over large areas. The structure of the receptor and supporting cells, which are, like the cells of the organ of Corti, mosaically located, is clearly visible. The auditory nerve is subdivided into cochlear and vestibular branches (Fig. 49c).

At the 20th stage of development, the structure of the vestibular apparatus continues to grow in size. The differentiation of the sensory epithelium into the ampullar crista of the horizontal semicircular canal is marked. The differentiation of the receptor epithelium of the saccular macula follows the differentiation of the receptor epithelium of the utricular macula.

In the pre- and postnatal development of a Steller sea lion, as in the majority of terrestrial and some semi-aquatic mammals, the size of the vestibular apparatus is twice the size of the cochlear part. In absolute hydrobionts, on the contrary, the size of the cochlea surpasses the size of the vestibular part of the inner ear, and in such semi-aquatic species as phocids, the sizes of cochlear and vestibular parts are similar.

Despite the fact that in most species the vestibular apparatus includes the same components, in various species specific features in their structure are marked. This is especially clearly shown in the location, form and size of the vestibular structures.

Thus, in a Steller sea lion, at the early stages of development, the anatomic structures of the cochlear and vestibular parts are simultaneously separated from each other and reveal common basic structural features found in the majority of mammals [Solntseva, 1998 a]. In a Steller sea lion, as well as in other species of mammals, in the first half of an early prefetal period, the outer and middle ears have similar features in development of separate structures. The species-specific features in the structural organization of the hearing and equilibrium organs are formed in the second half of an early prefetal period, depending on the ecological specialization of each species.

In the vestibular apparatus of a Steller sea lion, as well as in terrestrial mammals, in comparison with the sacculus (organ of vibration), the greatest development is reached by the utriculus (organ of gravitation). The earlier differentiation of the sensory epithelium into the receptor and supporting cells is marked in the utricular macula. It indicates that, compared to other semi-aquatic mammals, a Steller sea lion spends most of its time on a firm substratum. Therefore, in a Steller sea lion, as well as in terrestrial species, among structures of the inner ear, the gravitation organ has a paramount value and is vitally important in comparison with the organ of vibration, the development of which is marked in aquatic mammals (cetaceans).

2.2.4. Walrus (*Pinnipedia: Odobenidae-Odobenus rosmarus divergens*)

According to ecological specialization, a walrus is ascribed to the group of semi-aquatic mammals whose way of life is connected with both aquatic and terrestrial habitat conditions. However, unlike other pinnipeds, a walrus spends most of its time on land, and this has substantially affected the structure of its peripheral auditory system and vestibular apparatus, which are similar in structure and development to those found in terrestrial mammals.

The vestibular apparatus of a walrus, as well as that of other mammals, includes membranaceous saccules: a round sacculus, or the organ of vibration, and an oval utriculus, or the organ of gravitation, and also three semicircular canals, which react to angular acceleration changes. In the widened parts of the semicircular canals – the ampullae – the receptor structures, the ampullar cristae, are located. The ampullae of the semicircular canals are connected with the base of the utriculus. The receptor structures of the sacculus and utriculus are represented by the auditory spots, or maculae: a *macula utriculi*, which is located at the base of the utriculus, a *macula neglecta*, located on the medial wall of the utriculus, and a *macula sacculi*.

In the walrus, as in other mammals, the acoustic vesicle is formed at the stage of 20 pairs of somites, or the stage of a forelimb bud. The subdivision of the acoustic vesicle into superior and inferior saccules is marked at the 14th to 15th stages of development. From the inferior saccule, the cochlear canal and sacculus are formed, and from the superior saccule, the utriculus and the semicircular canals develop. Both parts are surrounded by a condensed mesenchyme.

In the walrus, at the 16th stage of development, the cartilaginous part of the external auditory meatus starts to form from the lateral walls of the first branchial cleft that is typical of the representatives of different ecological groups of mammals. In the bud of the auditory ossicles, represented by a mesenchymal condensation, the contours of the malleus, incus and stapes, which are continuously connected with each other, are formed. The cochlear canal twists spirally, forming a basal cochlea's turn surrounded with an aural capsule, which consists of a condensed mesenchyme.

In the vestibular apparatus of the walrus, as well as in other mammals, the top part of the wall of the superior saccule thickens, forming flat recesses, the opposite walls of which adjoin each other. Further along, these places of adhesion resolve, and from the external parts of the recesses, the semicircular canals are formed [Burlet, 1934]. Both vertical canals develop from a common anlage, and their posterior ends fall into the middle part of the utriculus (Fig. 50a). The other ends of the semicircular canals fall directly into the utriculus; therefore, the widenings (ampullae) are formed.

The vestibular apparatus of a walrus includes structures typical of other mammals. Three semicircular canals (two vertical and one horizontal) are located in three mutually perpendicular planes. The size of the vestibular apparatus of a walrus is twice the size of the cochlea, whereas in absolute hydrobionts (*Stenella attenuata*), the size of the vestibular apparatus is one-half the size of the cochlea [Solntseva, 1993]. Some phocids (*Erignathus barbatus*) were found to have a cochlea and vestibular parts approximately equivalent in size.

In the vestibular apparatus of a walrus, at the 17th stage of development, the lumens of the semicircular canals are increased. As has been shown in terrestrial mammals [Anson, Black, 1934], the connection between the utriculus and sacculus occurs with the help of a narrow canal – the ductus utriculosaccularis – into which the ductus endolymphaticus opens. The utriculus is connected to the sacculus by means of a sacculo-endolymphatic canal [Brunner, 1934]. Further along, between the sacculo-endolymphatic canal and the sacculus, a medial dorsal fold is formed. The same contact between the sacculus and utriculus has been found in a walrus at the 17th stage of development (Fig. 50a).

At the 18th stage of development, the anatomic formation of the utriculus has basically ended, whereas the formation of the sacculus continues (Fig. 50b). The receptor spot of the utricular macula adopts a more horizontal position relative to the receptor spot of the saccular macula, which, as in other mammals, is located almost vertically. Relative to each other, both spots form a right angle. The differentiation of the receptor formations is marked neither in the saccular macula, nor in the utricular macula.

In the widened parts of the semicircular canals, the ampullar cristae are marked. The structure of the cristae of the semicircular canals in a walrus is similar to the structure of the cristae found in terrestrial mammals. The ampullar crista of the posterior vertical semicircular canal forms much earlier than the crista of the horizontal semicircular canal. In the fetal period, the horizontal and anterior vertical semicircular canals form completely, while the anterior vertical semicircular canal completes development only at the birth [Gagloeva, 1973]. The author believes that the anterior vertical semicircular canal is very important to the vital functions of a developing fetus, as it is this structure that is most important in fixing a body in the vertical position.

At the 19th stage, the semicircular canals, sacculus, utriculus, ampullar cristae and ampullar parts of the semicircular canals form and increase in size (Fig. 50c). In terrestrial mammals, the differentiation of the receptor epithelium occurs simultaneously in several sites of maculae and cristae, spreading over large areas of the organ. In a walrus, a similar differentiation of the sensory epithelium is marked. The receptor and supporting cells form a mosaic distribution pattern, which is also typical of the location of the cells of the organ of Corti. The vestibular ganglion is well developed. Its neurons contain large, oval nuclei with expressed nucleoli.

At the 20th stage, the vestibular apparatus of a walrus is twice the size of the cochlea. The sizes of all structures of the vestibular apparatus are increased. In a walrus, as well as in terrestrial mammals, the initial differentiation of the receptor epithelium of the utricular macula and the ampullar crista of the horizontal semicircular canal is marked. The differentiation of the sensory epithelium of the saccular macula follows the differentiation of the structures mentioned above.

Thus, in the early prefetal period of a walrus, the cochlear and vestibular parts are simultaneously separated from each other, and the vestibular apparatus consists of the same components found in other mammals. The formation of the auditory and vestibular structures in a walrus, as well as in the majority of mammals, is extended in time that is explained by the presence of heterochrony in development of the inner ear. In a walrus, the size of the utriculus considerably surpasses the size of the sacculus, and the initial cellular differentiation of the sensory epithelium occurs in the utricular macula, which is typical of terrestrial mammals. These features of the structure and development of the utriculus indicate the vital necessity of the organ of gravitation in comparison with the organ of vibration in mammals, whose way of life is connected with the land to a greater degree [Solntseva, 1997b; 1998c]. In other representatives of pinnipeds, particularly in phocids (*Erignathus barbatus*), similarity in the size of the sacculus and utriculus is evident, as well as the simultaneous differentiation of the sensory epithelium of maculae on the receptor and supporting cells, which must be connected with the necessity of the functioning of the structures of the vestibular apparatus in both air and water.

2.3. Aquatic mammals

2.3.1. Odontocetes (*Cetacea: Odontoceti*)

Spotted dolphin (*Stenella attenuata*), White whale (*Delphinapterus leucas*)

In representatives of absolute hydrobionts – the spotted dolphin (stenella) and the white whale (beluga) – the auditory pit is formed at the stage of 14 to 15 pairs of somites. From this auditory pit, an acoustic vesicle develops at the stage of 20 pairs of somites (the stage of a forelimb bud, stage 13). The subdivision of the acoustic vesicle into superior and inferior saccules is marked at the 14th to 15th stages. The cochlear canal and sacculus are formed from the inferior saccule, and the utriculus and semicircular canals develop from the superior saccule. Both parts are surrounded by a condensed mesenchyme (Fig. 51a; 52a).

At the 16th stage of development, the sites of the upper part of the wall of the superior saccule thicken, and from them the flat recesses are formed, the opposite walls of which adjoin each other. Further along, these places of adhesion resolve, and from the external parts of the recesses, the semicircular canals are formed. Both vertical canals develop from a common anlage, and their posterior ends fall into the middle part of the utriculus. The other ends of the semicircular canals fall directly into the utriculus, forming the widenings (ampullae).

The vestibular apparatus of *Stenella attenuata* and *Delphinapterus leucas* includes structures that are typical of terrestrial mammals. Three semicircular canals (two vertical and one horizontal) are located in three mutually perpendicular planes. The size of the vestibular apparatus of the dolphin is one-half the size of the cochlea, while in terrestrial forms, the size of the vestibular apparatus is twice the size of the cochlea. Cellular differentiation of the receptor structures is absent.

In the cochlear canal, both the base formed by the columnar epithelium and the roof consisting of cuboidal epithelium are discernible. At this stage, the cochlear canal starts spiralling and the basal turn of the cochlea is formed; it is surrounded with the aural capsule consisting of condensed mesenchyme.

At the 17th stage of development, the lumens of the semicircular canals increase in size. The differentiation of the sensory epithelium into the receptor and supporting cells is marked only in the saccular macula. Further along, the cellular elements differentiate into the utricular macula and, later, in the ampullar cristae (Fig. 51b). In the anterior vertical semicircular canal, the formation of the ampulla has finished. The formation of the posterior vertical semicircular canal together with the ampulla continues, as well as the formation of the horizontal semicircular canal, the lumen of which considerably increases and acquires a rounded form. The cochlea's medial turn is formed. The differentiation of receptor cells of the organ of Corti is absent.

At the 18th stage of development, the size of the semicircular canals continues to increase. The receptor spot of the utricular macula adopts a more horizontal position relative to the receptor spot of the saccular macula, which is located almost vertically. As a result, both receptor spots form almost a right angle relative to each other. The maculae represent receptor formations covered by a columnar epithelium. In terrestrial mammals, the central and peripheral parts of the receptor formations differ in their functional properties. Despite the structural similarity, each of the maculae performs a specific function and is an independent organ.

In the ampullae of the semicircular canals, the ampullar cristae are marked. In structure, their receptor epithelium is similar to the receptor epithelium of the maculae (Fig. 51b). Above the surface of the cristae's receptor epithelium, a small gelatinous cupula, or otolithic membrane, is located. In *Stenella attenuata,* the crista of the anterior vertical semicircular canal is large, practically reaching the opposite wall of the canal. The crista is located perpendicularly to the surface of the semicircular canal. The cristae's crests can be flat or steep. The sensory epithelium of the crista covers its whole crest and both lateral surfaces. In the posterior vertical semicircular canal, the ampulla is formed, while in the horizontal semicircular canal, this process still continues. In humans, the formation of the horizontal and posterior semicircular canals is complete by the 7th month of prenatal development, and the formation of the anterior vertical semicircular canal is complete only at birth [Gagloeva, 1973, 1976]. The author connects it with the fact that the anterior vertical semicircular canal, in comparison with the horizontal and posterior vertical semicircular canals, is vitally important for the developing fetus, as this structure takes part in fixing a body in the vertical position.

The cochleas of a spotted dolphin and a white whale finish their anatomic formation with the formation of the apical turns and include 2.0 turns

(Fig. 52b). The cartilaginification of an aural capsule is marked. The elements of the organ of Corti are at the same stage of the cellular differentiation in all cochlea turns. At this stage of development, the columnar epithelium moves apart and two thickenings are formed (axial and lateral) from which the cells of the organ of Corti and the anatomic structures of the cochlear canal will further develop (Fig. 52c-d).

At the 19th stage of development, the size of the semicircular canals, sacculus, utriculus, ampullar cristae and ampullar parts of the semicircular canals increases (Fig. 52c). In the dolphin, differentiation of the saccular macula occurs mostly in the central site of the sensory epithelium (Fig. 51c), while in terrestrial mammals, the process of differentiation of the receptor epithelium occurs simultaneously at several sites of maculae and cristae, spreading over large areas of the organ [Titova, 1968]. In the dolphin, the different structures of the receptor cells and supporting cells are clearly visible. The receptor cells are gathered in one row in the upper part of the sensory epithelium. The body of a cell has a cylindrical form and contains a large, oval nucleus. The nuclei of the cells are light (Fig. 51d).

The supporting cells extend from the basal membrane to the apical surface of the receptor epithelium. The narrowed parts of the supporting cells are located between the receptor cells. The borders of the supporting cells are not visible. Their nuclei are small, dark and oval. The receptor and supporting cells, as well as the cells of the organ of Corti, form a mosaic distribution pattern. The vestibular ganglion is well developed. Its neurons contain large, oval nuclei with expressed nucleoli.

In the cochlea, the differentiation of the elements of the cochlear canal and the cells of the organ of Corti is marked. The flattening of the cells of the cuboidal epithelium, and the loosening of connective tissue adjacent to this epithelium, occur. At this location, the formation of the tympanic and vestibular scalae begins. The cells of the organ of Corti are differentiated from the cells of the columnar epithelium, which move apart relative to each other. The differentiation of the elements of the organ of Corti starts in the basal cochlea's turn and spreads gradually over the turns located above. As a result, in all turns of the cochlea, a different degree of anatomic and cellular differentiation is marked.

At the 20th stage of development, the size of the cochlea increases substantially, so much so that it becomes twice the size of the vestibular apparatus. The size of the semicircular canals, sacculus, utriculus and ampullar cristae increases. The differentiation of the sensory epithelium in the ampullar crista of the horizontal semicircular canal begins. However, the differentiation of the receptor epithelium of the cristae and utricular macula lags behind the differentiation of the sensory epithelium of the saccular macula.

At this stage, the cochlea is anatomically formed. In the cochlear canal, the differentiation of a spiral limb, a vascular stria, tectorial membrane and Reissner's membrane starts. The vascular stria is formed from an undifferentiated epithelium. The future spiral incisure consists of a multi-row, high columnar epithelium. The size of the neurons of the spiral ganglion increases. The vestibular and cochlear branches of the auditory nerve are well expressed. The differentiation of the elements of the cochlear canal and the cells of the organ of Corti continues.

Thus, during the early prefetal period in the family Delphinidae, the anatomic structures of the vestibular and cochlear apparatuses are simultaneously separated from each other, which has also been observed in terrestrial and semi-aquatic mammals. However, in aquatic forms, the formation of the auditory and vestibular structures is prolonged due to the presence of heterochrony in the development of the inner ear.

In a spotted dolphin and in a white whale, the earlier differentiation of the sensory epithelium into receptor and supporting cells is marked in saccular macula, while the cytological differentiation of the cells of the organ of Corti, utricular macula and ampullar cristae has not yet begun. In terrestrial mammals, the initial cellular differentiation occurs in utricular macula. The earlier differentiation of the sensory epithelium of the saccular macula in dolphins indicates that in aquatic mammals the organ of vibration carries out a more important function in comparison with the organ of gravitation, which in terrestrial forms, on the contrary, develops earlier and is vitally important for them. It can be supposed that in a spotted dolphin and in a white whale, the substantial increase of the cochlea's size relative to the size of the vestibular apparatus is connected with its paramount auditory function among structures of the inner ear, namely with the development of a special way of spatial orientation and communication – echolocation [Solntseva, 1996; 1999a].

2.3.2. Mysticetes (Cetacea: Mysticeti)

Minke whale (Cetacea: Mysticeti-*Balaenoptera acutorostrata*)

The vestibular apparatus of a minke whale, as well as that found in other mammals, consists of three semicircular canals (two vertical and one horizontal) located in three mutually perpendicular planes, and membranaceous saccules – a round sacculus and an oval utriculus. In the widened parts of the semicircular canals, the ampullae, the receptor structures, the ampullar cristae, are located. The ampullae of the semicircular canals are connected with the base of the utriculus. The receptor structures of the sacculus and utriculus are represented by the auditory spots, or maculae: a utricular macula, located at the base of the utriculus, a *macula neglecta*, located on the medial wall of the utriculus, and a saccular macula.

The subdivision of an acoustic vesicle into the superior and inferior saccules is marked at the 14th to 15th stages of development. The cochlear canal and sacculus are formed from the inferior saccule, and the semicircular canals and utriculus develop from the superior saccule (Fig. 53a).

The vestibular apparatus of the minke whale includes the same structures typical of other mammals. The size of the vestibular apparatus is 2 to 3 times smaller than the size of the cochlea that is found in representatives of odontocetes (*Stenella attenuata*, *Delphinapterus leucas*). The vertical semicircular canals develop from a common anlage, their posterior ends fall into the middle part of the utriculus. The other ends of the semicircular canals fall directly into the utriculus, forming the widenings (ampullae).

At the 17th stage, the lumens of the semicircular canals are very narrow (Fig. 53b). The initial differentiation of the sensory epithelium into receptor and supporting cells is marked in the macula sacculi. At this point, the formation of the anterior and posterior vertical semicircular canals has begun. In the anterior vertical semicircular canal, an ampulla forms. The lumen of the horizontal semicircular canal widens considerably. The sacculus and utriculus are anatomically formed, and the sacculus is larger than the utriculus. Cellular differentiation of the maculae is absent.

At the 18th stage, the size of the sacculus, utriculus and semicircular canals increases (Fig. 53c), and the utriculus surpasses the sacculus in size. The saccular macula acquires a horizontal position relative to the utricular macula; therefore, both maculae form a right angle relative to each other. Although the maculas have similar structures, each performs a very specific function and is an independent organ. The ampullar cristae are located in the ampullae of the semicircular canals. The receptor epithelium of the ampullar cristae is similar in structure to the receptor epithelium of the maculae. Above the surface of the cristae's receptor epithelium, a small gelatinous cupula (otolithic membrane) is located. As in odontocetes, the crista of the anterior vertical canal is large and is located perpendicularly to the ampulla of the semicircular canal. In the maculae, the initial cellular differentiation of the receptor epithelium is marked.

At the 19th stage, in the vestibular apparatus of a minke whale, earlier differentiation of the sensory epithelium occurs in the macula sacculi. As in terrestrial mammals, it is carried out simultaneously in several sites of maculae and cristae. The structures of the receptor cells and supporting cells, which have a mosaic distribution, are clearly visible. The vestibular and cochlear branches of the auditory nerve are equally well developed and look like narrow, compact bundles.

At the 20th to 21st stages, the structure of the vestibular apparatus grows considerably compared to the previous stage of development.

In contrast to terrestrial mammals, in which the vestibular apparatus is twice the size of the cochlea, the cochlea in a minke whale is 2 to 3 times bigger than the vestibular apparatus (Fig. 53a, b, c). In the ampullar crista of the horizontal semicircular canal, which forms after the vertical canals, the sensory epithelium is differentiated into hair cells and supporting cells. The differentiation of the sensory epithelium into the ampullar cristae and the utricular macula lags behind the differentiation of this epithelium in the saccular macula.

Thus, in this representative of aquatic mammals – the minke whale – the auditory and vestibular structures are simultaneously separated from each other at the early stages of development, as occurs in other mammals. Unlike the cochlear apparatus (the organ of Corti), earlier differentiation in the vestibular apparatus of the sensory epithelium of maculae into receptor and supporting cells is noted. However, by the moment of birth, the vestibular and cochlear structures are equally ready for functioning, beginning from an early postnatal ontogenesis [Solntseva, 1998 b]. In terrestrial mammals, the initial cellular differentiation occurs in the utricular macula, while in a minke whale, as well as in representatives of odontocetes, the initial differentiation of the sensory epithelium is observed in the saccular macula. This indicates that in aquatic mammals the organ of vibration (*sacculus*) carries out a more important function in comparison with the organ of gravitation (*utriculus*), which in terrestrial species, on the contrary, is vitally important. In a minke whale, as in a spotted dolphin and in a white whale, the size of the cochlea considerably surpasses the size of the vestibular apparatus. This enables us to assume that, among the functions of the inner ear, the auditory function has a paramount value, as it is with the help of the sound signals produced and perceived by aquatic mammals that cetaceans can communicate and orient themselves in the ocean.

2.3.3. Comparative analysis of the development of the vestibular apparatus in mammals with different ecologies

The question of the evolutionary origin of a labyrinth in vertebrates still remains unanswered, despite existing hypotheses that explain its evolution beginning with a lancelet up to a mammal. A well-known hypothesis is that the labyrinth has appeared on the basis of lateral line organs, all located openly on the surface of the animal's body and all in direct contact with the environment. A complication of the structures and functions of the lateral line organ caused the appearance of a new structural formation – the vestibular apparatus. However, no researcher has been able to explain how the evolution progressed from an open labyrinth to a closed one located deep in the cranium.

Although the present research does not directly answer this question, it contributes to its solution and provides an important comparative analysis of the early embryogenesis of the vestibular and cochlear parts of the inner ear in terrestrial, semi-aquatic and aquatic mammals. The sensory systems, in particular, demonstrate the range of evolutionary and adaptive transformations that have appeared in mammals during their transition from a terrestrial to an aquatic way of life.

In this chapter, we have generalized the results of a comparative analysis of the prenatal development of the vestibular apparatus and the cochlear apparatus in some representatives of terrestrial, semi-aquatic and aquatic mammals [Solntseva, 2000 a, b, 2001 b, 2002b].

The vestibular apparatus of the investigated species consists of a system of membranaceous saccules and closed canals filled with endolymph. This system is referred to as a membranaceous labyrinth and includes a round sacculus, an oval utriculus and three semicircular canals located in three mutually perpendicular planes. In each of the semicircular canals, there are widenings (ampullae) that form connections with the utriculus. In the ampullae, the receptor structures – the ampullar cristae – are located. The receptor structures of the sacculus and utriculus are the auditory spots (maculae): the saccular macula, located on the lateral wall of the sacculus; the utricular macula, located at the base of the utriculus; and the *macula neglecta*, absent in many species of mammals, located in the inner ear on the medial wall of the utriculus. In terrestrial, semi-aquatic and aquatic mammals, all structures of the membranaceous labyrinth differ among themselves in their location in the inner ear, as well as in their size and form. However, for all species, the presence in the maculae of the otolithic membrane of a gelatinous consistence is typical, as well as the presence of the gelatinous cupula on the tops of the ampullar cristae. The macula is a receptor formation, consisting of the sensory cells, which are covered with the otolithic membrane and small crystals – otoconias – plunged into the otolithic membrane. The ampullar cristae have similar structures, but, unlike the maculae, the surface layer of the cristae's receptor epithelium is covered with a gelatinous cupula.

Another receptor formation of the inner ear is the *papilla basilaris*, which in higher reptilians and birds develops into the auditory organ (*papilla acustica basilaris*), and in mammals into the organ of Corti, which is located in the closed cochlear canal, involuted into a spiral cochlea.

Relative to each other, the receptor spots of the saccular and utricular maculae form a right angle. Although some researchers have posited that there are no basic distinctions in the structures of these maculae [Burlet, 1934], we have been able to determine, through use of the electron microscope, that the sensory epithelium of the organ of equilibrium consists of two types of receptor hair cells. The cells of the first type have a jug-like form, and the cells of the second type have a cylindrical form. More significant distinctions between both types of sensory cells are revealed in connection with the features of their innervation [Wersall et al., 1965]. Studies of other authors have shown that through evolutionary processes the receptor cells of the first type appeared in the inner ear of mammals in connection with the change of position of the animal's body in a gravitational field after their coming out on land [Titova, 1968].

As the comparative analysis of the development of the inner ear has shown, an acoustic vesicle develops in most mammals at the stage of 20 pairs of somites (forelimb bud, stage 13).

In terrestrial mammals, pinnipeds and cetaceans, the acoustic vesicle divides into the superior and inferior saccules at the 14th to 15th stages of development. From the inferior saccule, the sacculus and a cochlear canal are formed, and from the superior saccule, the utriculus and semicircular canals develop. Both parts are surrounded by a condensed mesenchyme. The wall of the acoustic vesicle consists of a single-layer epithelium. The epithelial thickening of the medial wall of the acoustic vesicle is a anlage of terminal organs of a labyrinth – *macula communis*. The macula and the acoustic vesicle increase in size and are simultaneously divided into superior and inferior parts. By means of an epithelial bridge, these parts are temporarily connected to each other. Later, the epithelial bridge is replaced by an indifferent epithelium, and two neuroepithelial spots are formed, one of which is located in the pars superior and the other in the *pars inferior*. The anlage of terminal organs is located in the *pars superior* and gives rise to the development of the macula utriculi and crista ampullaris (ampullar cristae) of the anterior vertical and horizontal semicircular canals. The anlage of terminal organs in the *pars inferior* forms a process inward and backward in the ampulla of the posterior vertical semicircular canal, forming the ampullar crista.

The other part of this anlage grows in length and is divided into two anlages: a small top and a large bottom. Out of the top anlage, the saccular macula is formed; the bottom anlage develops further, forming the anlage of the organ of Corti (Alexander, 1904).

At the 16th stage, the cochlear canal twists spirally, forming a lower, or basal, turn of the cochlea, which is surrounded with an aural capsule consisting of a condensed mesenchyme. At this stage, the formation of the cochlea in terrestrial and semi-aquatic species follows the formation of the equilibrium organ. In most investigated species, the semicircular canals are very narrow in diameter. The sacculus and utriculus are small size and round or oval.

At this same stage of development in representatives of different ecological groups, the sites of the upper part of the wall of the superior saccule thicken, and flat recesses are formed from them, the opposite walls of which adjoin each other. Later, these places of adhesion resolve, and from the external parts of the recesses, the semicircular canals are formed. The inferior and posterior vertical semicircular canals develop from a common anlage, and their back ends fall into the middle part of the utriculus. The other ends of the semicircular canals fall directly into the utriculus, forming the widenings (ampullae).

At the 17th stage of development, the lumens of the semicircular canals, sacculus, utriculus and ampullar cristae increase in size. In otariids, as well as in terrestrial species, the initial cellular differentiation of the sensory epithelium into receptor cells and supporting cells is marked in the utricular macula. In odontocetes and mysticetes, earlier differentiation of the sensory epithelium is marked in the saccular macula, while in phocids and in the walrus, cellular differentiation is marked neither in the saccular macula nor in the utricular macula.

At this stage of development, the cochlear apparatus in all investigated species is a cochlear canal of a slit-like form, whose base, formed by a columnar epithelium, and roof, consisting of cells of a cuboidal epithelium, are well discernible.

In terrestrial species, otariids and the walrus, the size of the vestibular apparatus is twice the size of the cochlear part of the inner ear, the lumens of the semicircular canals are wide, the utriculus has an oval form and the sacculus has a rounded form. In phocids, the size of the cochlear and vestibular parts of the inner ear is similar, while the form and size of the sacculus, utriculus and semicircular canals are similar to those found in terrestrial species. In cetaceans, the vestibular apparatus is extraordinarily small. The utriculus and sacculus are connected by means of a narrow canal (ductus utriculosaccularis), which opens into the ductus endolymphaticus. The utriculus is connected to the sacculus by means of a sacculo-endolymphatic canal. In the cochlear part of the inner ear, the medial turn of the cochlea is formed. The structures of the cochlear canal are not formed, and cellular differentiation of the organ of Corti is absent.

At the 18th stage of development, in all studied species of mammals, an increase in the size of the structures of the vestibular apparatus occurs proportionally to the growth of the prefetal body. In otariids, as well as in other species whose way of life is to a great degree connected with land, the size increase of all structures of the vestibular apparatus occurs proportionally to the size increase of the cochlea. In absolute hydrobionts (cetaceans), the increase of the cochlea's size considerably outstrips the growth of the structures of the organ of equilibrium. The receptor spot of the utricular macula acquires a more horizontal position in relation to the receptor spot of the saccular macula, which lies almost vertically. As a result, both spots form a right angle relative to each other. The maculas represent the receptor formations covered by a columnar epithelium. Each of them performs a very specific function.

In cetaceans and pinnipeds, as well as in terrestrial species, the ampullar cristae are located in the formed ampullae of the semicircular canals. The receptor epithelium of the ampullar cristae is structurally similar to the receptor epithelium of the maculae. Above the surface of the sensory epithelium, the otolithic membrane, which in aquatic species is much thinner than in terrestrial and semi-aquatic mammals, is located. In cetaceans, the ampullar crista is large and occupies a significant part of the ampullar space of the semicircular canals. The receptor epithelium covers the whole surface of the crista. At this stage, unlike the formation of the anterior and posterior vertical semicircular canals, the formation of the horizontal semicircular canal continues in all studied species. In humans, the growth of the horizontal and posterior vertical semicircular canals comes to an end by the 7th month of prenatal development, while the growth of the anterior vertical semicircular canal stops only at birth [Gagloeva, 1973]. The author connects it to the fact that in comparison with the posterior vertical and horizontal semicircular canals, the anterior vertical semicircular canal is vitally important to the developing fetus, as this structure takes part in the fixation of a body in a vertical position.

The differentiation of the sensory epithelium of the maculae and cristae is noted in all investigated animals. In a walrus, as well as in terrestrial species, the initial cellular differentiation is observed in the utricular macula. Phocids are characterized by the simultaneous differentiation of the sensory epithelium in the utricular and saccular maculae. In cetaceans, the initial cellular differentiation of the sensory epithelium is marked only in one site of the saccular macula.

At this stage, the formation of the cochlea finishes with the formation of the last apical turn. In representatives of odontocetes (*Stenella attenuata*, *Delphinapterus leucas*), the height of the cochlea amounts to 2.0 turns, and in pinnipeds (*Eumetopias jubatus*, *Erignathus barbatus*, *Pusa hispida*, *Odobenus rosmarus divergens*) and in representatives of mysticetes (*Balaenoptera acutorostrata*) to 2.5 turns. In terrestrial species (*Sus scrofa domestica*, *Rattus norvegicus*), the cochlea is formed by 3 turns, and in bats (*Rhinolophus ferrumequinum*) by 3.5 turns. In some terrestrial mammals, the height of the cochlea reaches 4.5 turns (*Cavia porcellus*). The elements of the cochlear canal are not formed, and the cells of the organ of Corti are at the same stage of differentiation in all turns of the cochlea. From the columnar epithelium of the cochlear canal, two thickenings are formed – axial and lateral. From them, the structures of the cochlear canal and the cells of the organ of Corti develop.

At the 19th stage of development, in the saccular macula, the differentiation of the sensory epithelium into receptor cells and supporting cells occurs simultaneously in several sites of maculae and cristae, spreading over most of their surface. The structure of the cells, like the cells of the organ of Corti, form a mosaic distribution pattern and are well discernible. The neurons of the vestibular ganglion contain large, oval nuclei with expressed nucleoli. The size of all the structures of the organ of equilibrium is considerably increased. In otariids and in a walrus, as well as in terrestrial mammals, the size of the utriculus surpasses the size of the sacculus. In cetaceans, the utriculus and sacculus are similar in size to those in phocids.

In the cochlea, the formation of the elements of the cochlear canal is observed. The flattening of the cells of a cuboidal epithelium, and the loosening of connective tissue adjacent to this epithelium, occurs. In this location, the tympanic and vestibular scalae are formed. Reissner's membrane is also formed. The differentiation of the cells of the organ of Corti begins in the basal turn of the cochlea at the moment when the cells of a columnar epithelium start to move apart, and is spread gradually over the turns located above. As a result, in all turns of the cochlea, a different degree of anatomic and cellular differentiation is observed.

At the 20th stage of development in otariids and in the walrus, as in terrestrial mammals, the vestibular apparatus is twice the size of the cochlear part. These apparatuses are similarly sized in phocids. In cetaceans, the vestibular apparatus is one-half the size of the cochlea. The cellular differentiation of the sensory epithelium of the maculae, cristae and the organ of Corti continues. The cochlear and vestibular branches of the auditory nerve form.

In the cochlear canal, a spiral limb, vascular stria and tectorial membrane are formed. The differentiation of the cells of the organ of Corti continues. The size of the cochlea increases. At this stage, the basic process of anatomic formation of the structures of the inner ear has ended.

Both in phylogenesis and in ontogenesis, the inner ear is the first to form, as it is the core, phylogenetically most-ancient formation of the peripheral auditory system. As the development of the inner ear continues, the other parts of the peripheral auditory system of different evolutionary age start to form. The most evolutionary young among them is the outer ear.

The comparative analysis of the development of the auditory and vestibular structures in representatives of terrestrial, semi-aquatic and aquatic mammals has shown that the formation of these structures occurs in an early prefetal period and is extended as a result of the presence of the heterochrony in the development of the inner ear.

In an early embryogenesis in representatives of different ecological groups, the auditory and vestibular structures are simultaneously separated from each other and reveal similar structural features. In the first half of an early prefetal period (stages 13-15) in most mammals, both auditory and vestibular formations exhibit common structural features. Specific features in the structural organization of the hearing and equilibrium organs are formed in the second half of an early prefetal period (stages 16-20) at similar stages of development and in a certain sequence. In immature-born species, the anatomic formation of the structures of the inner ear ends in an early prefetal period, while the cellular differentiation of the sensory epithelium of the cochlea, maculae and cristae continues up to the early stages of postnatal ontogenesis. In mature-born species (cetaceans), the differentiation of structures of the inner ear completes at birth [Solntseva, 2002b].

In studied groups of mammals, the features connected with the stages of differentiation of the sensory epithelium of the maculae and cristae into the receptor cells and supporting cells are revealed. In terrestrial and semi-aquatic mammals (otariids, walruses) whose way of life is connected to a great degree with the stay on a firm substratum, the initial cellular differentiation of the sensory epithelium occurs in the utricular macula, indicating the vital role of the organ of gravitation in the activity of these mammals. The simultaneous cellular differentiation of the sensory epithelium in the saccular and utricular maculae, as well as the similarity in size of the cochlear and vestibular parts of the inner ear in phocids, indicates that for these species the organs of gravitation and vibration are equally important. Each of these organs is adapted for functioning in a habitat with certain physical properties. In absolute hydrobionts (cetaceans), the initial cellular differentiation of the saccular macula indicates that in aquatic mammals the organ of vibration carries out a more important function than the organ of gravitation.

All parts of the peripheral auditory system are multicomponent formations. Unlike the outer and middle ears, which are characterized by diverse structural variations and a wide range of adaptable transformations connected with peculiarities of the species ecology, the inner ear in representatives of different ecological groups, in spite of its variety of functions, maintains a monotonous structural organization. Usually, both in the cochlear and vestibular analyzers, the topography, form and size of separate components vary.

In echolocating mammals, the substantial growth of the cochlea's size compared to the size of the vestibular apparatus, as well as other features in the structure of the cochlear canal and the cells of the organ of Corti, serves as the adaptation of the cochlea to the perception of frequencies of a wide range, including ultrasounds (dolphins, bats). At the same time, a large cochlea and extraordinarily small vestibular apparatus in absolute hydrobionts with a various orientation of hearing can serve as the adaptations of the inner ear to an aqueous medium, as hearing in the aquatic mammals dominates among distant analyzers, thus providing the survival of these animals in conditions of constant dwelling in aquatic environment.

The comparative study of the peripheral auditory system's development revealed the general regularities of its formation in ontogenesis of representatives of various ecological groups: (1) in most mammals at the early stages of development (stages 13 to 16), the peripheral auditory system has common basic features in structure; (2) species-specific features in the structural organization of the peripheral auditory system are formed in the early prefetal period [according to the periodization of G.A. Schmidt, 1968], depending on the frequency tuning of the auditory system in each species; (3) these structural features are caused by habitat peculiarities and develop simultaneously from the homologous rudiments of the peripheral auditory system in phylogenetically distant and close forms; (4) in mammals, the morphological features of the peripheral auditory system, which were formed in an early prefetal period, continue to develop in a fetal period and during the whole period of a postnatal development.

Abbreviations (for Figures 1-27)

a - *pinna (auris)*
b - *external auditory meatus*
at - *tympanic ring*
c - *cartilago auriculae*
Ceg - *glandulae ceruminosae*
HF - *hair follicle*
ch - *cochlea*
cs - *cochlear canal*
ct - *tympanum*
d - *corium*
ep - *epidermis*
fo - *fenestra ovalis*
i - *incus*
lf - *ligamentum fibrosum*
lm - *ligamentum mallei*
m - *malleus,*
md - *modiolus*
mm - *musculus*
ms - *musculus stapedius*
mt - *tympanic membrane*
mbm - *handle of malleus*
bs - *footplate of stapes*
pl - *lenticular arm*
pb - *short arm of incus*
pm - *muscular arm*
B - *basal turn*
M - *middle turn*
A - *apical turn*
e - *orifice of the auditory meatus*
k - *distal part of the auditory meatus*
f - *proximal part of the auditory meatus*
z - *the blocking of the lumen in auditory meatus, the auditory meatus obliteration area*
mtk - *pars tensa*
mtp - *pars flacida*
mmt - *musculus tensor tympani*
n. 8 - *nervus vestibulocochlearis*
opr - *periotic bone*
opt - *tympanic bone*
och - *cochlear bone*
kb - *bone beam*
pg - *processus gracilis*
sg - *processus sigmoideus*
s - *stapes*
st - *tympanic scala*
svt - *vestibular scala*
sv - *venous sinus*
IHC - *inner hair cells*
OHC - *outer hair cells*
CH - *Hensen cells*
icc - *inner columnar cells*
occ - *outer columnar cells*
va - *vestibular apparatus*
chl - *cochlear branch of n. acousticus*
vb - *vestibular branch of n. acousticus*
mb - *basilar membrane*
sk - *cells of spiral ganglion*
ds - *dendrites of spiral ganglion*
rm - *Reissner's membrane*
l - *limb*
ls - *spiral ligament*
t - *tectorial membrane*
sp - *stria vascularis*
T - *columns and tunnel Corti*
KC - *Claudius cells*
DC - *Deiters cells*

ABBREVIATIONS (FOR FIGURES 28-53)

1 - *cochlear canal*
2 - *vestibular apparatus*
3 - *cerebrum*
4 - *cochlea*
5 - *sacculus*
6 - *utriculus*
7 - *ampulla of semicircular canal*
8 - *semicircular canal*
9 - *stapes*
10 - *tympanic membrane*
11 - *handle of malleus*
12 - *pinna*
13 - *musculus stapedius*
14 - *vestibular scala*
15 - *tympanic scala*
16 - *cochlear branch of n. acousticus*
17 - *vestibular branch of n. acousticus*
18 - *crista ampullaris*
19 - *vestibular ganglion*
20 - *utricular macula*
21 - *saccular macula*
22 - *receptor cells of saccular macula*
23 - *auditory capsule*
24 - *malleus*
25 - *cochlear ganglion*
26 - *incus*
27 - *external auditory meatus*
28 - *n. acousticus*
29 - *sacculo-endolymphatic canal*
30 - *basal turn of the cochlea*
31 - *middle turn of the cochlea*
32 - *apical turn of the cochlea*
33 - *auditory tube*
34 - *tympanum*
35 - *musculus tensor tympani*
36 - *ligamentum fibrosum*
37 - *tympanic bone*
38 - *long arm of mallei*
39 - *lenticular arm*
40 - *peribullar sinus*
41 - *venous sinus*
42 - *cavernous plexus*
43 - *epithelial cells in distal part of the auditory meatus*
44 - *cutaneous deepening behind an eye from which the outer ear is formed*
45 - *ossification foci*
46 - *prominences localised around cutaneous deepening*
47 - *modiolus*
48 - *Reissner's membrane*
49 - *fenestra ovalis*
50 - *axial thickening*
51 - *lateral thickening*
52 - *long arm of incus*

FIGURES

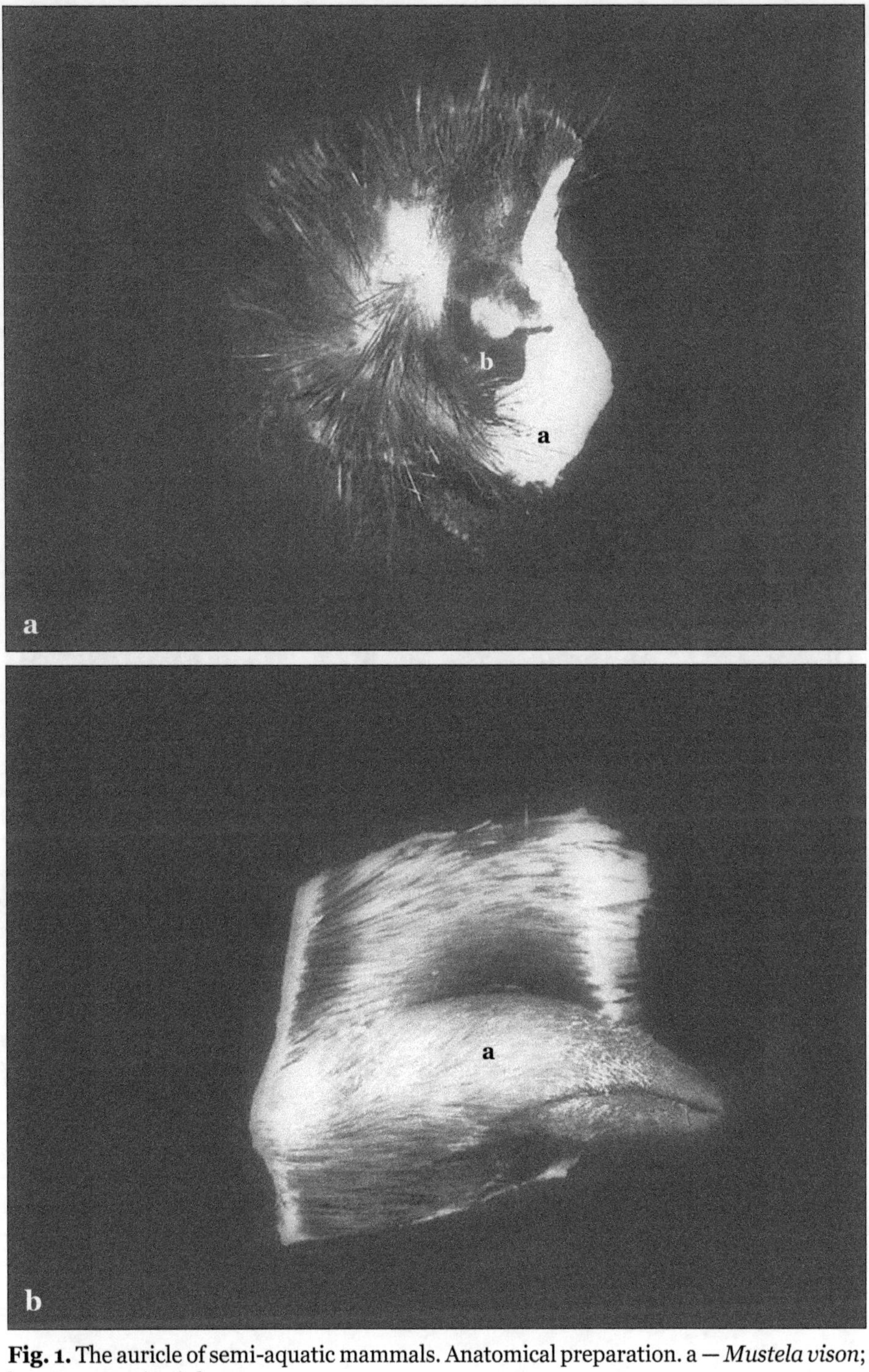

Fig. 1. The auricle of semi-aquatic mammals. Anatomical preparation. a — *Mustela vison*; b — *Eumetopias jubatus*.

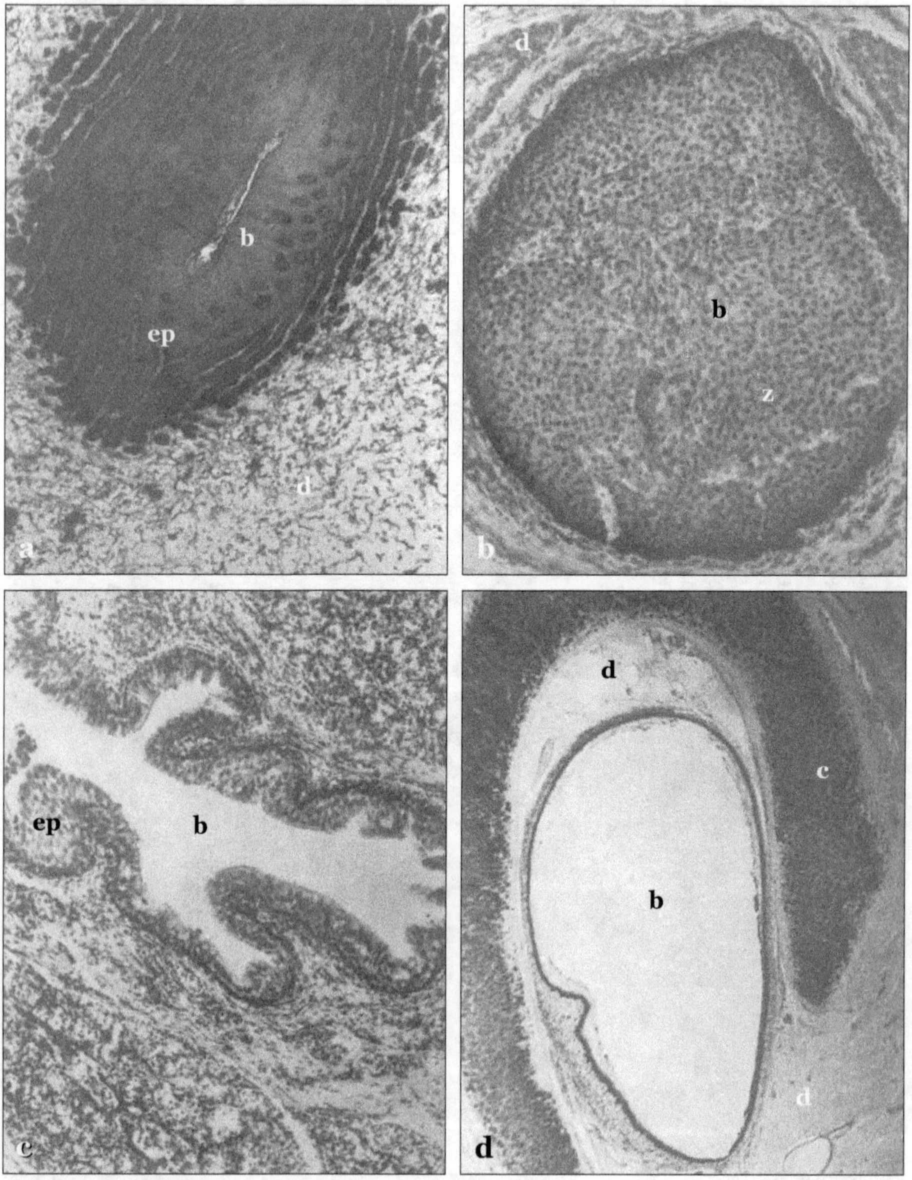

Fig. 2. The external auditory meatus of *Tursiops truncatus*. Cross sections. Staining with hematoxylin-eosin. a — orifice area; b — obliteration area of the auditory meatus lumen; c — middle part of the auditory meatus; d — end part of the auditory meatus.

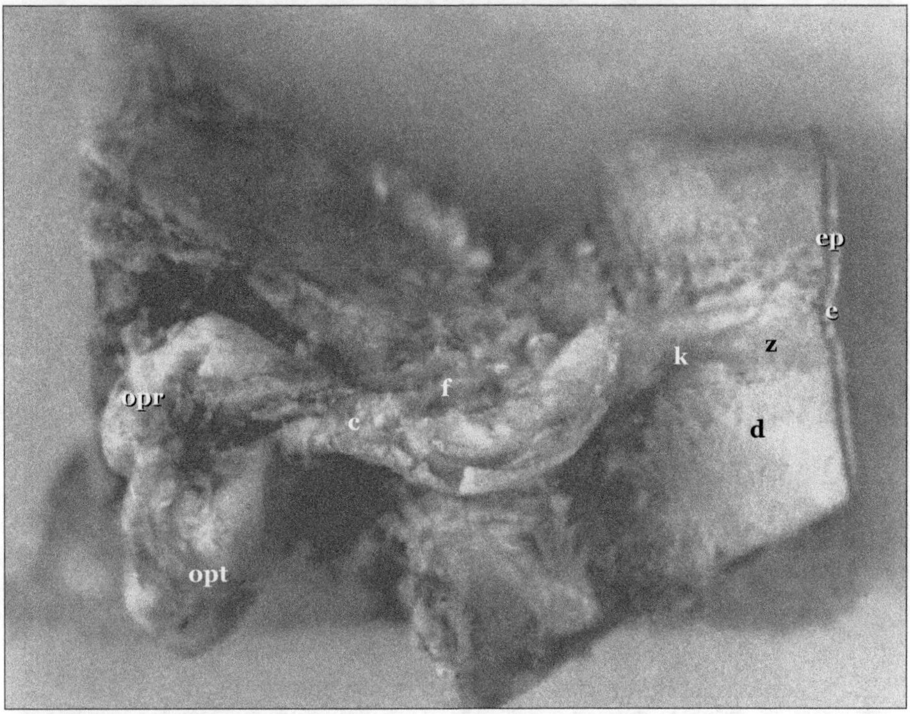

Fig. 3. The external auditory meatus and bulla tympanica of *Tursiops truncatus*. Anatomical preparation.

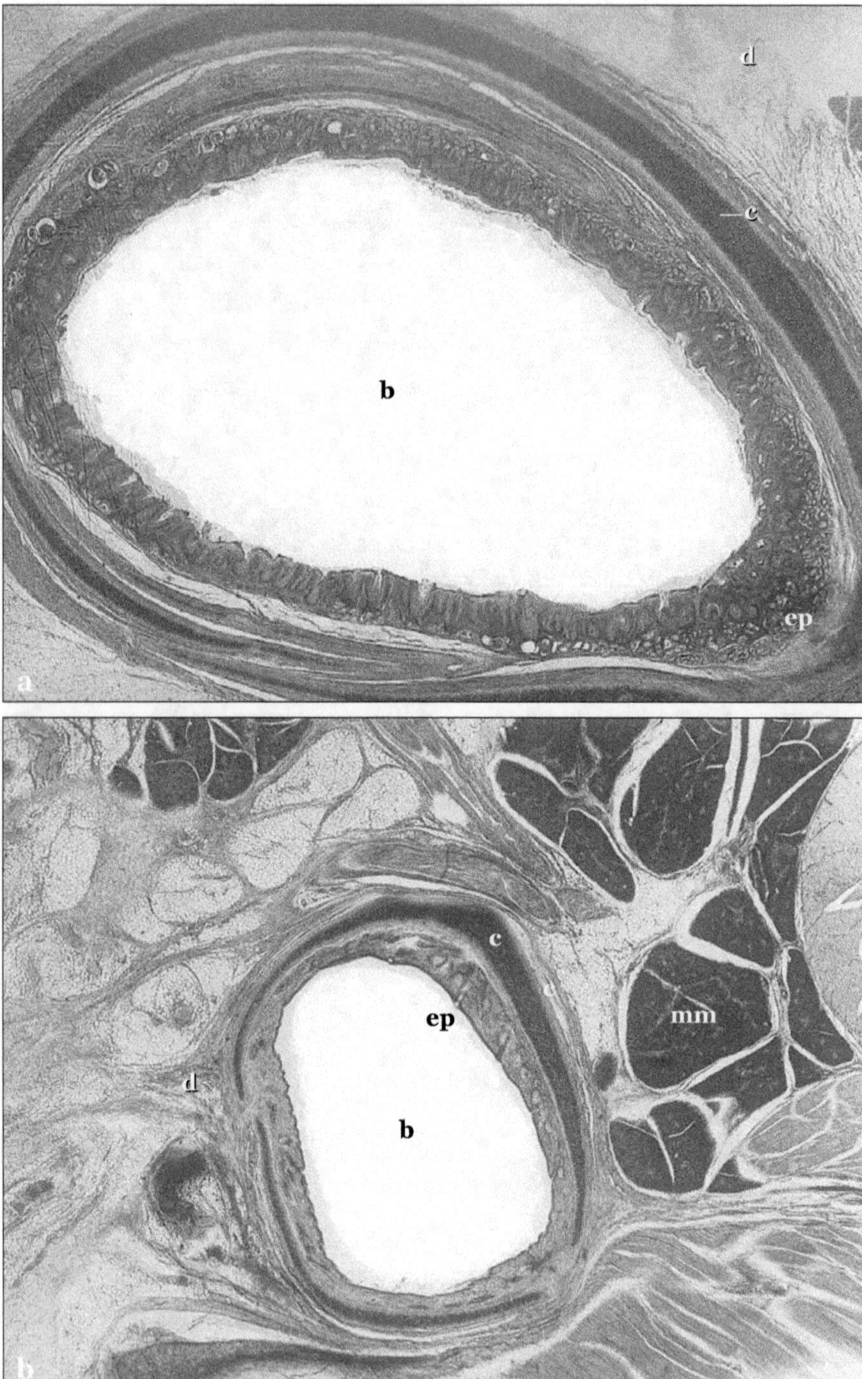

Fig. 4. The external auditory meatus in some Carnivora. a — *Vulpes vulpes*; b — *Mustela vison*. Cross sections. Staining with hematoxylin-eosin.

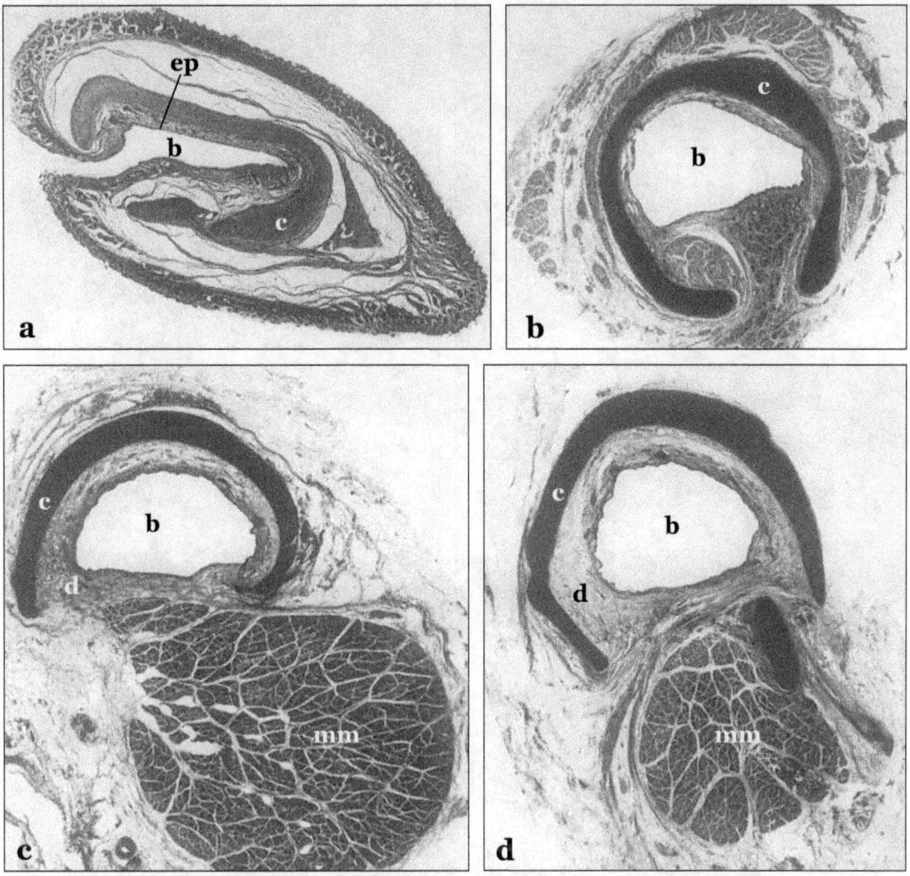

Fig. 5. The external auditory meatus of *Callorchinus ursinus*. Cross sections. Staining with hematoxylin-eosin. a — cross sections across the basis of an auricle; b — orifice area of the auditory meatus; c — middle part of the auditory meatus; d — end part of the auditory meatus.

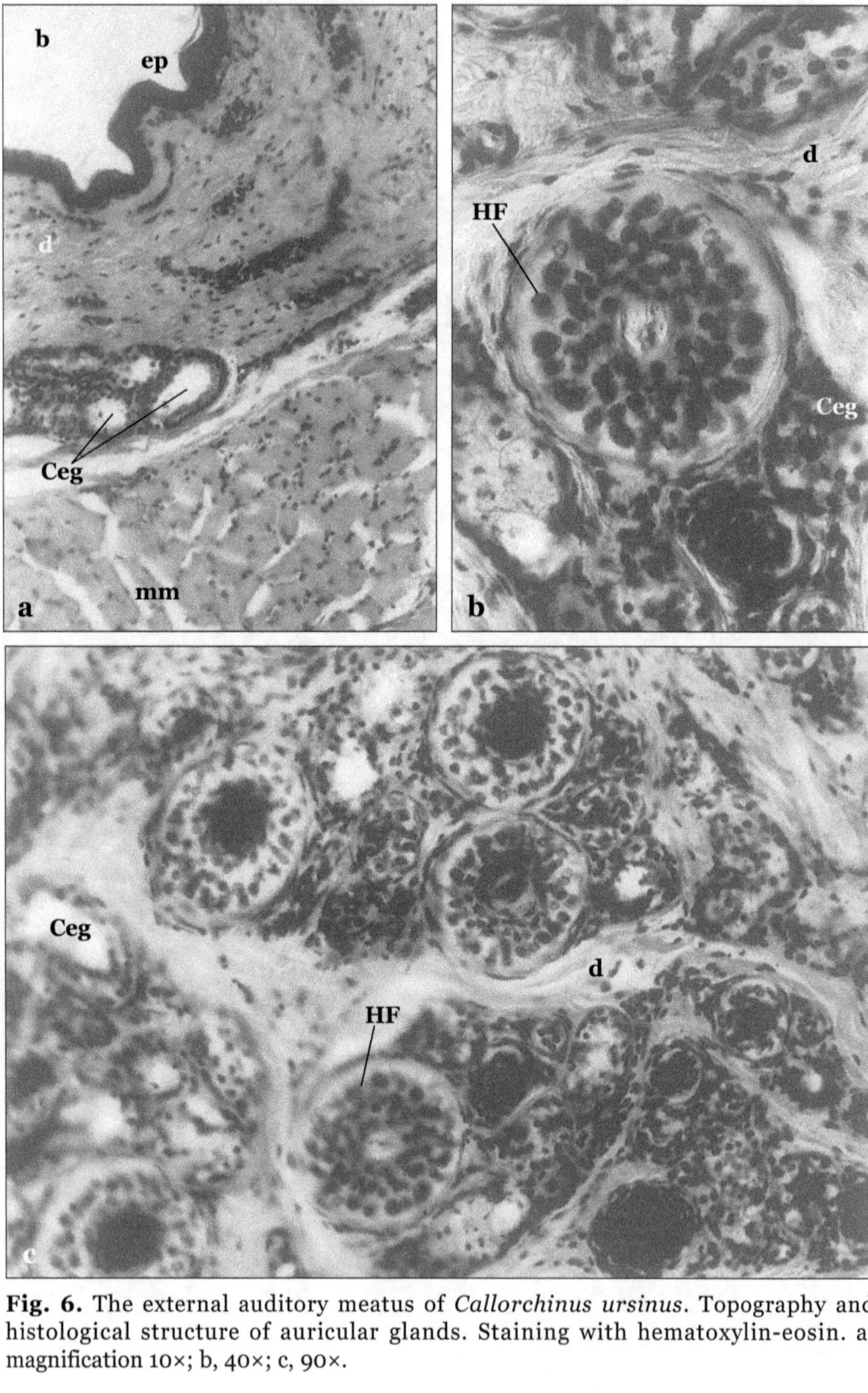

Fig. 6. The external auditory meatus of *Callorchinus ursinus*. Topography and histological structure of auricular glands. Staining with hematoxylin-eosin. a, magnification 10×; b, 40×; c, 90×.

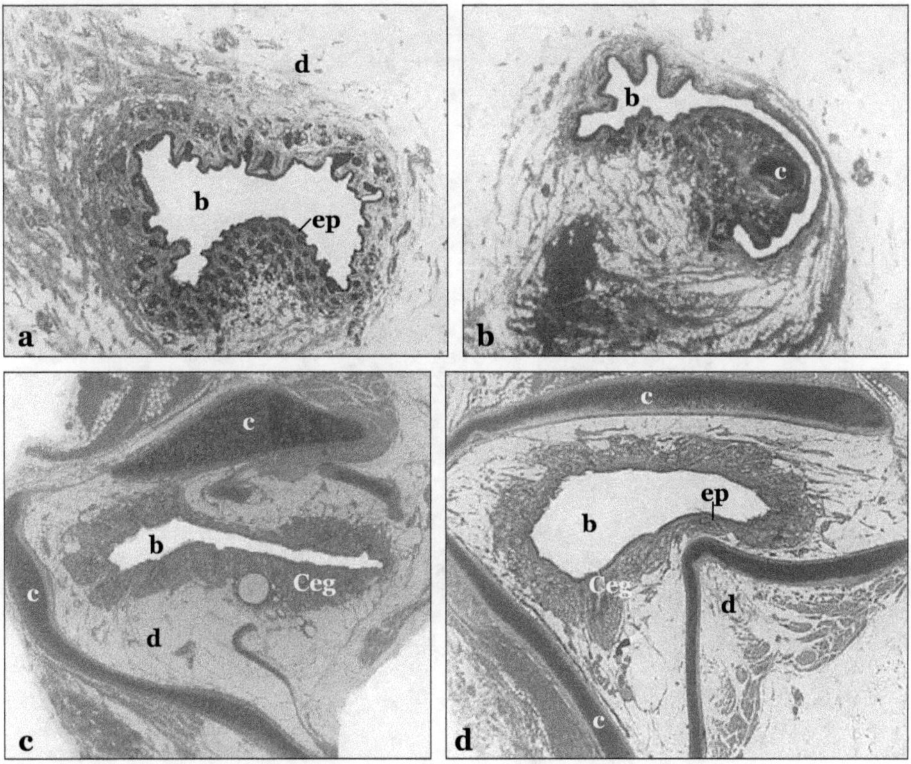

Fig. 7. The external auditory meatus of *Pusa caspica*. Cross sections. Staining with hematoxylin-eosin. a — orifice area of the auditory meatus; b — S-shaped bend area of the auditory meatus; c — auditory meatus cavity surrounded by cartilaginous plates flexibly connected to each other, with the auditory meatus lumen alteration in the area of its middle part; d — venous sinuses localized in dense connective tissue; e–g — end part of the auditory meatus.

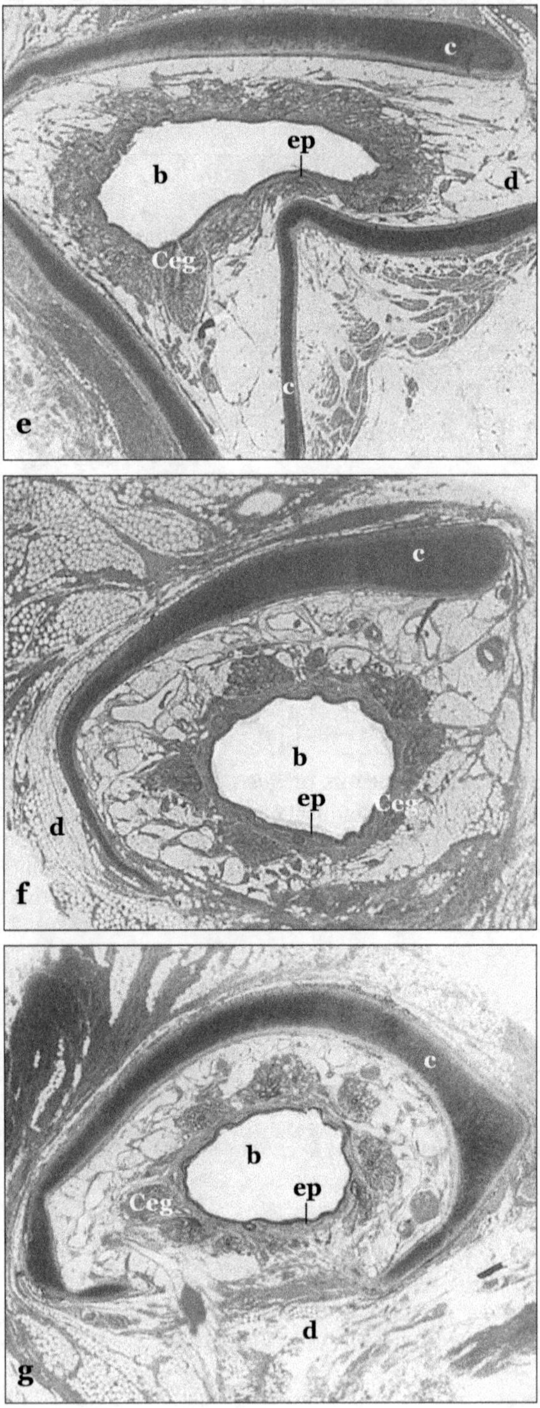

Fig. 7. Continued.

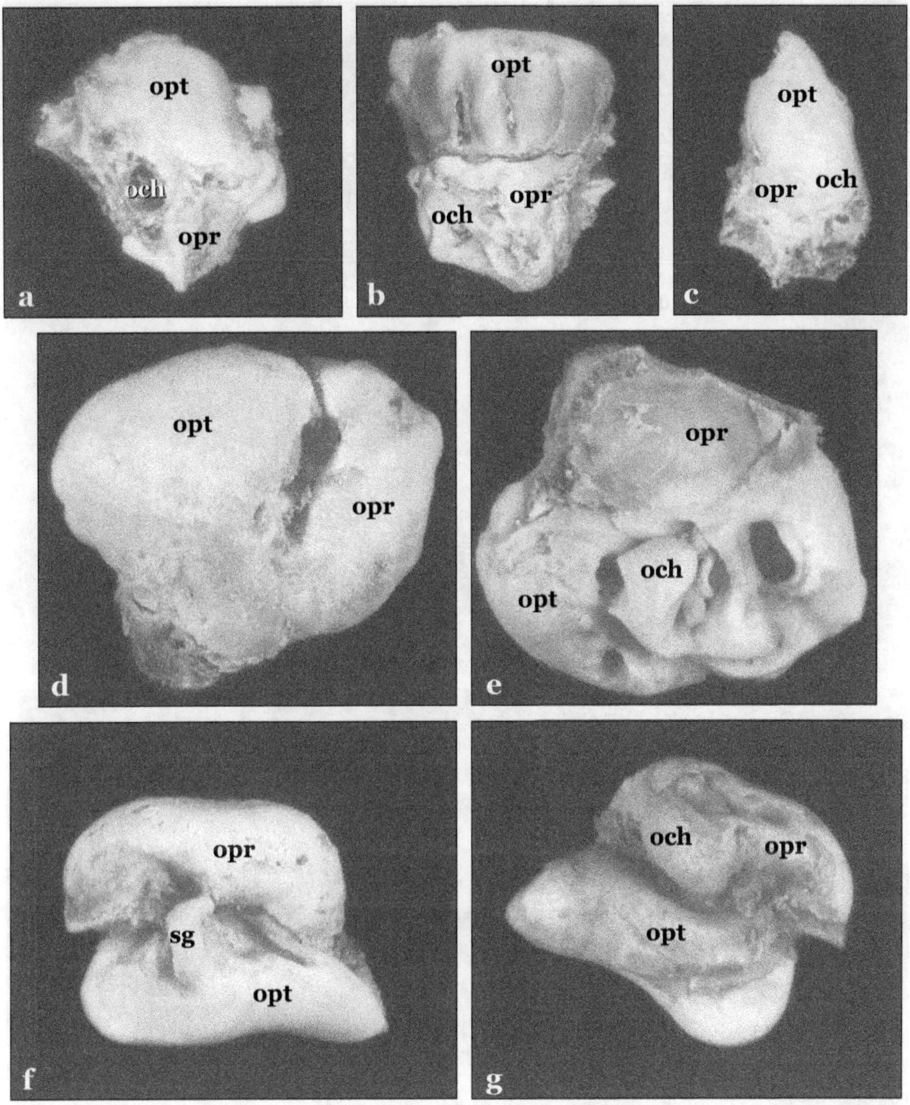

Fig. 8. The bulla tympanica in different species of mammals. Anatomical preparation. a — *Vulpes vulpes*; b — *Ovis ammon*; c — *Myocastor coypus*; d, e — *Pagophilus groenlandicus*; f, g — *Tursiops truncatus*; h — *Phoca vitulina*; i — *Balaenoptera physalus*.

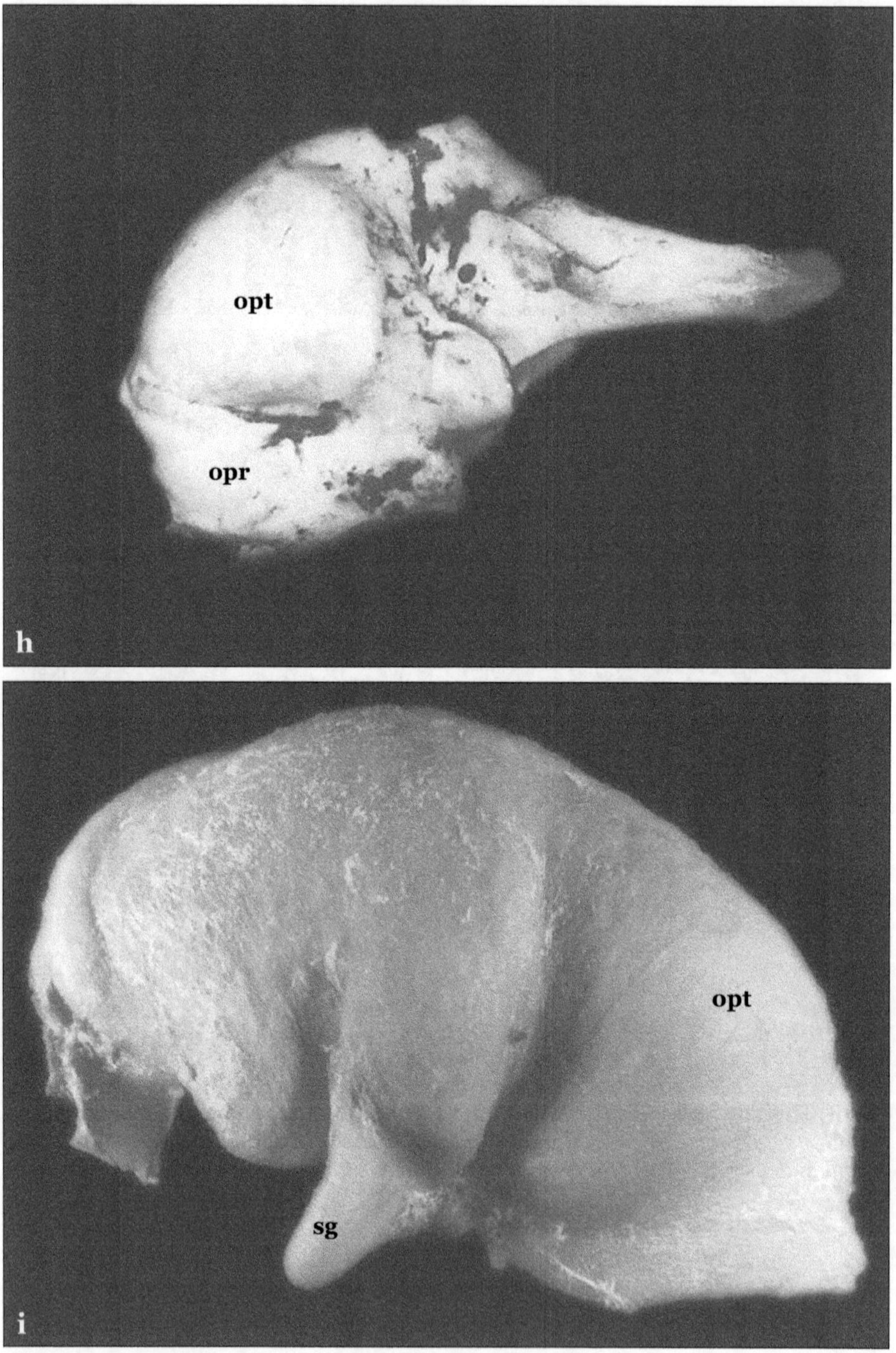

Fig. 8. Continued.

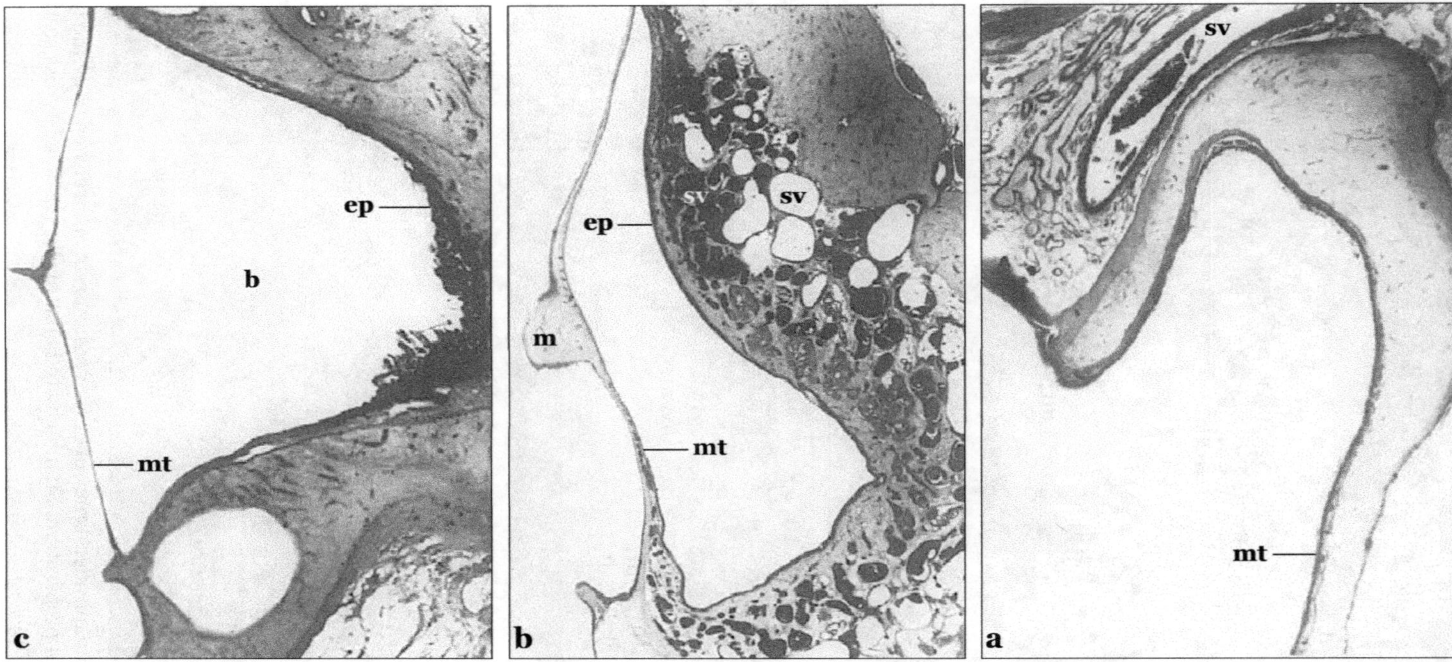

Fig. 9. The tympanic membrane in different species of mammals. Histological preparations. Staining with hematoxylin-eosin. Magnification 25×. a — *Myocastor coypus*; b — *Pagophilus groenlandicus*; c — *Tursiops truncatus*; d — *Pipistrellus pipistrellus*.

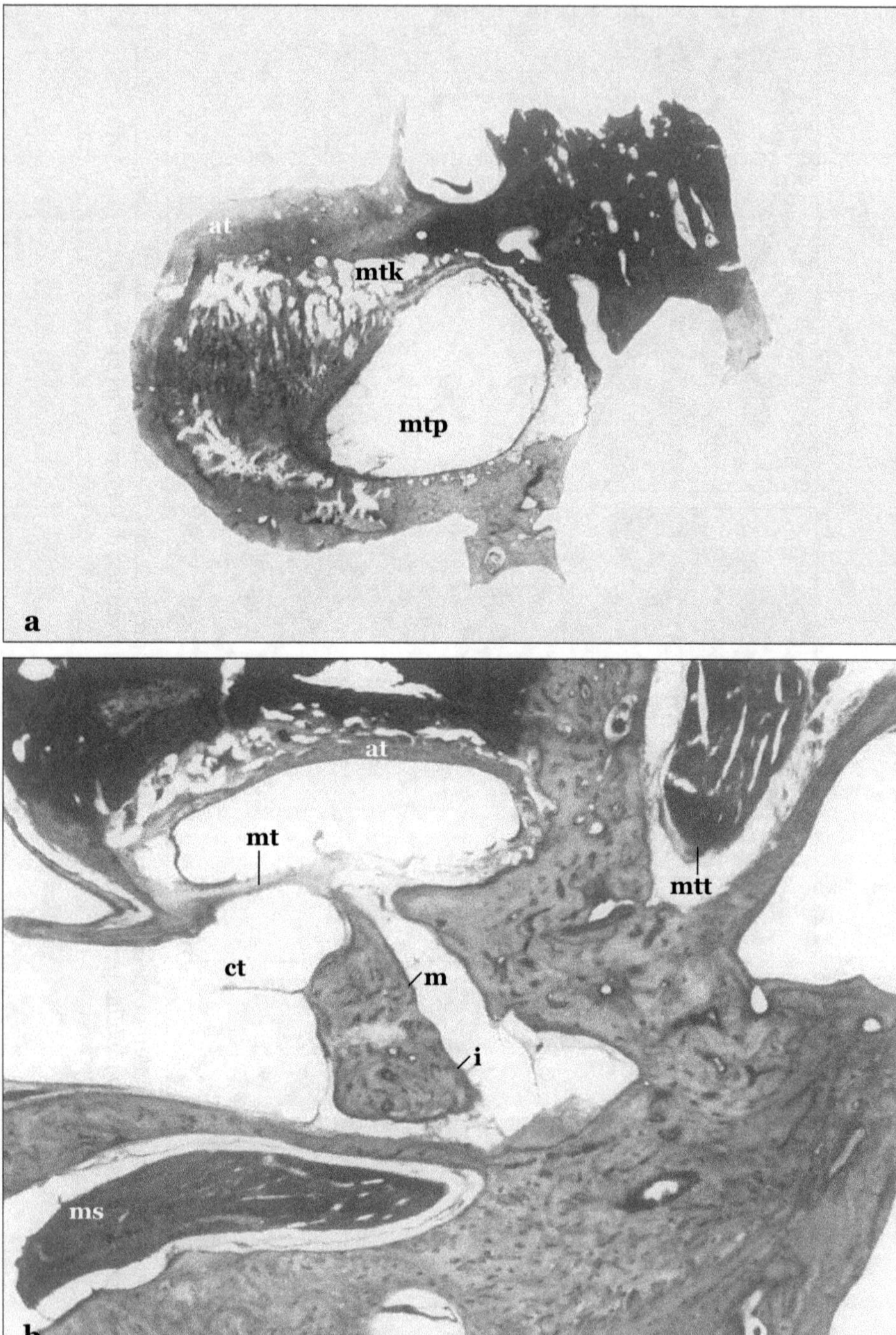

Fig. 10. The tympanic membrane of *Callorchinus ursinus*, stretched upon the tympanic ring. Histological preparation. Staining with hematoxylin-eosin. a — cross section; b — longitudinal section.

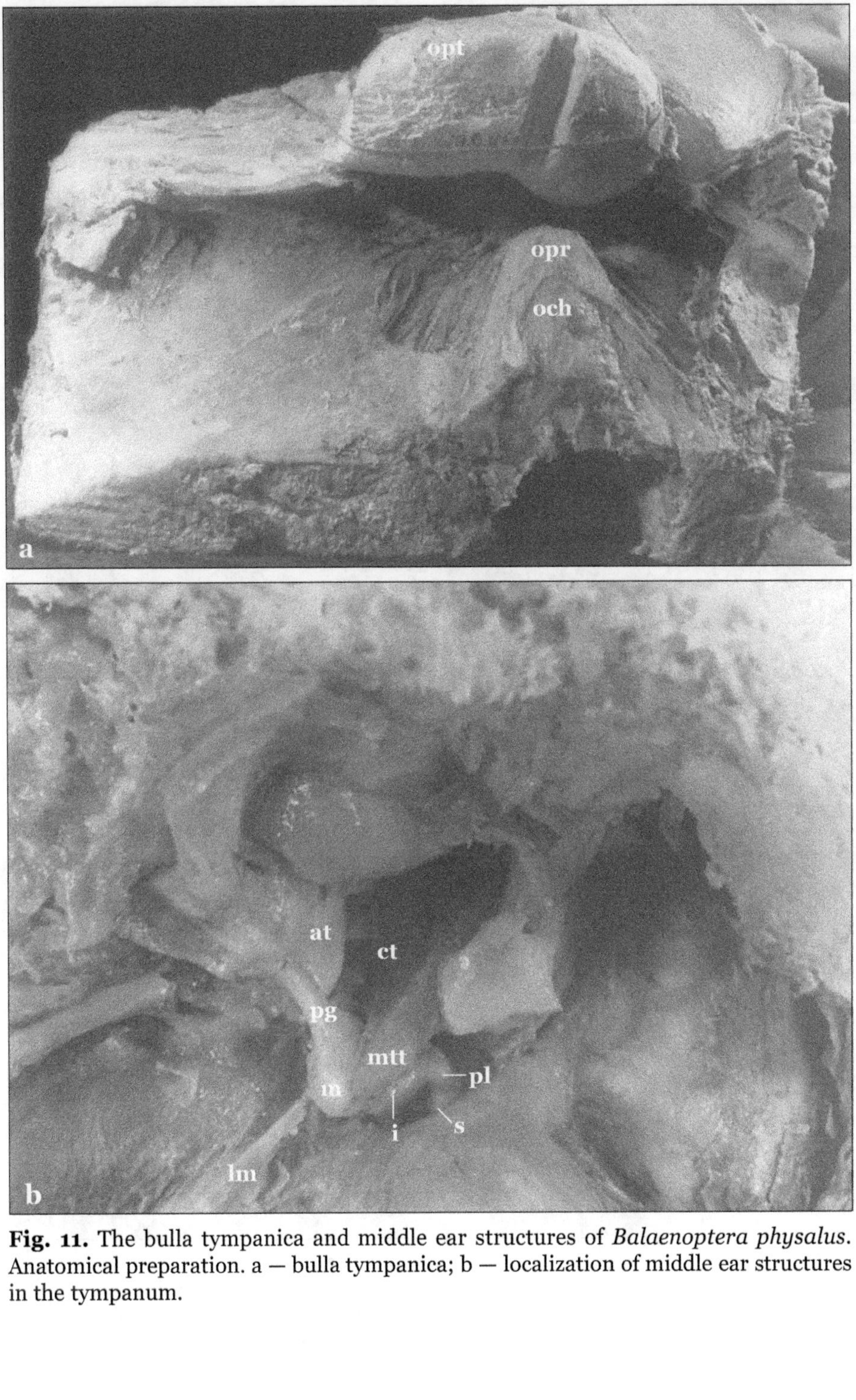

Fig. 11. The bulla tympanica and middle ear structures of *Balaenoptera physalus*. Anatomical preparation. a — bulla tympanica; b — localization of middle ear structures in the tympanum.

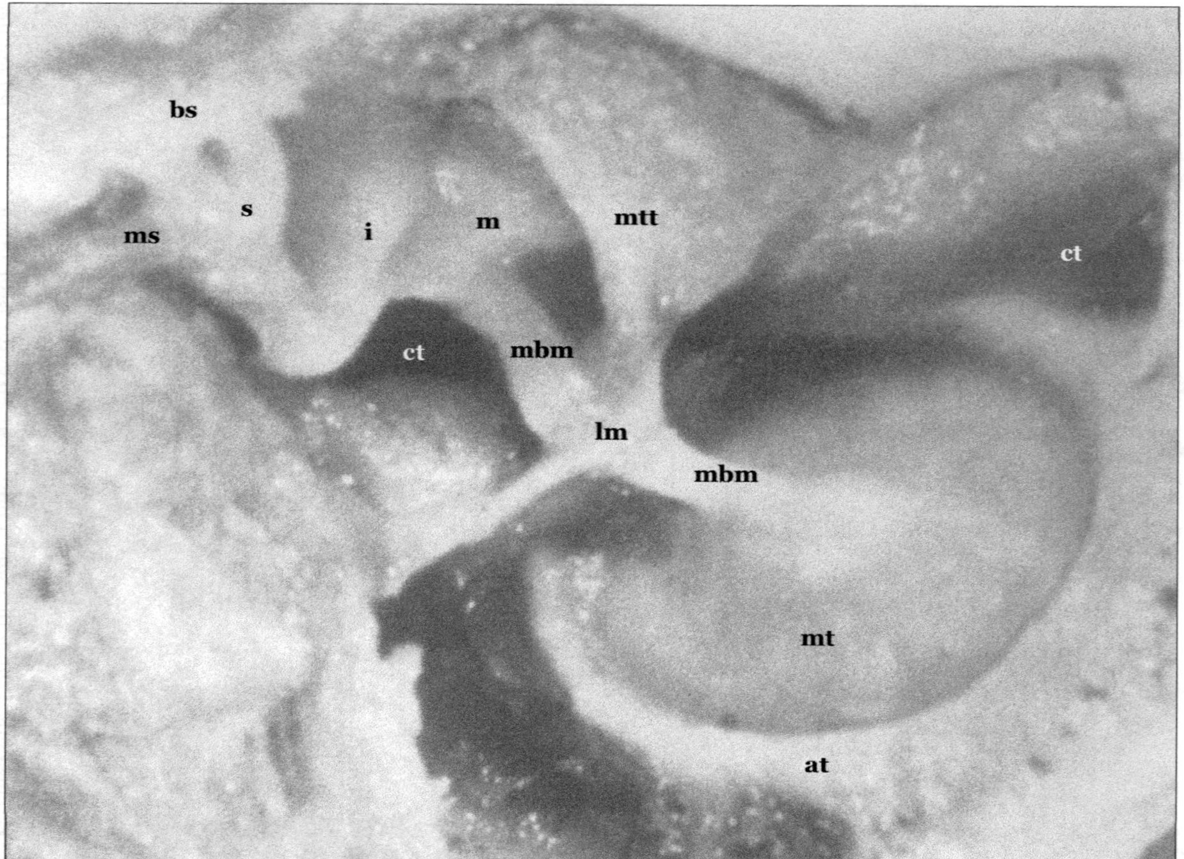

Fig. 12. Arrangement of middle ear structures in the tympanum of *Pagophilus groenlandicus*. Anatomical preparation.

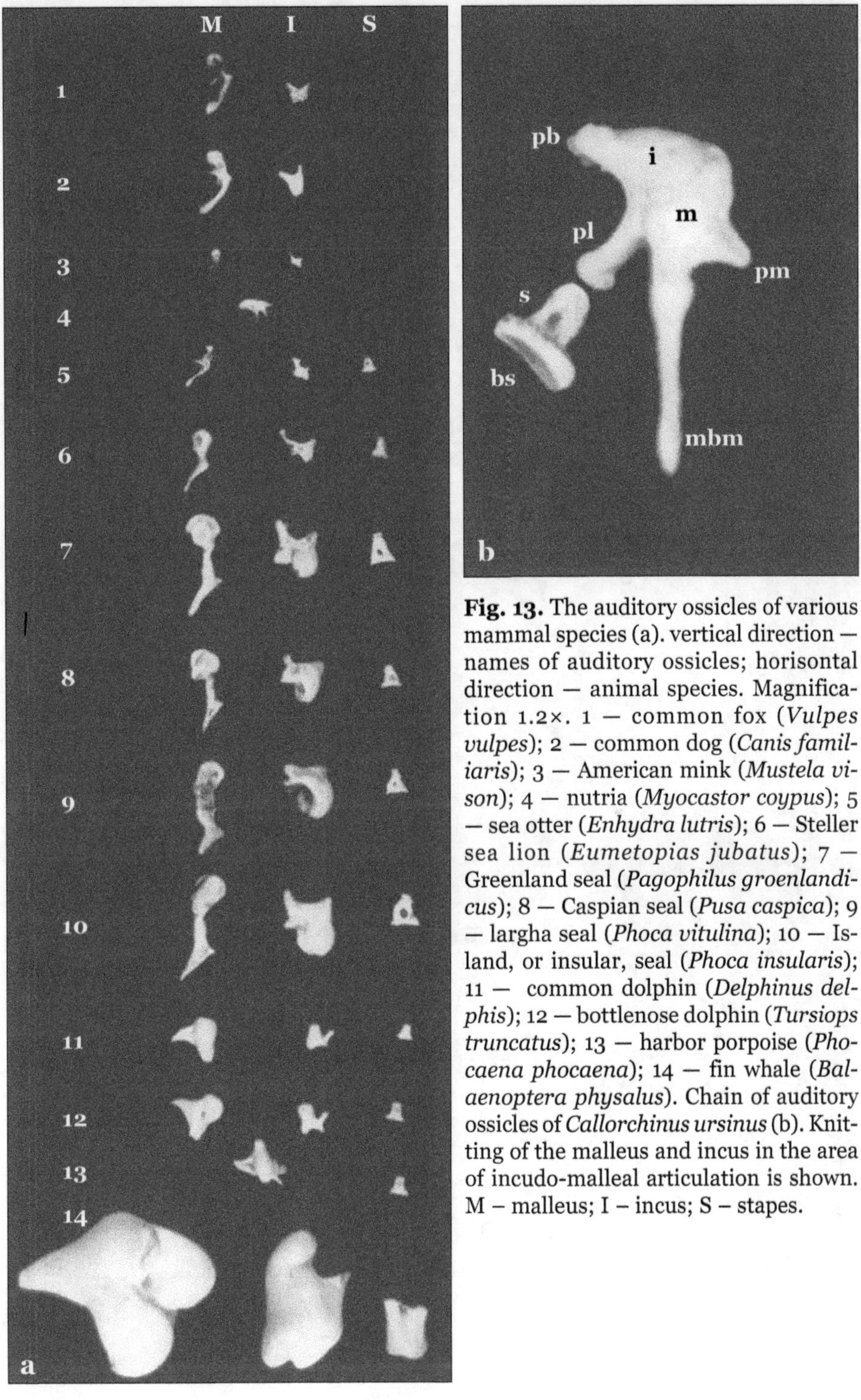

Fig. 13. The auditory ossicles of various mammal species (a). vertical direction – names of auditory ossicles; horisontal direction – animal species. Magnification 1.2×. 1 – common fox (*Vulpes vulpes*); 2 – common dog (*Canis familiaris*); 3 – American mink (*Mustela vison*); 4 – nutria (*Myocastor coypus*); 5 – sea otter (*Enhydra lutris*); 6 – Steller sea lion (*Eumetopias jubatus*); 7 – Greenland seal (*Pagophilus groenlandicus*); 8 – Caspian seal (*Pusa caspica*); 9 – largha seal (*Phoca vitulina*); 10 – Island, or insular, seal (*Phoca insularis*); 11 – common dolphin (*Delphinus delphis*); 12 – bottlenose dolphin (*Tursiops truncatus*); 13 – harbor porpoise (*Phocaena phocaena*); 14 – fin whale (*Balaenoptera physalus*). Chain of auditory ossicles of *Callorchinus ursinus* (b). Knitting of the malleus and incus in the area of incudo-malleal articulation is shown. M – malleus; I – incus; S – stapes.

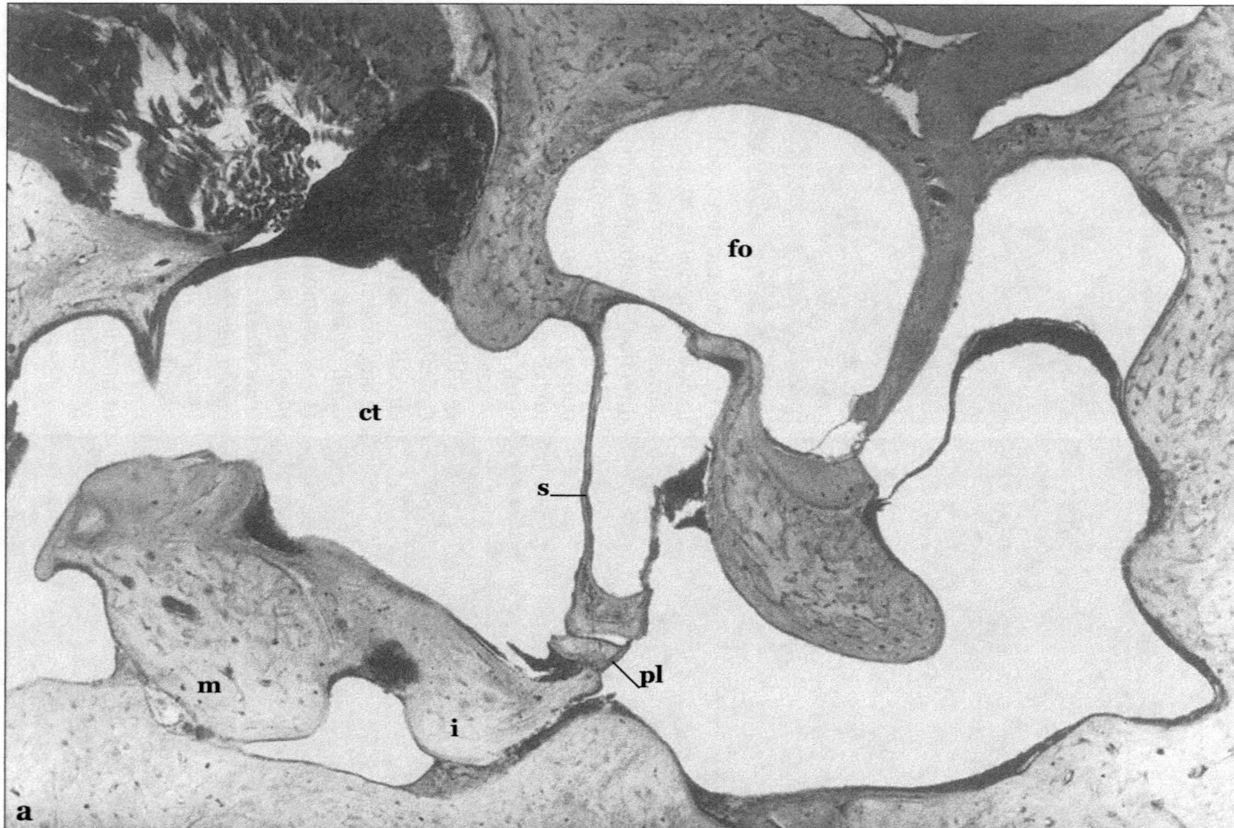

Fig. 14. Histological preparations of the middle ear in various species of mammals. Staining with hematoxylin-eosin. Magnification 25×. a — *Vulpes vulpes*; b — *Sorex araneus*; c — *Mustela vison*; d, e — *Callorchinus ursinus*; f — *Pagophilus groenlandicus*; g — *Tursiops truncatus*.

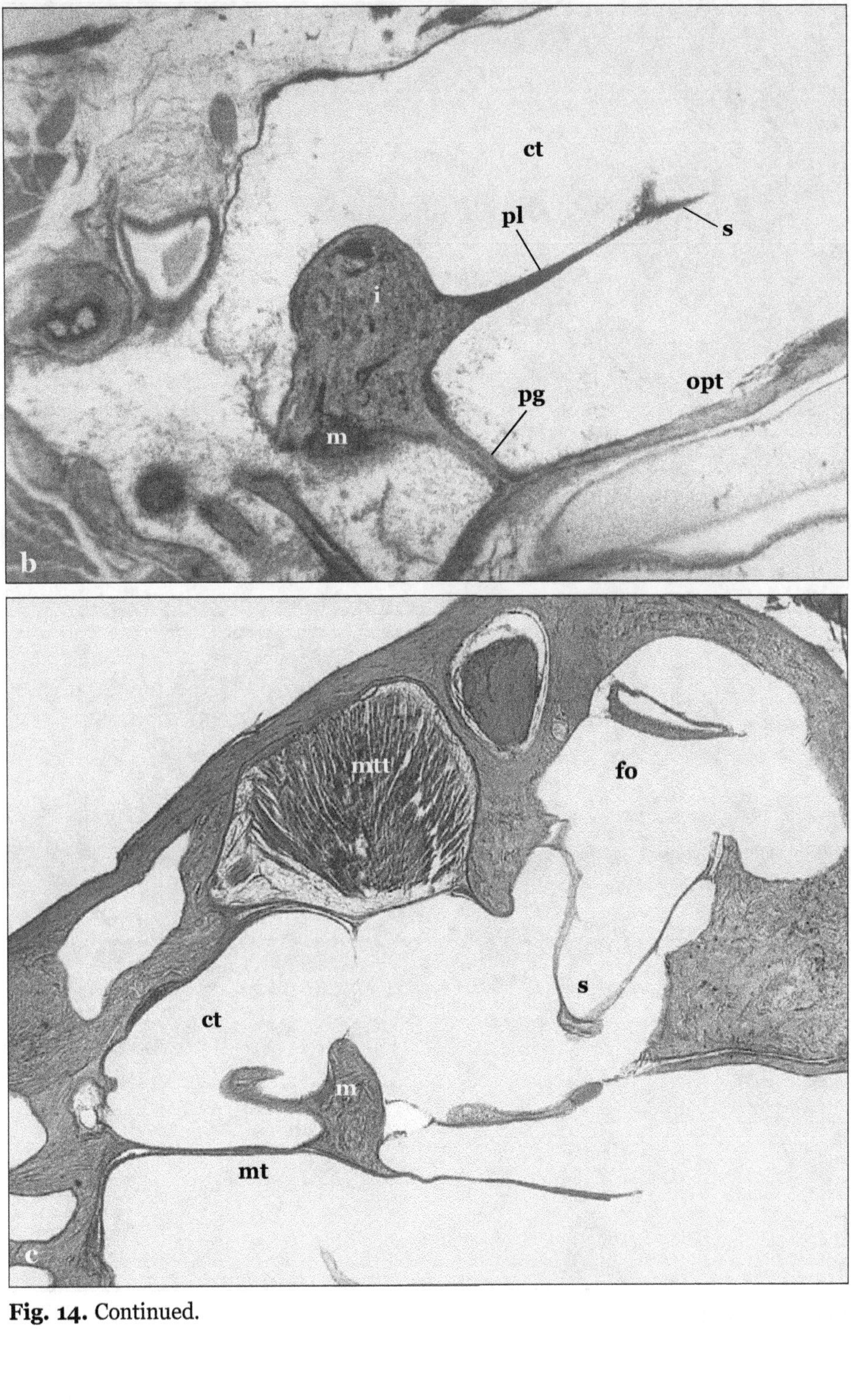

Fig. 14. Continued.

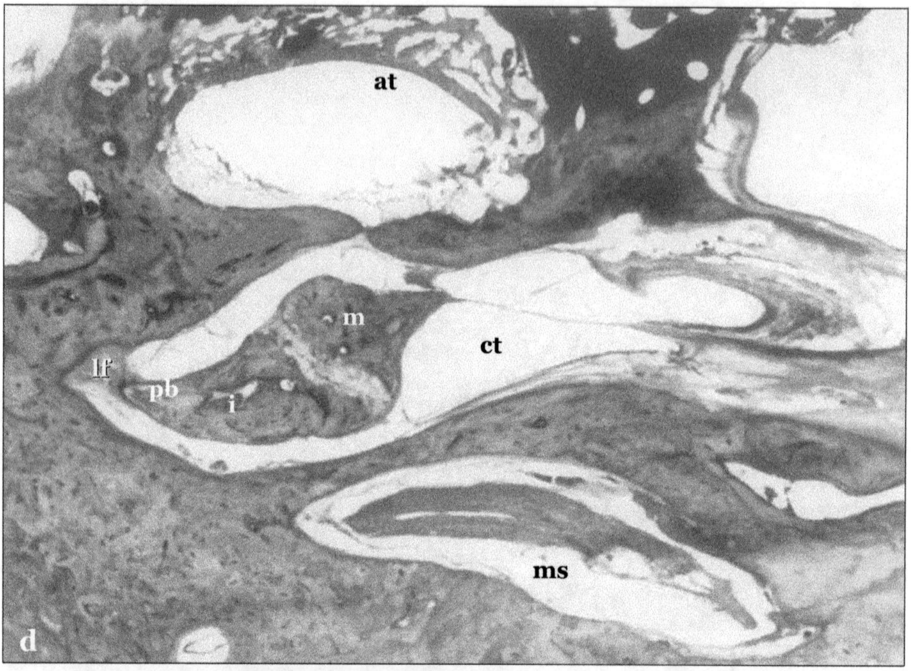

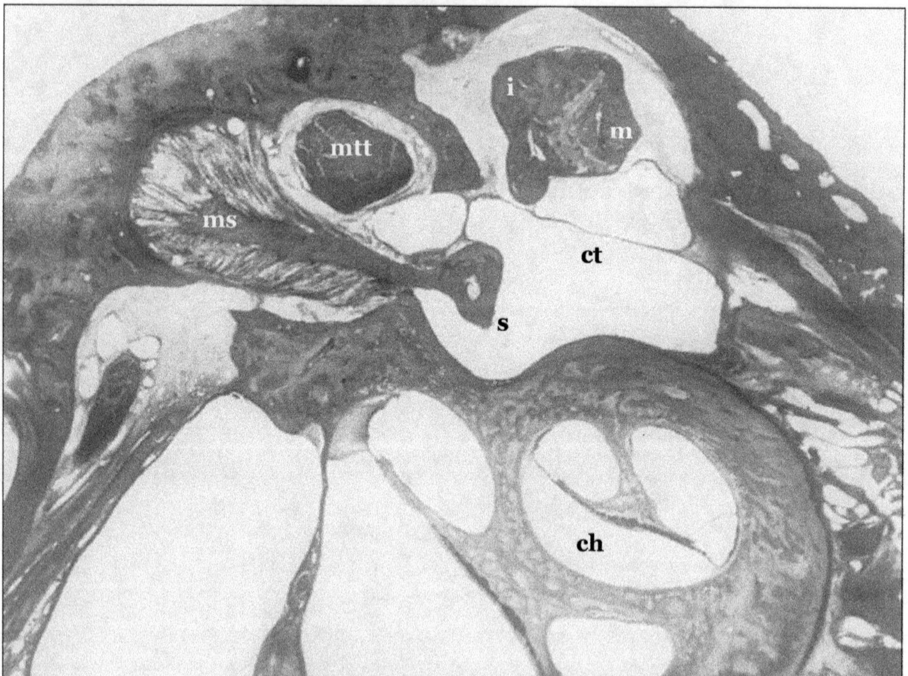

Fig. 14. Continued.

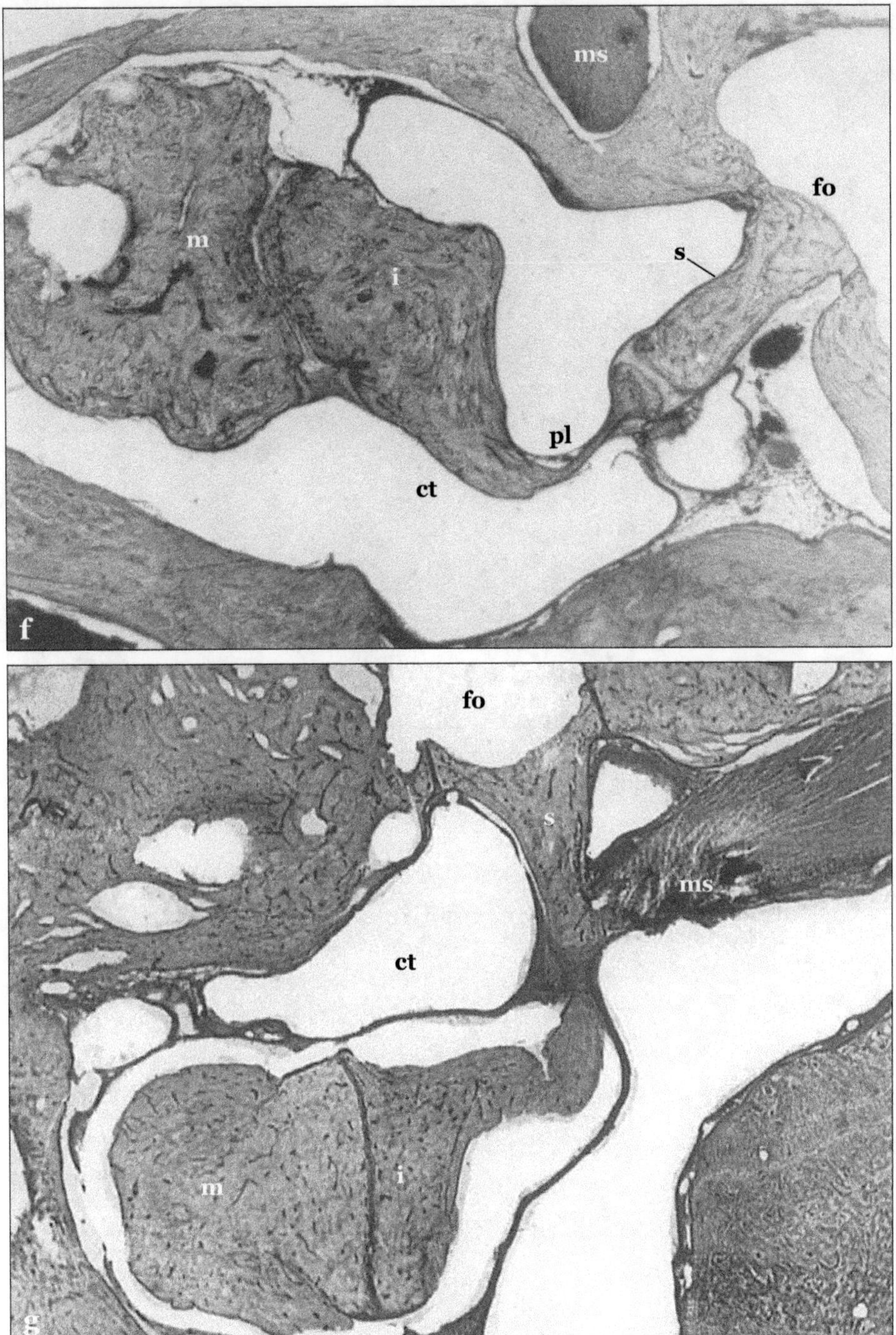

Fig. 14. Continued.

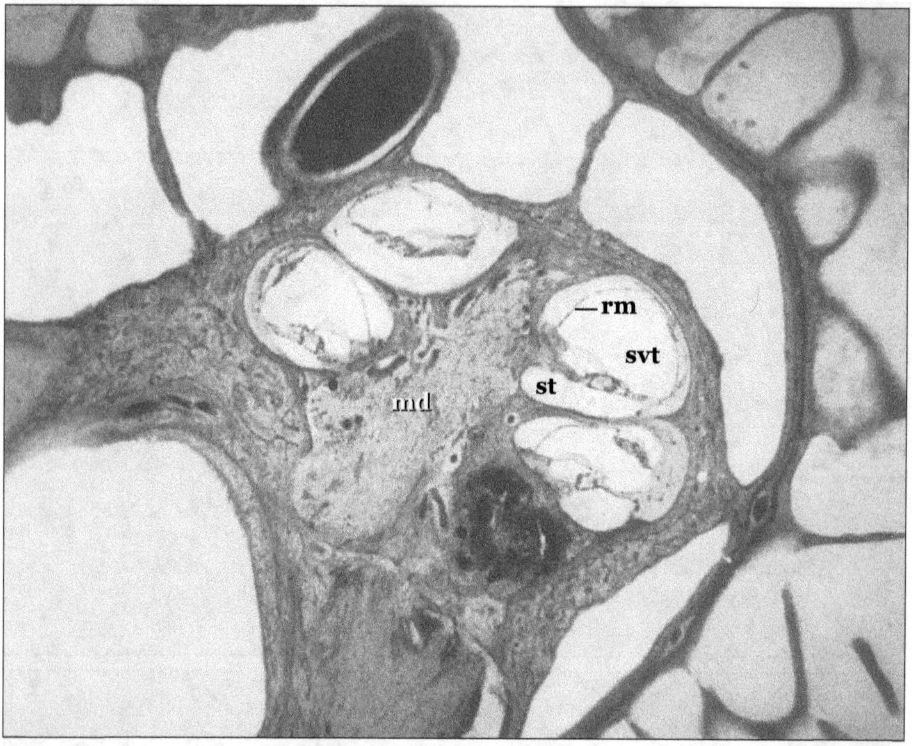

Fig. 15. The cochlea of *Talpa europaea*. Slit through the cochlea axis. Two turns of cochlea are shown. Staining with hematoxylin-eosin.

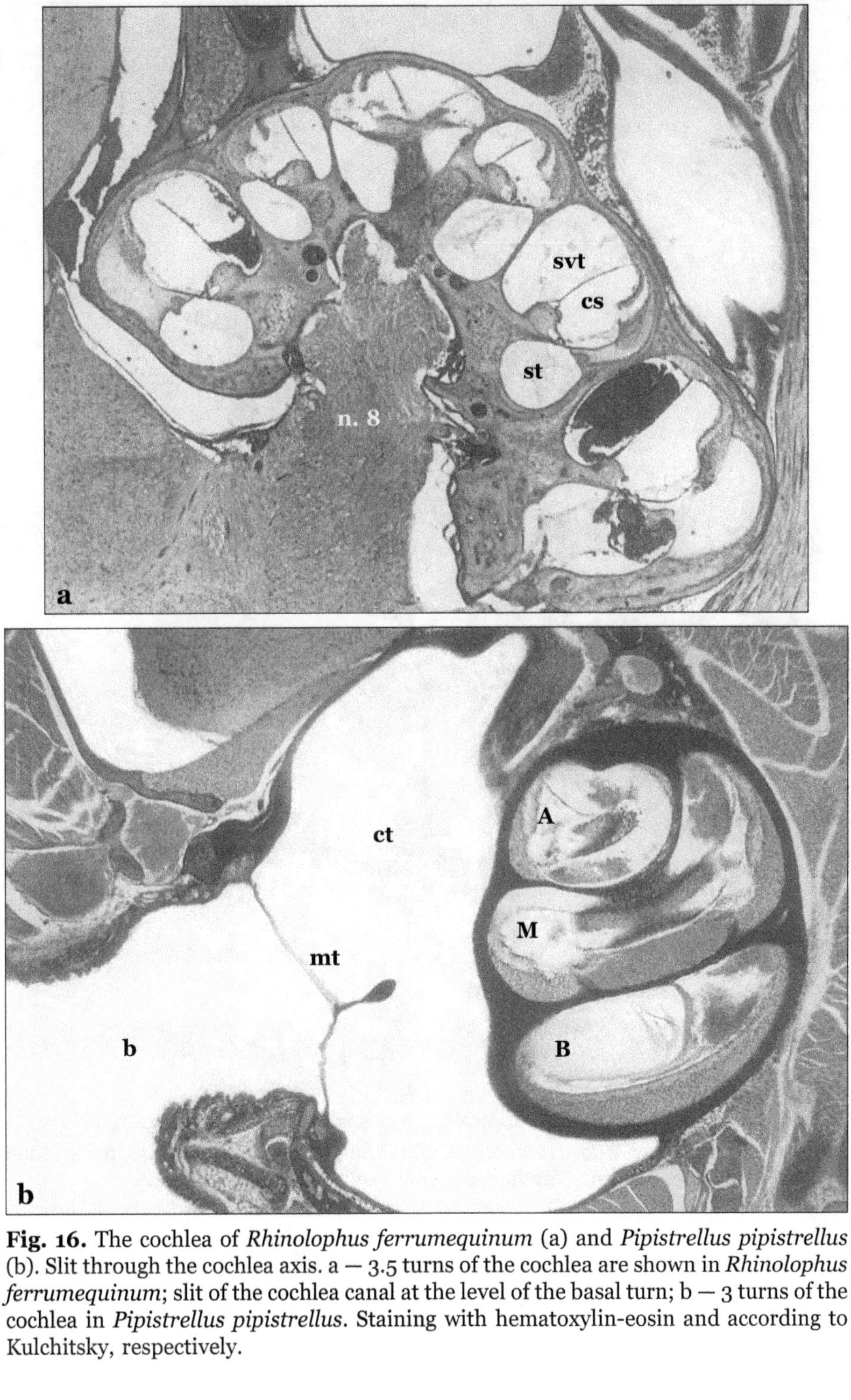

Fig. 16. The cochlea of *Rhinolophus ferrumequinum* (a) and *Pipistrellus pipistrellus* (b). Slit through the cochlea axis. a — 3.5 turns of the cochlea are shown in *Rhinolophus ferrumequinum*; slit of the cochlea canal at the level of the basal turn; b — 3 turns of the cochlea in *Pipistrellus pipistrellus*. Staining with hematoxylin-eosin and according to Kulchitsky, respectively.

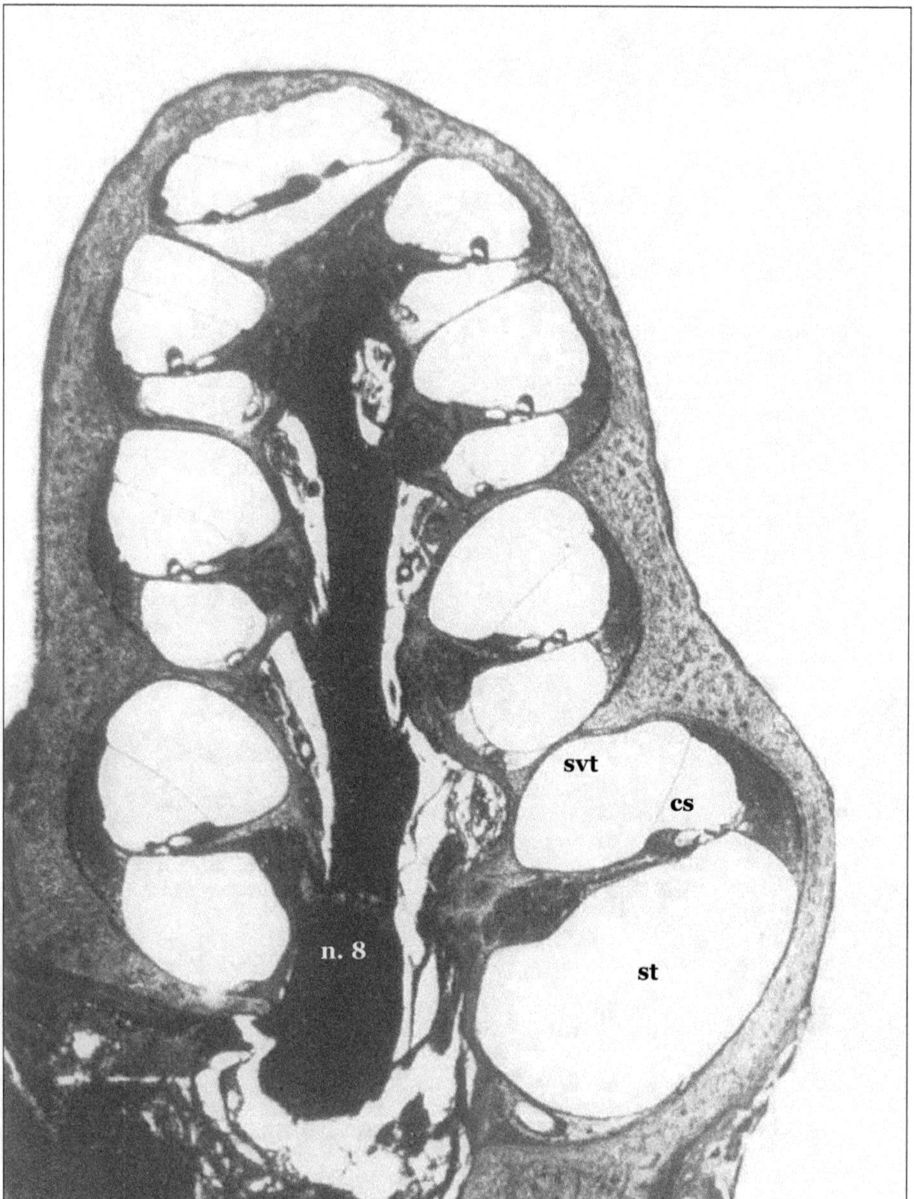

Fig. 17. The cochlea of *Myocastor coypus*. Slit through the cochlea axis. 4.5 turns of the cochlea are shown. Staining with hematoxylin-eosin. Magnification 25×.

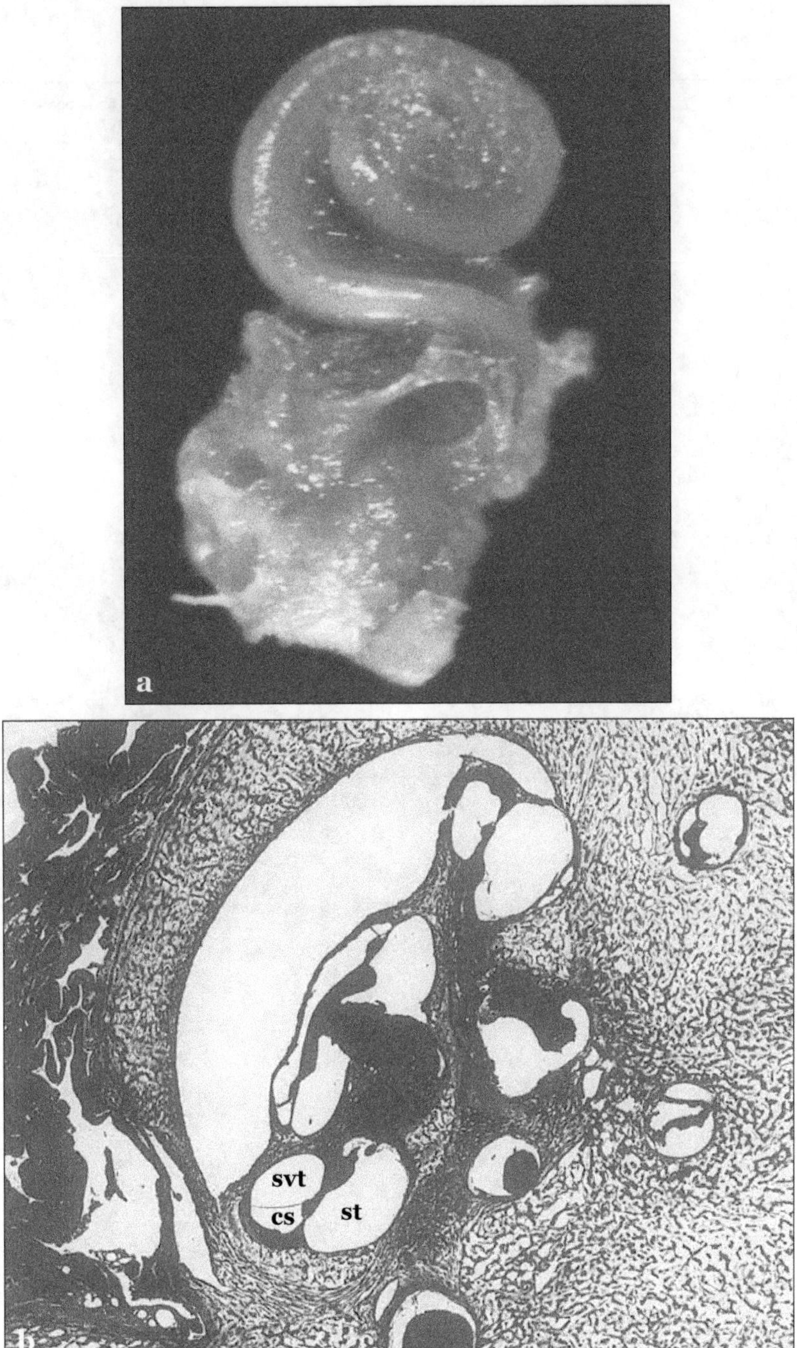

Fig. 18. The cochlea of *Delphinus delphis*. 1.5 turns of the cochlea are shown. Staining according to Kampas, magnification 25×. a — cochlea, anatomical preparation; b — slit through the axis of the cochlea.

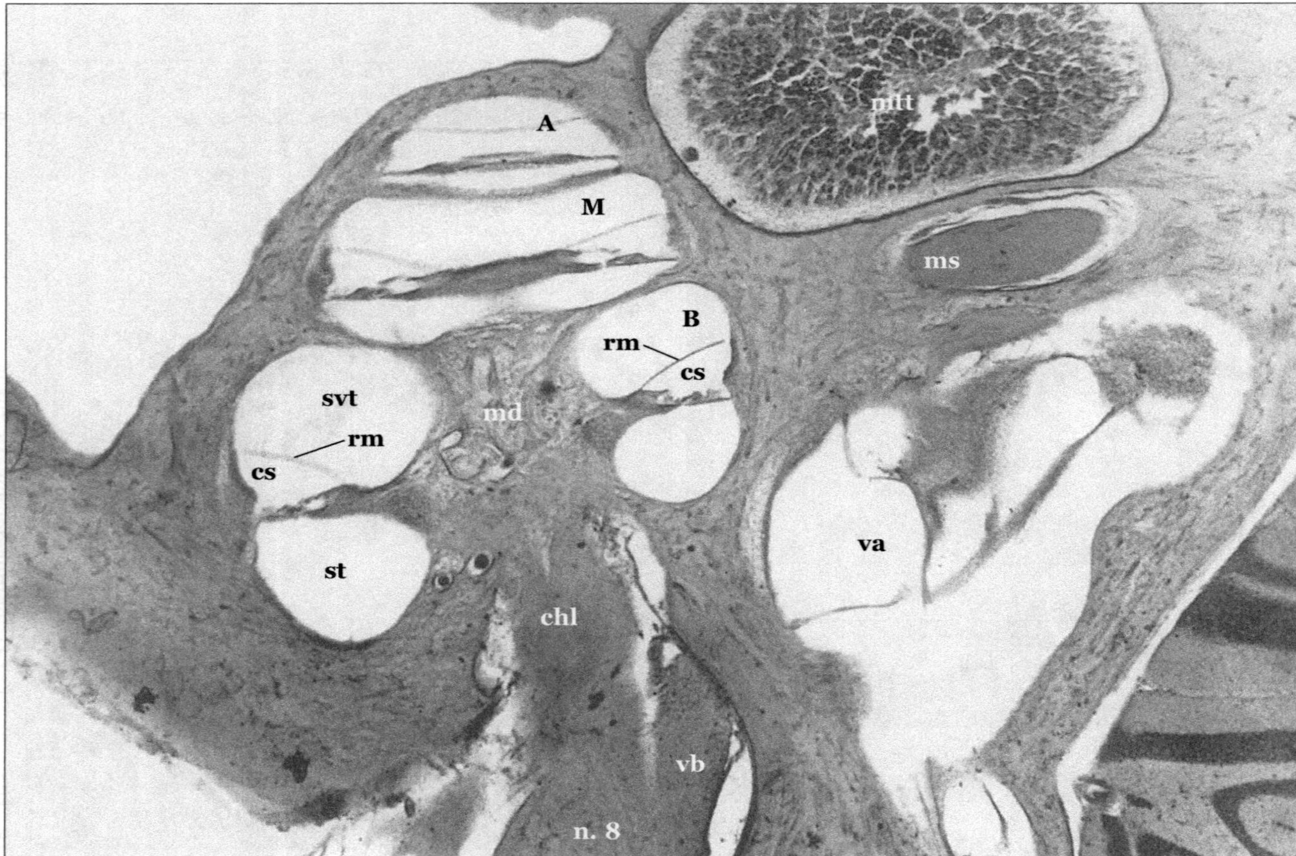

Fig. 19. The cochlea of *Enhydra lutris*. Slit through the cochlea axis. 3 turns of the cochlea are shown. Staining with hematoxylin-eosin.

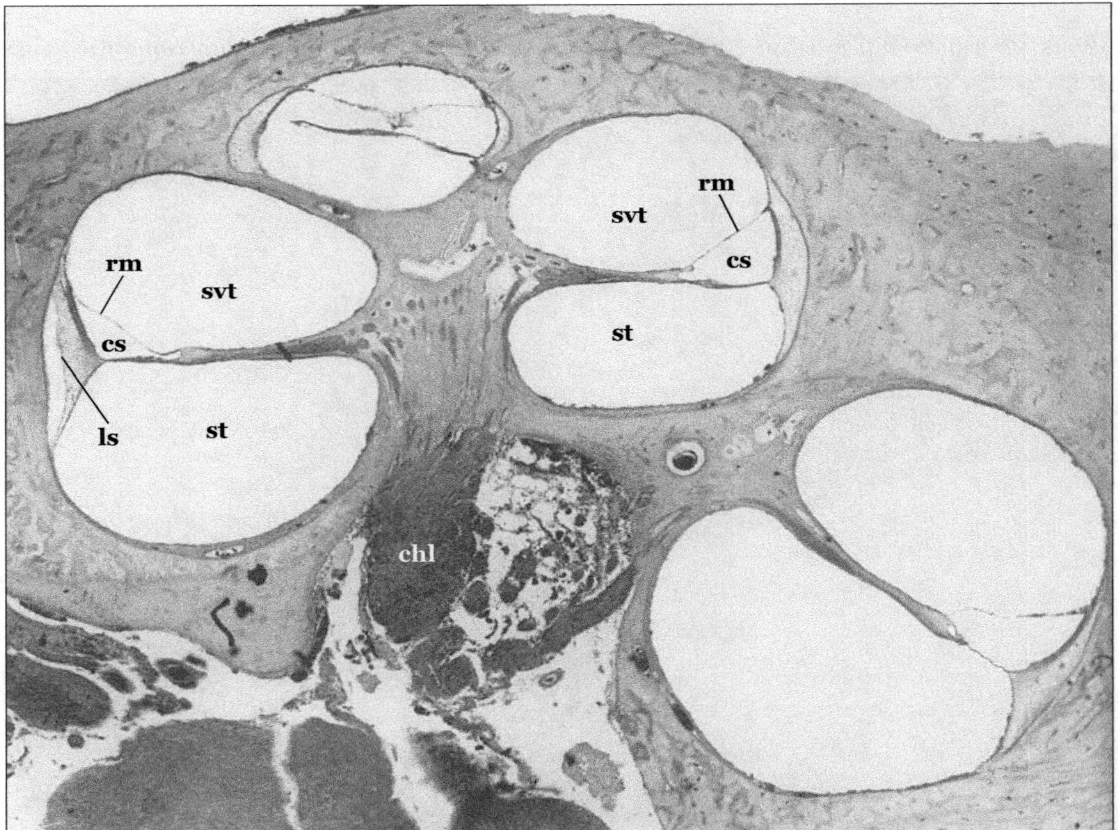

Fig. 20. The cochlea of *Pagophilus groenlandicus*. Slit through the cochlea axis. 2.5 turns of the cochlea are shown. Staining with hematoxylin-eosin. Magnification 25×.

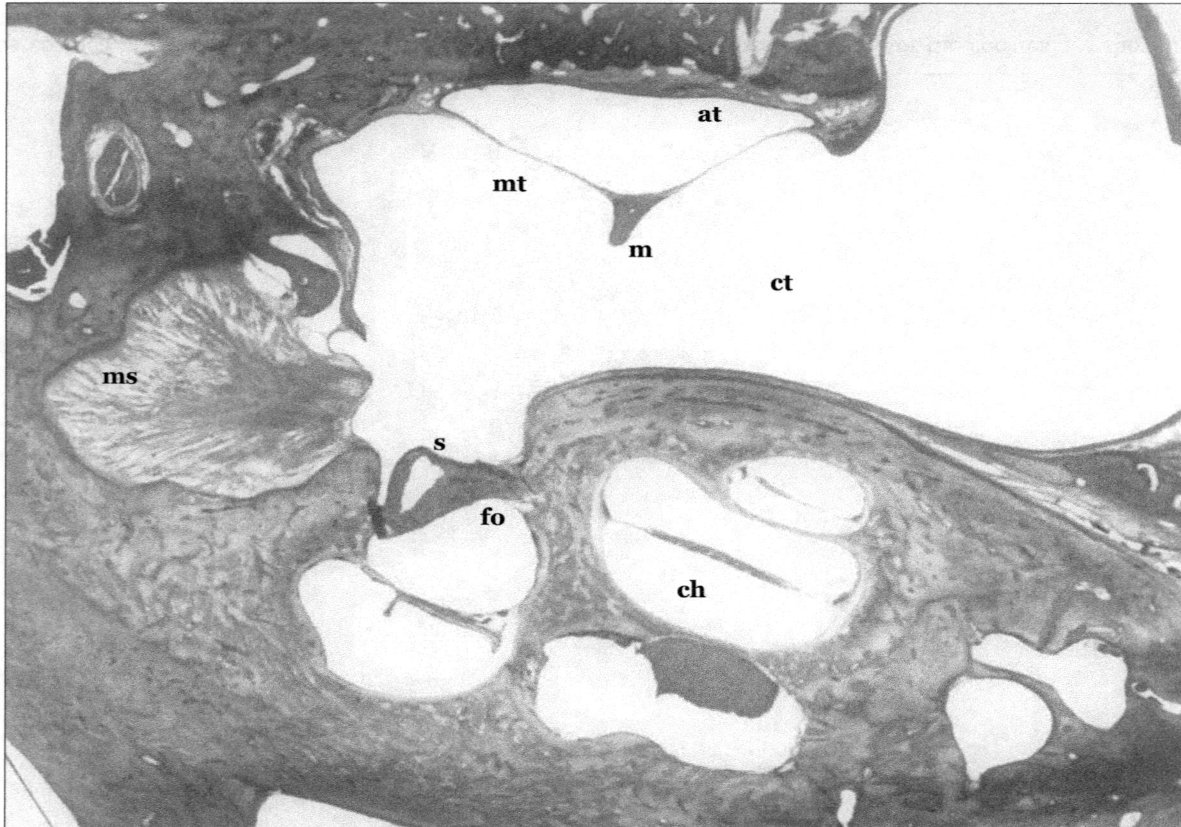

Fig. 21. The cochlea of *Callorchinus ursinus*. Slit through the cochlea axis. 2.5 turns of the cochlea are shown. Staining with hematoxylin-eosin.

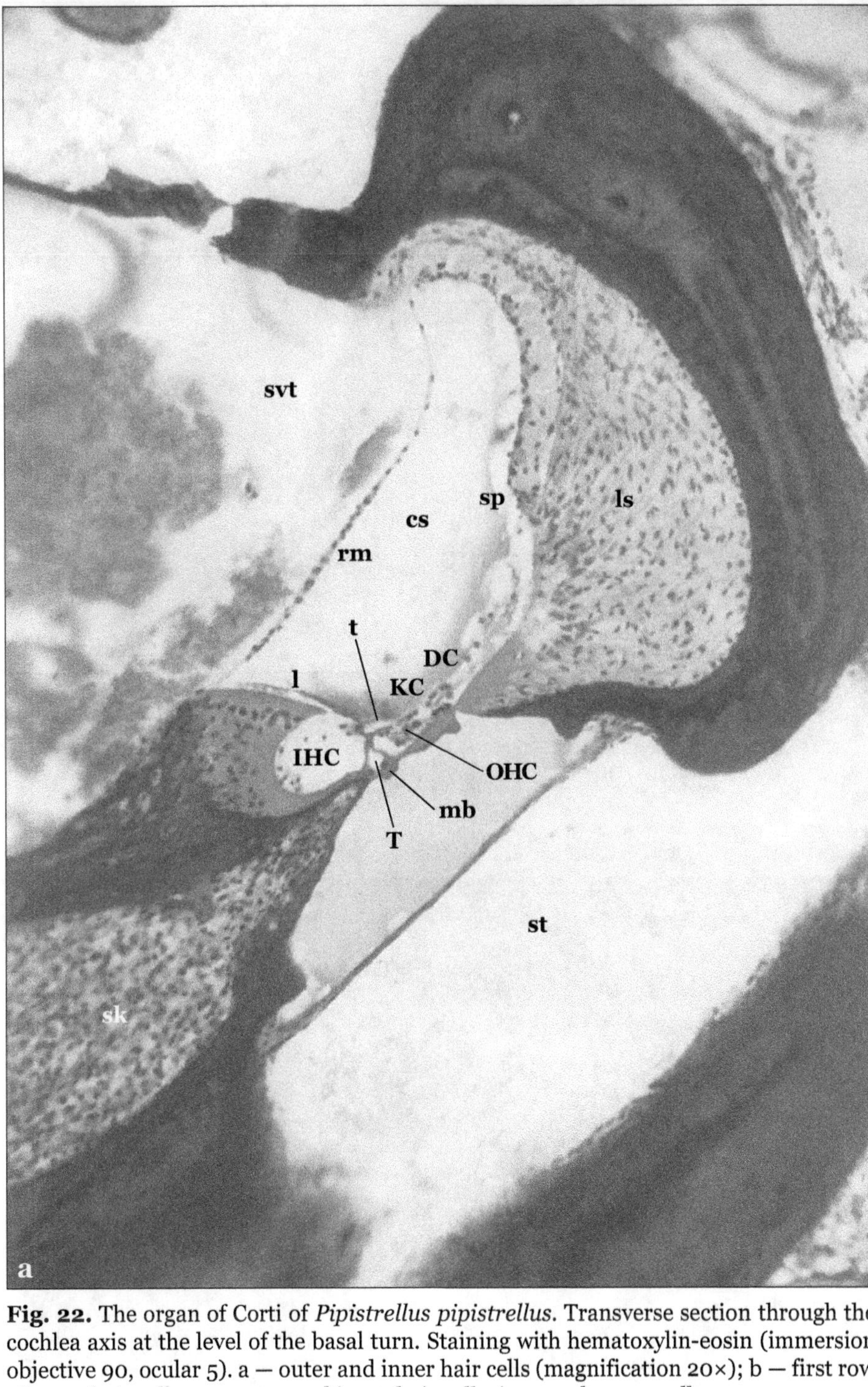

Fig. 22. The organ of Corti of *Pipistrellus pipistrellus*. Transverse section through the cochlea axis at the level of the basal turn. Staining with hematoxylin-eosin (immersion objective 90, ocular 5). a — outer and inner hair cells (magnification 20×); b — first row of outer hair cells; c — outer and inner hair cells, inner columnar cells.

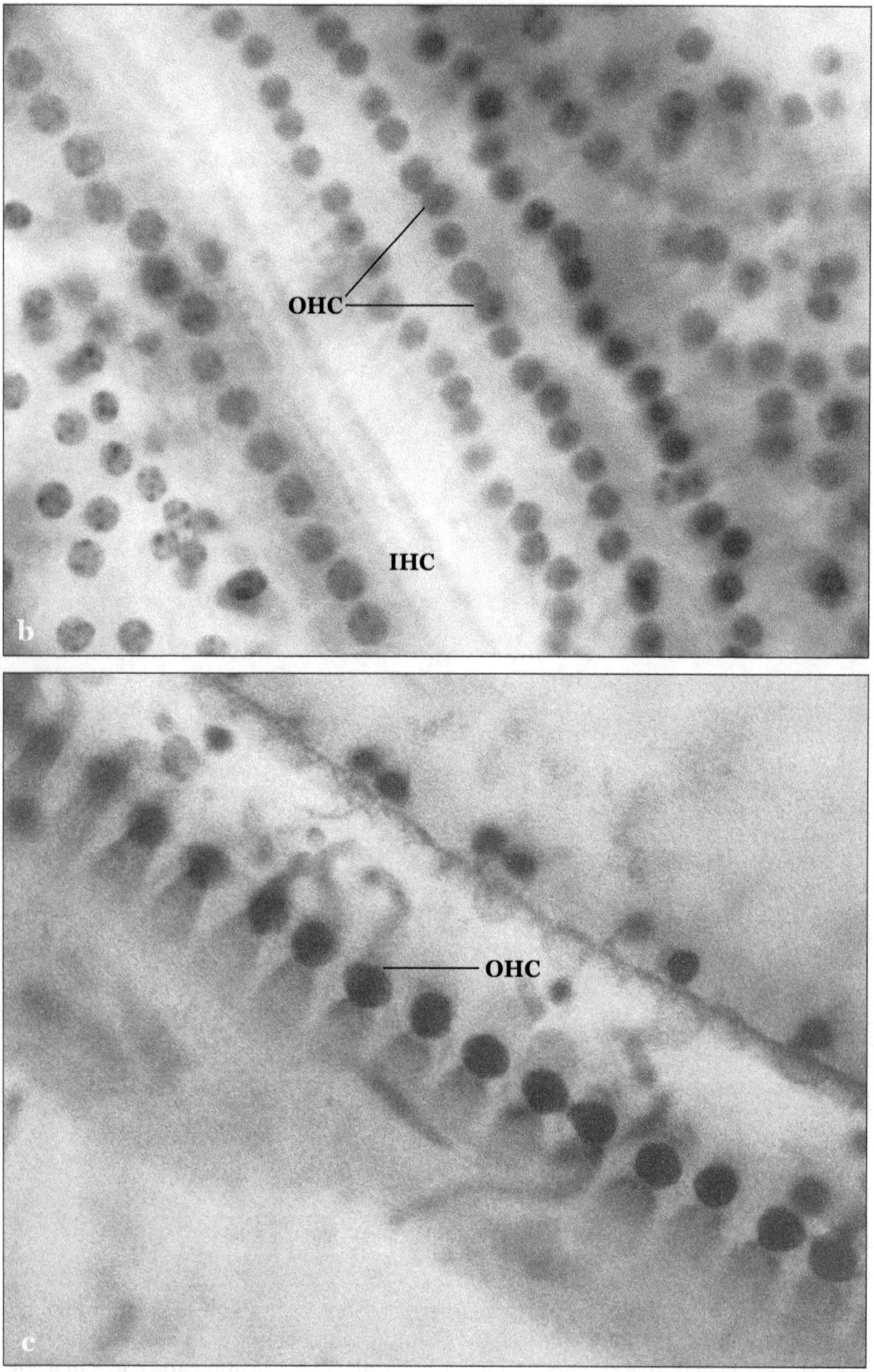

Fig. 22. Continued.

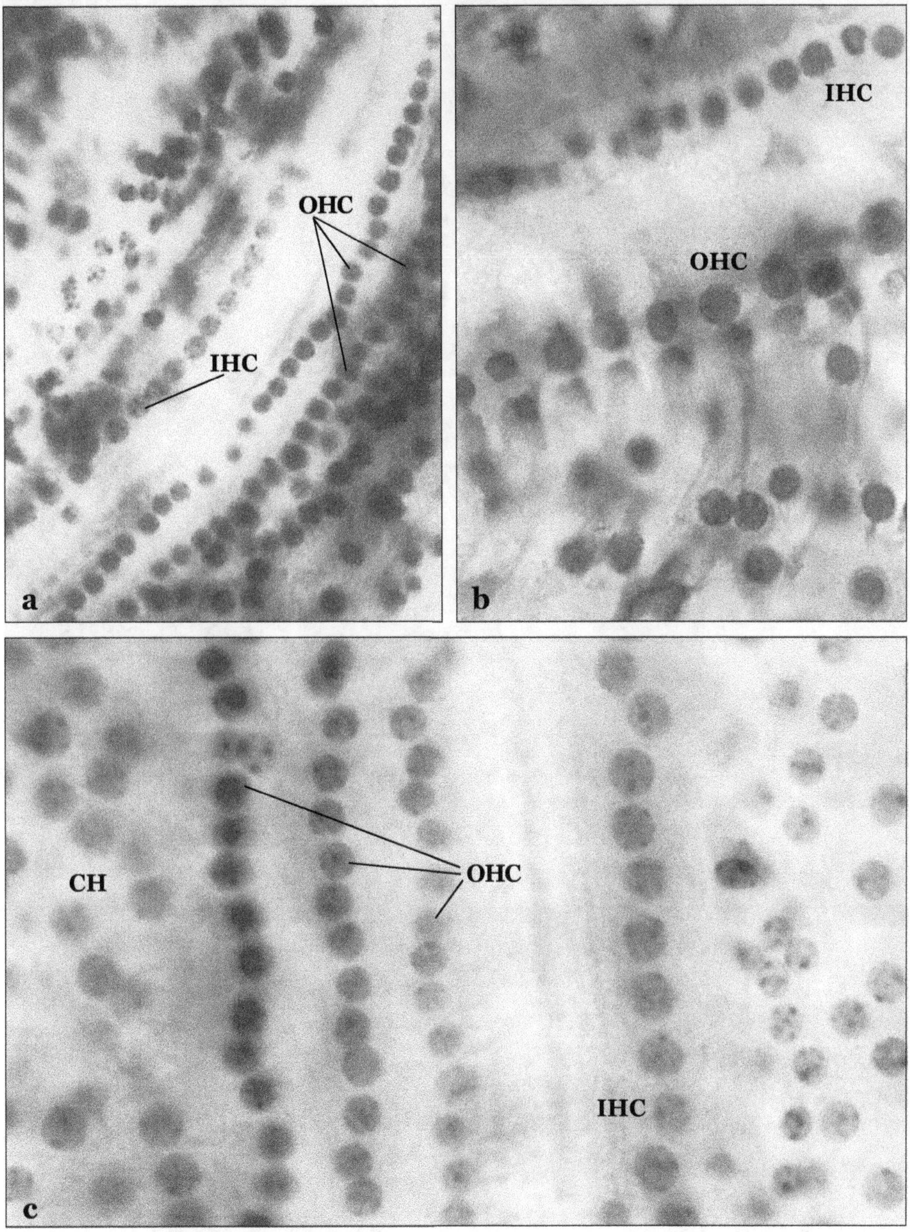

Fig. 23. The organ of Corti of *Pagophilus groenlandicus*. Slit through the cochlea axis at the level of the basal turn. Staining with hematoxylin-eosin (immersion objective 90, ocular 5). a — general view (objective 5, ocular 5); b, c — outer and inner hair cells; d, e – outer and inner columnar cells.

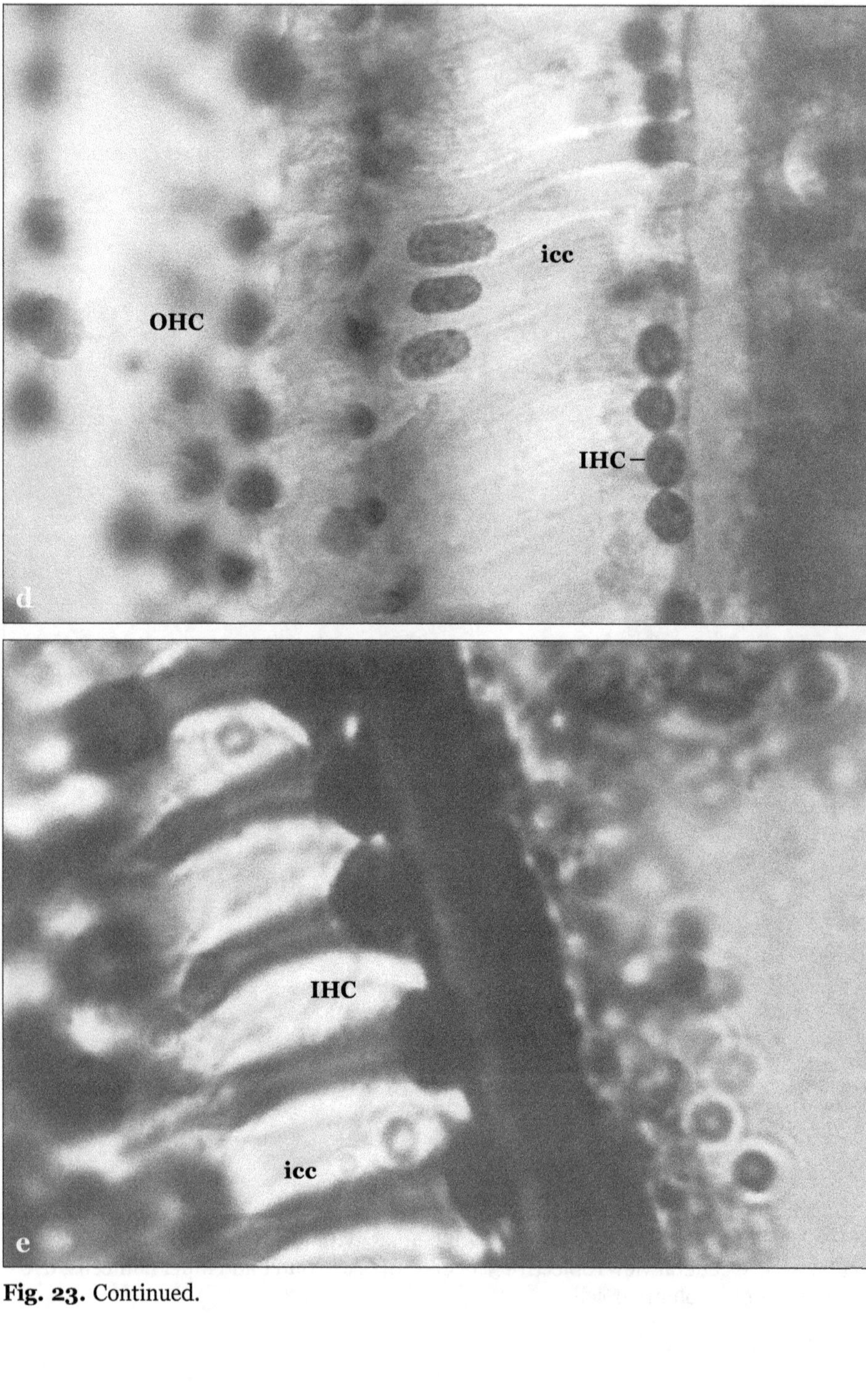

Fig. 23. Continued.

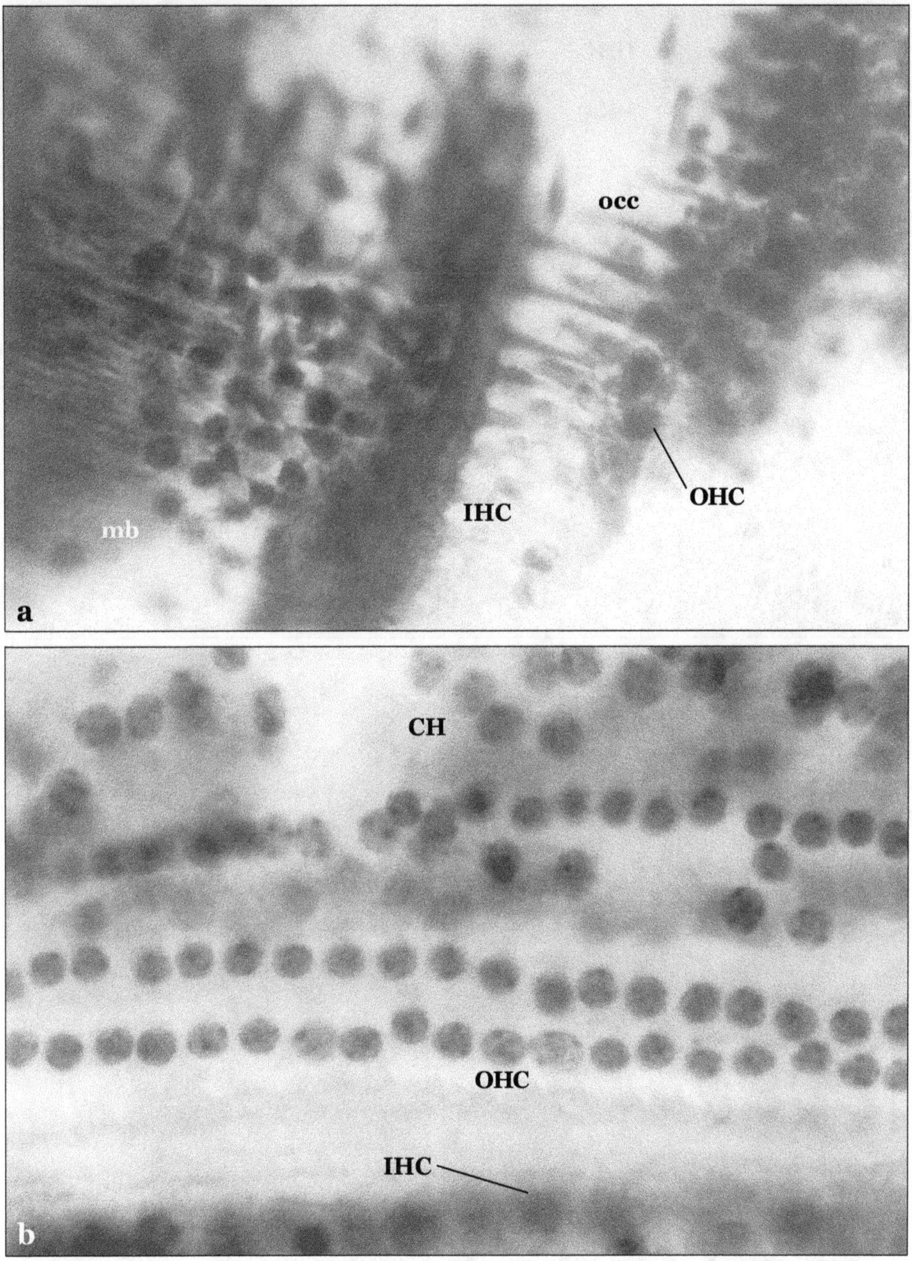

Fig. 24. The organ of Corti of *Callorchinus ursinus*. Slit through the cochlea axis at the level of the basal turn. Staining according to Kulchitsky (immersion objective 90, ocular 7). a — general view; b — outer and inner hair cells.

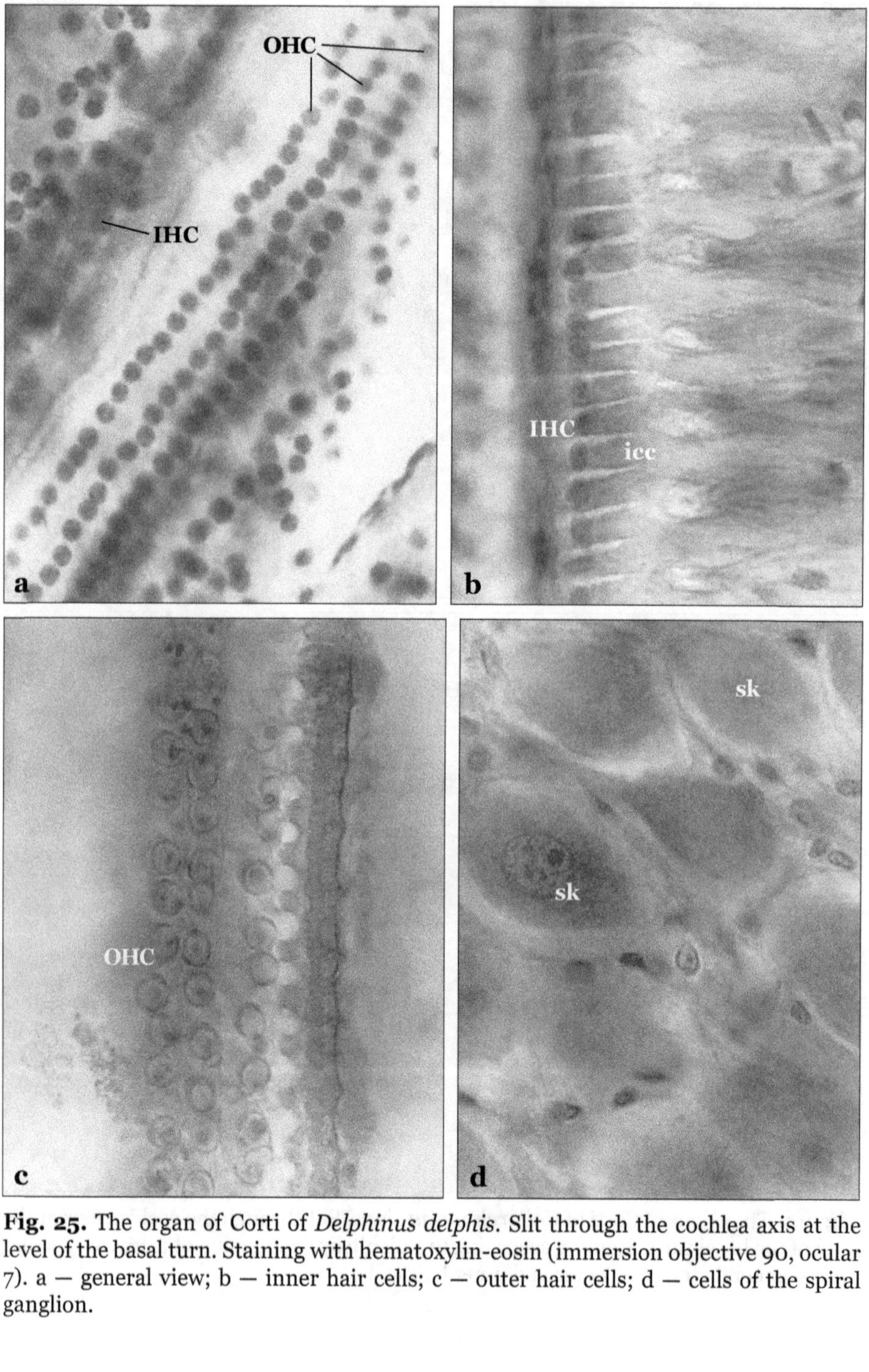

Fig. 25. The organ of Corti of *Delphinus delphis*. Slit through the cochlea axis at the level of the basal turn. Staining with hematoxylin-eosin (immersion objective 90, ocular 7). a — general view; b — inner hair cells; c — outer hair cells; d — cells of the spiral ganglion.

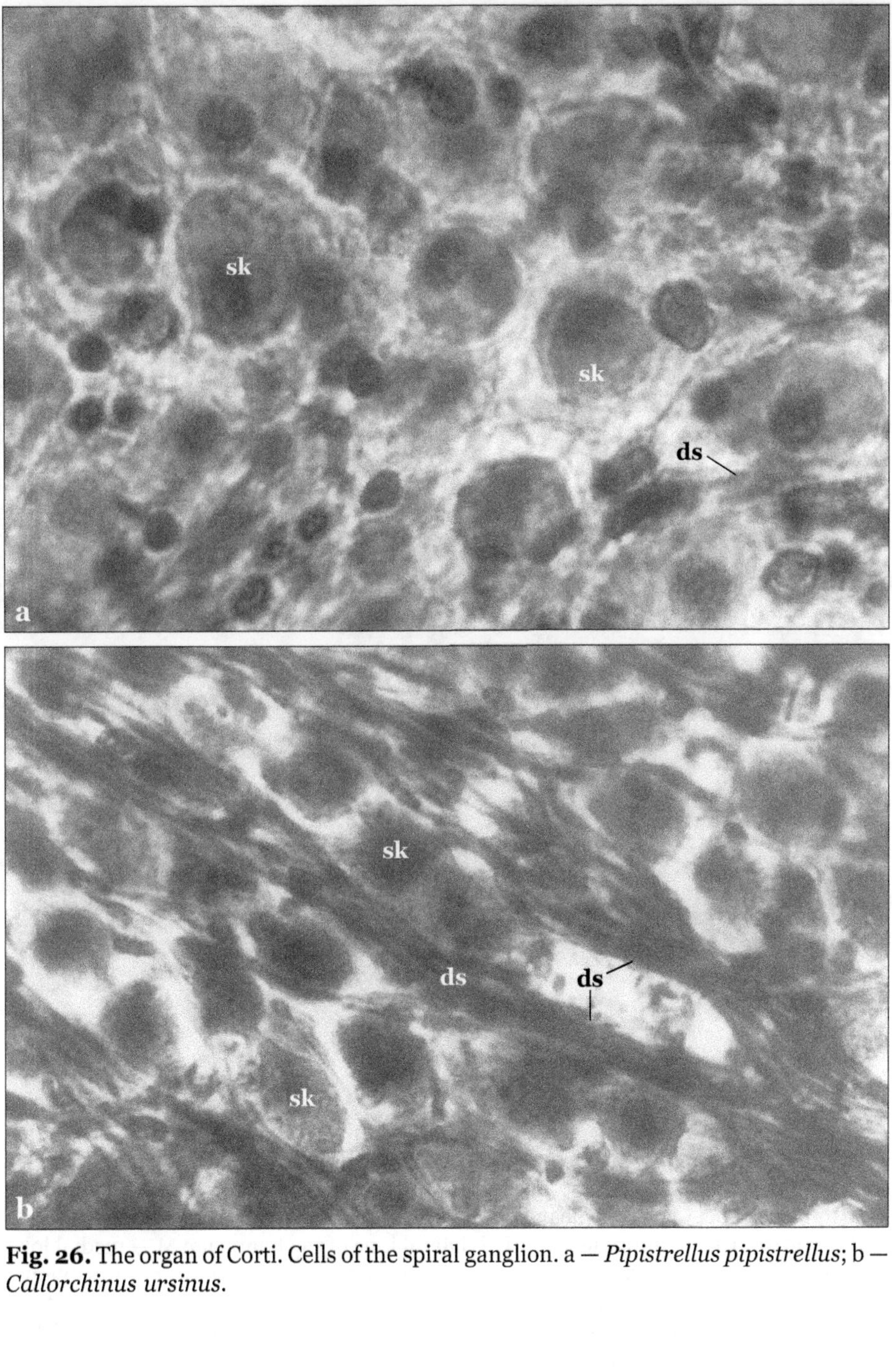

Fig. 26. The organ of Corti. Cells of the spiral ganglion. a — *Pipistrellus pipistrellus*; b — *Callorchinus ursinus*.

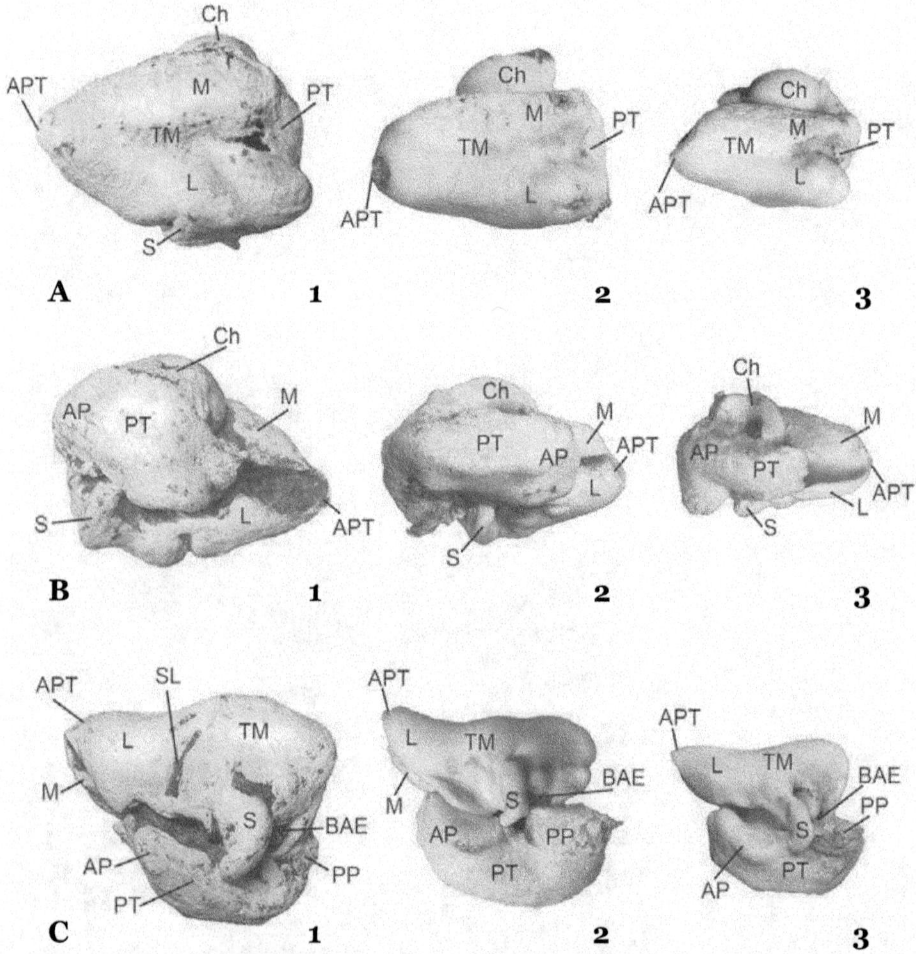

Fig. 27. The bulla tympanica of dolphins, (1) *Inia geofrensis*, (2) *Tursiops truncatus*, and (3) *Delphinus delphis* in 6 planes: A — ventral; B — dorsal; C, D — parasagittal; E — anterior; F — posterior. PT — periotic; AP — anterior process of the periotic; Ch — cochlear part of the periotic; Chc — cochlear canal; TM — tympanic; s — pr. sigmoid; BAE — bony part of the external auditory meatus; L — lateral lobe of the tympanic; M — medial lobe of the tympanic; PTB — pr. tubarius; EF — elliptical foramen; FS — fossa for the stapedial muscle; FC — fenestra cochleae; CD — lateral part of the ductus cochlearis; VD — lateral part of the ductus vestubularis; CT — crista transversalis; SL — sulcus lateralis; APT — anterior process of the tympanic; PP — posterior process of the tympanic.

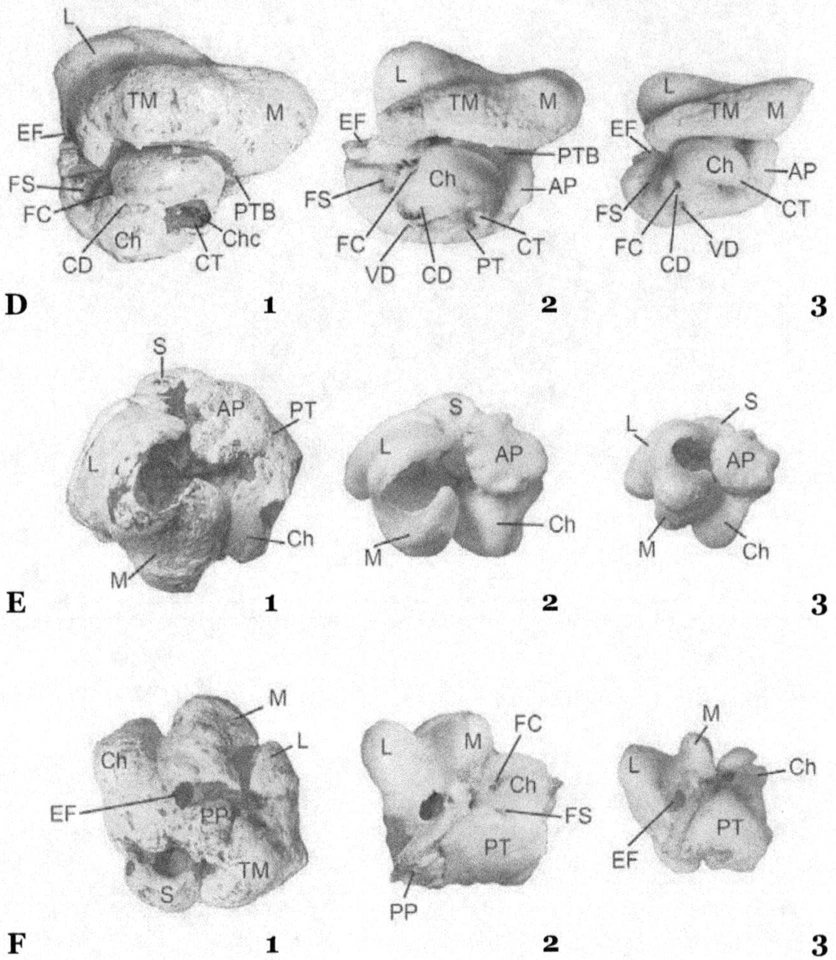

Fig. 27. Continued.

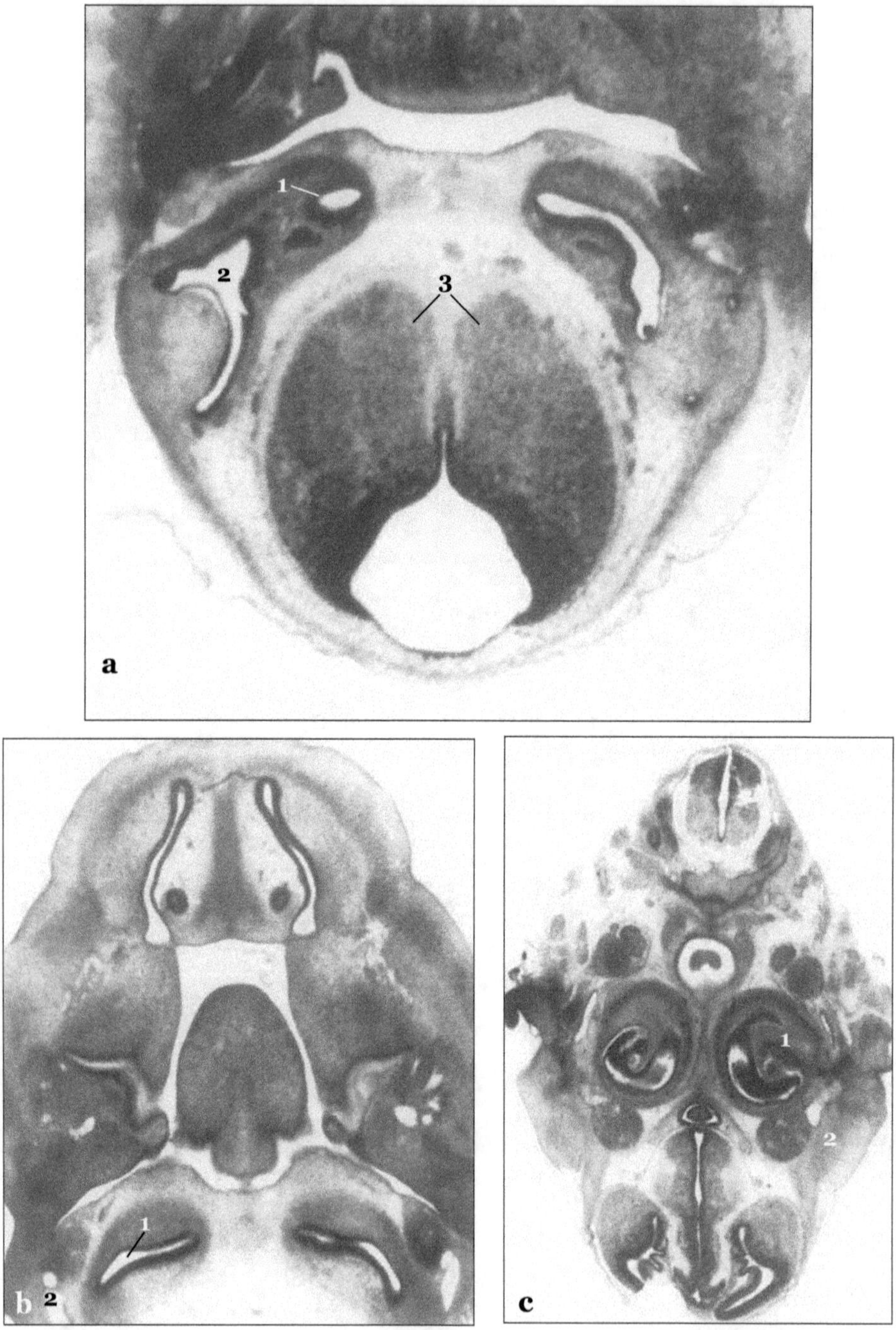

Fig. 28. Histotopography of the peripheral auditory system in dorsoventral sections of prefetal heads, stages 14, 15. a — *Rattus norvegicus*; b — *Cavia porcellus*; c — *Rhinolophus ferrumequinum*. Subdivision of the auditory vesicle into the cochlea and vestibular parts is shown.

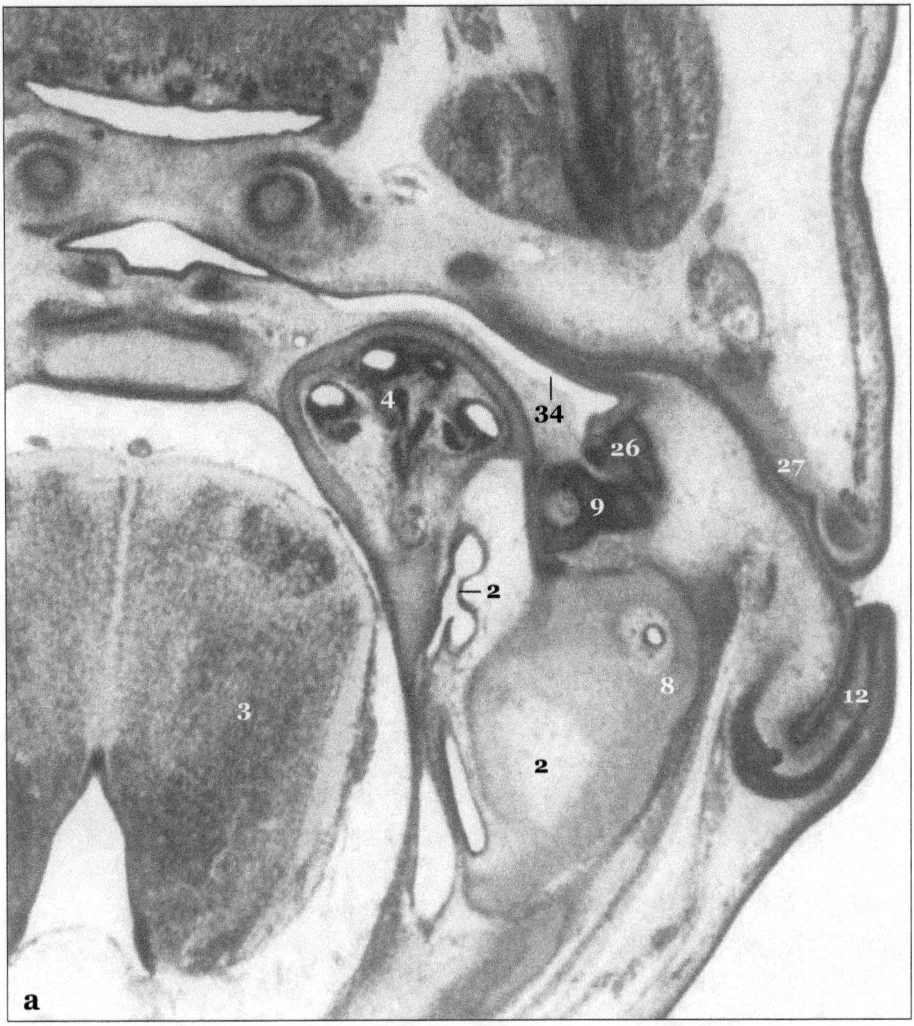

Fig. 29. Histotopography of the peripheral auditory system in dorsoventral sections of prefetal heads, stage 17. a — *Rattus norvegicus*. Contours of the auricle are shown, a medial turn of the auricle is formed, auditory ossicles are differentiated and represented by independent formations; b — *Sus scrofa domestica*; c — *Cavia porcellus*.

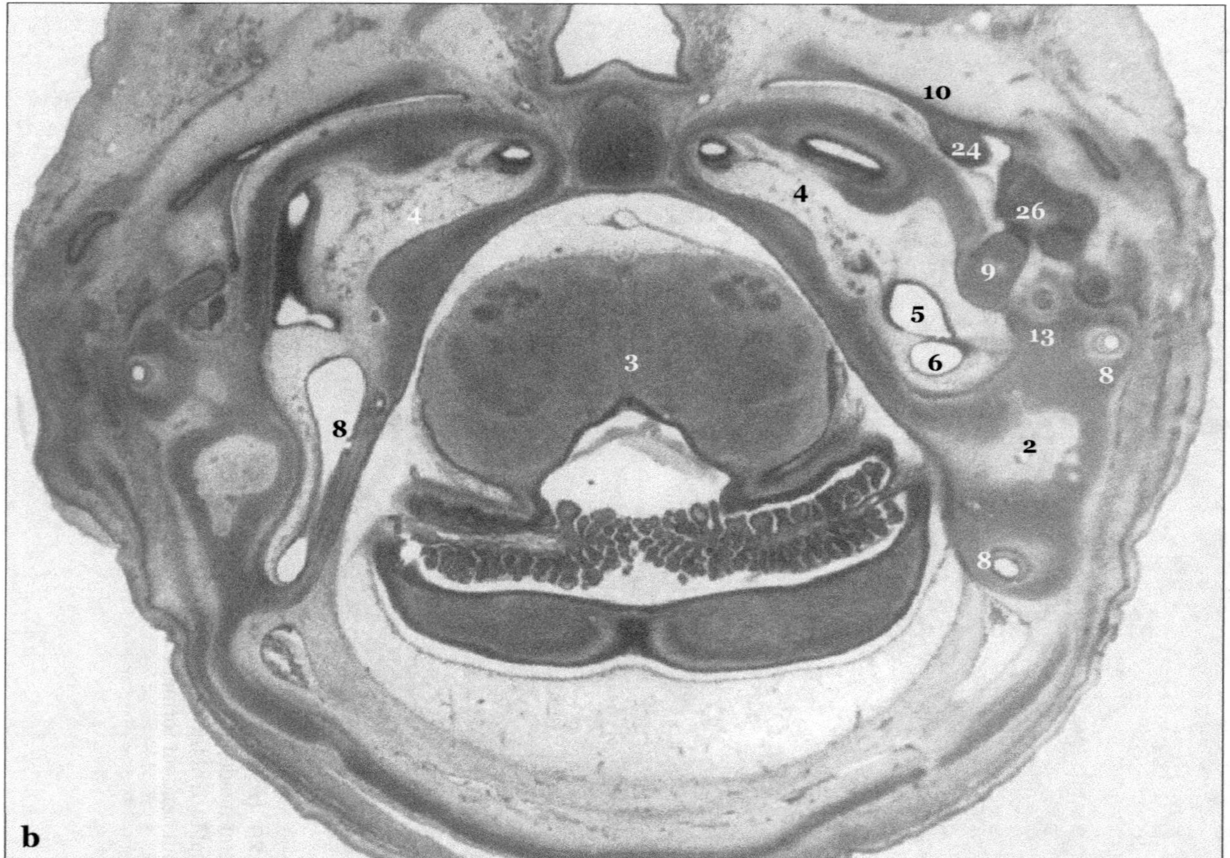

Fig. 29. Continued.

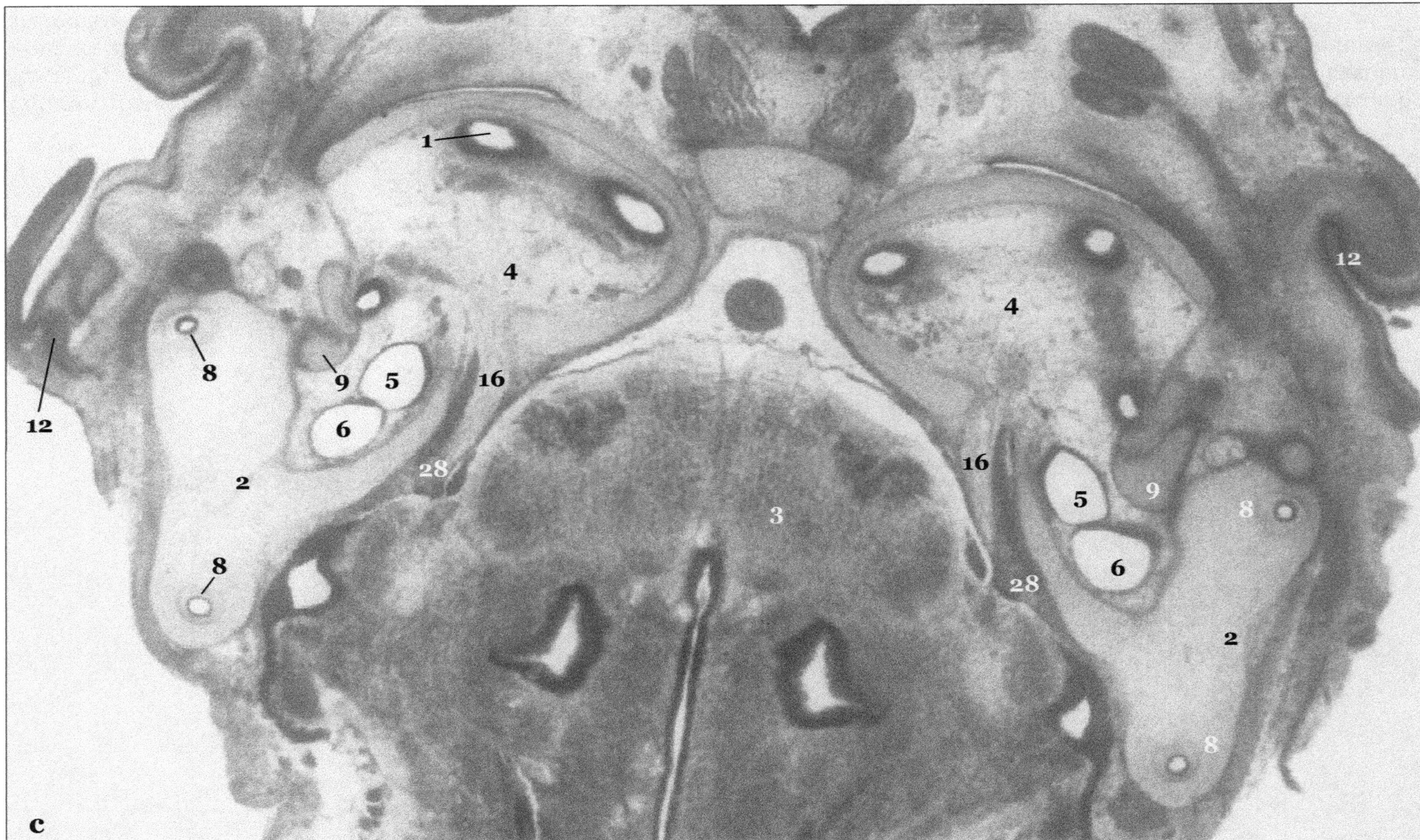

Fig. 29. Continued.

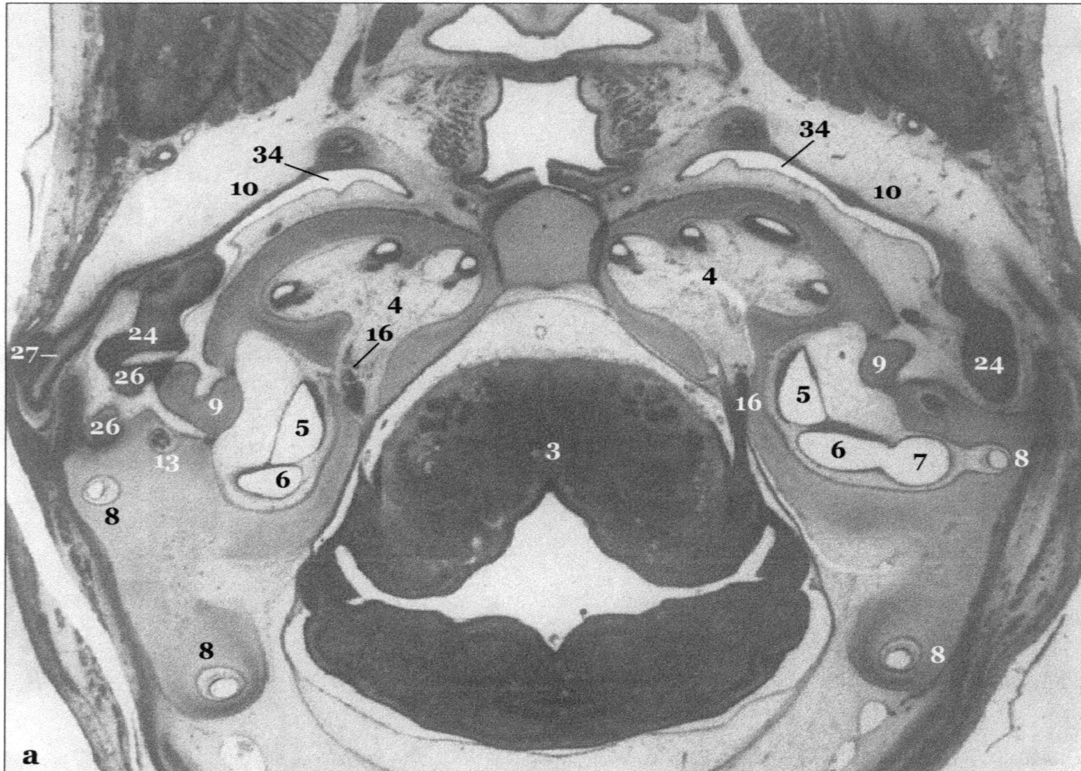

Fig. 30. Histotopography of the peripheral auditory system in dorsoventral sections of prefetal heads of *Sus scrofa domestica*. a, b — stages 18, 19. The auditory meatus is filled with epithelial cells; the manubrium of the malleus is connected to the tympanic membrane along its entire length, as in other terrestrial species; c — stages 19, 20. The cochlea is formed by 3 turns, as in definitive forms; an anatomical formation of the vestibular apparatus is completed.

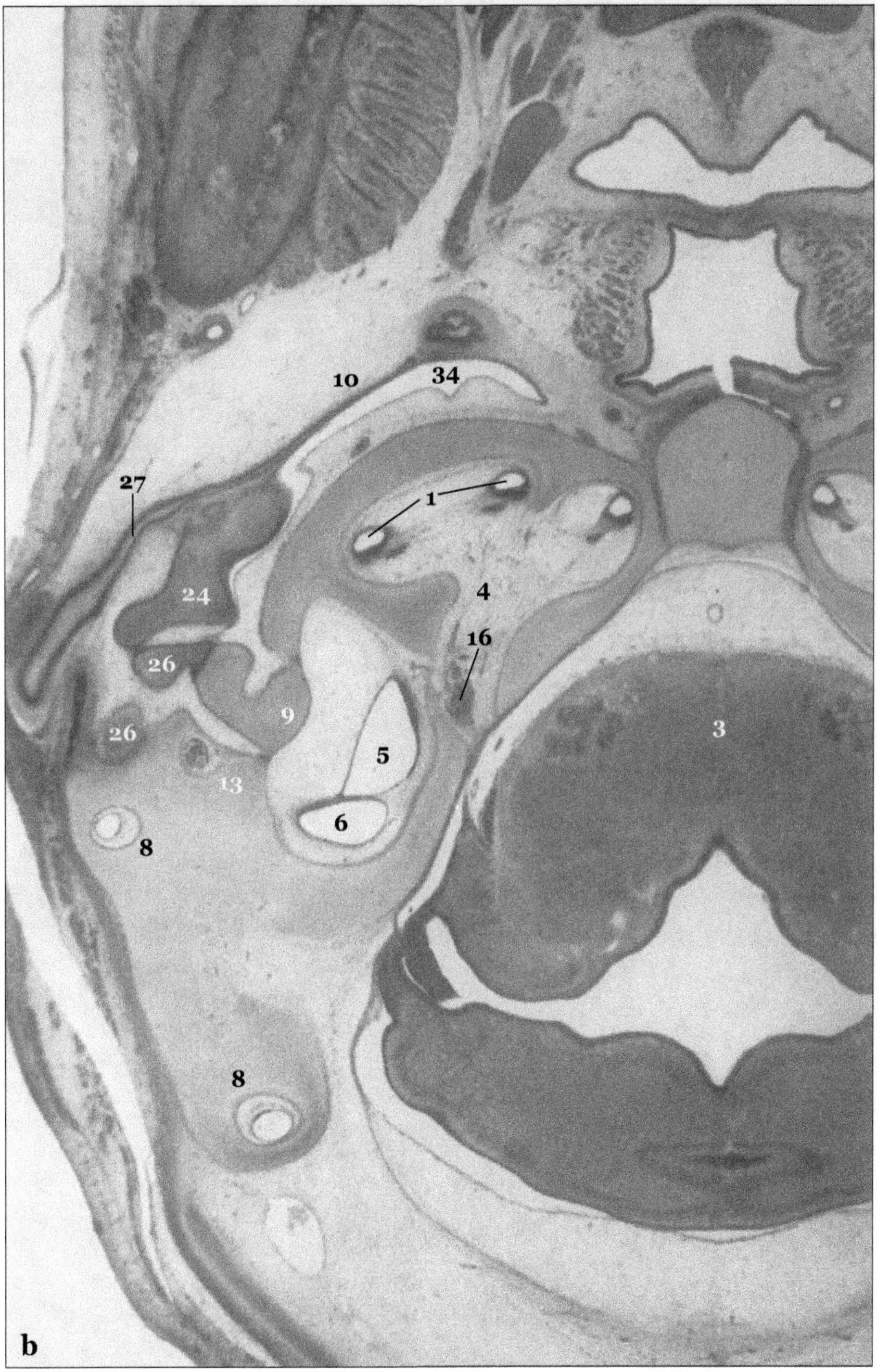

Fig. 30. Continued.

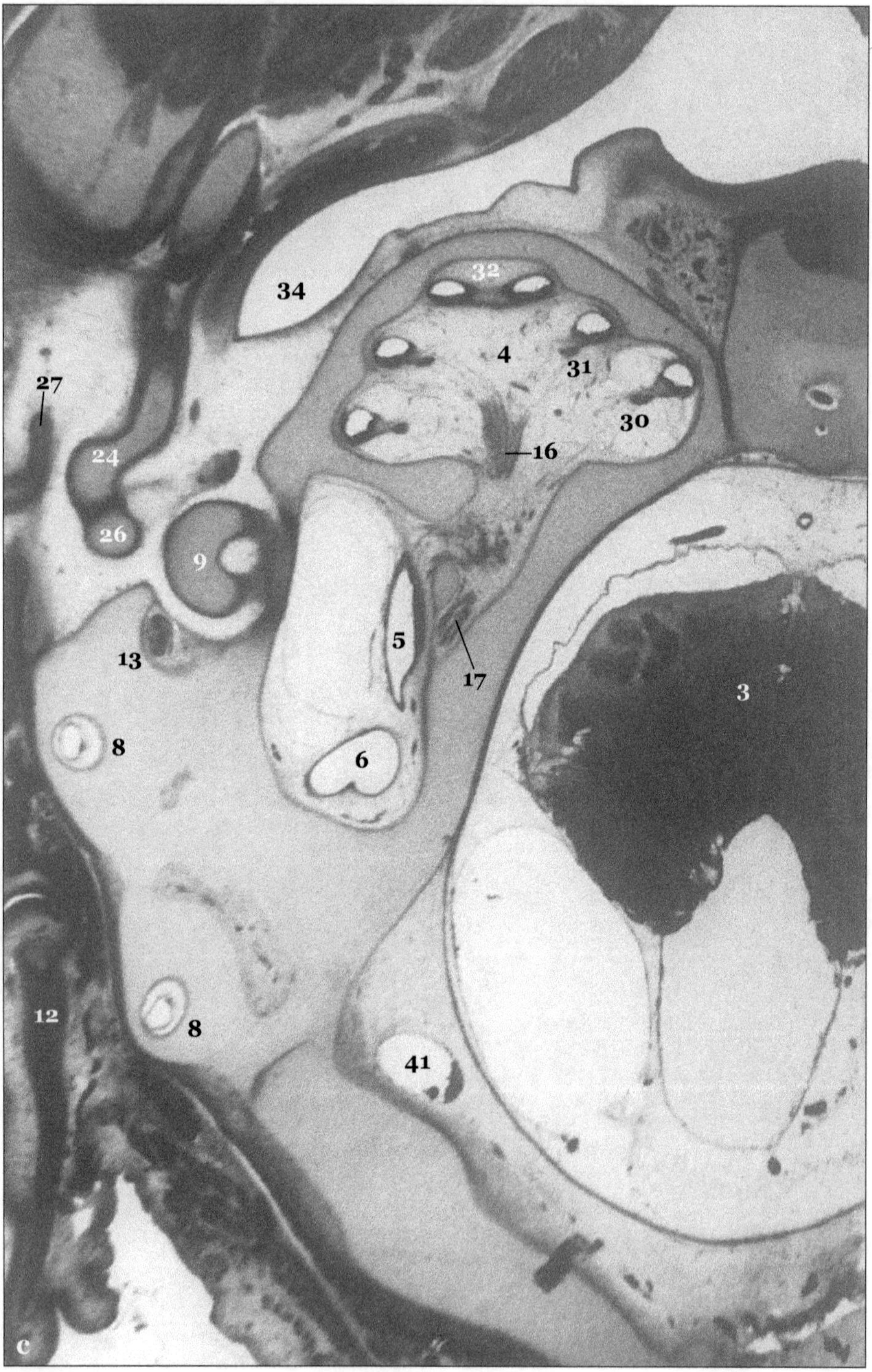

Fig. 30. Continued.

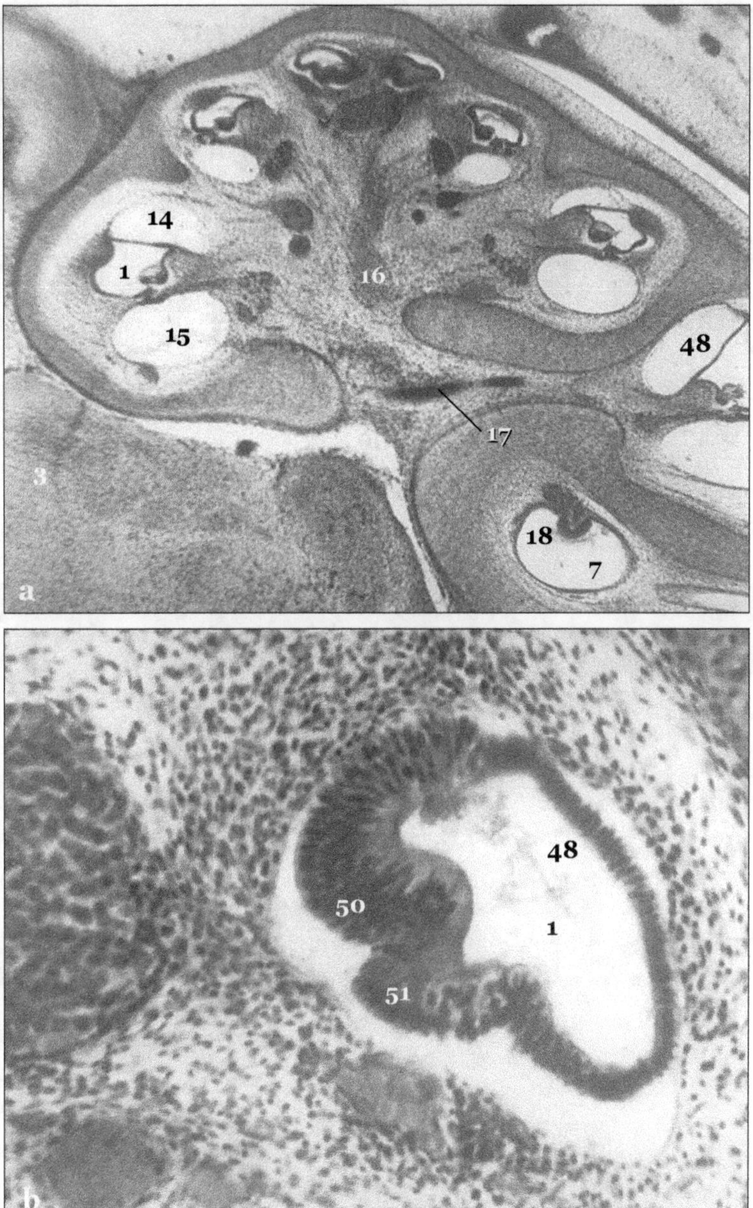

Fig. 31. The cochlea of *Rhinolophus ferrumequinum* prefetus, stages 20, 21 (a–d), and *Pipistrellus pipistrellus*, stages 22, 23 (e–g). Histological preparations. Staining according to Kampas. a — *Rhinolophus ferrumequinum*; general view of the cochlea; the different stages of cochlear canal elements and organ of Corti differentiation in the turns of the cochlea are shown; differentiation of the elements of the cochlear canal starts with the basal turn of the cochlea, gradually spreading onto the medial and then apical turns; b — apical turn; c — medial turn; d — basal turn; e–g — *Pipistrellus pipistrellus*; an anatomical formation of the cochlear and vestibular apparatuses is completed.

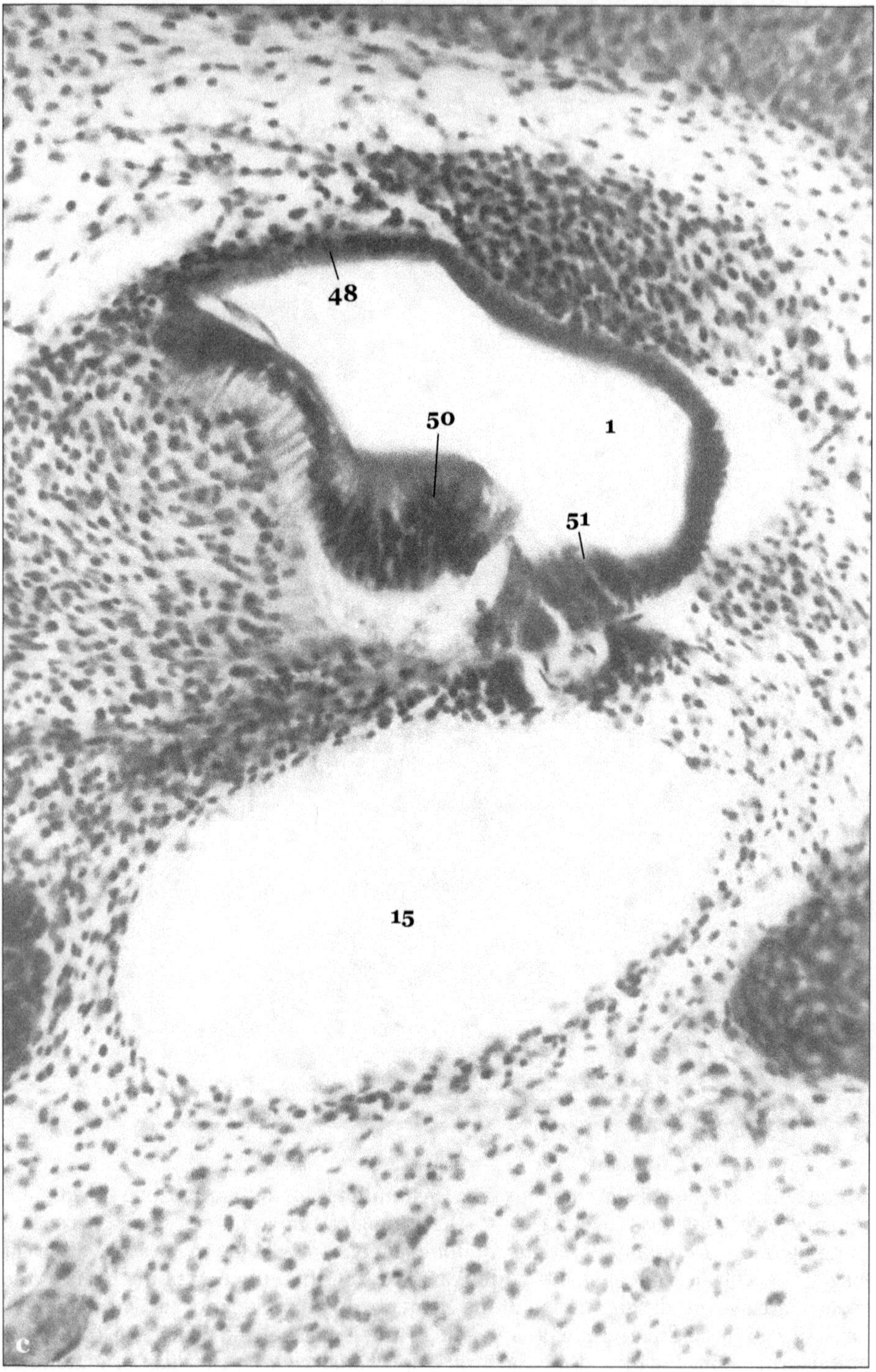

Fig. 31. Continued.

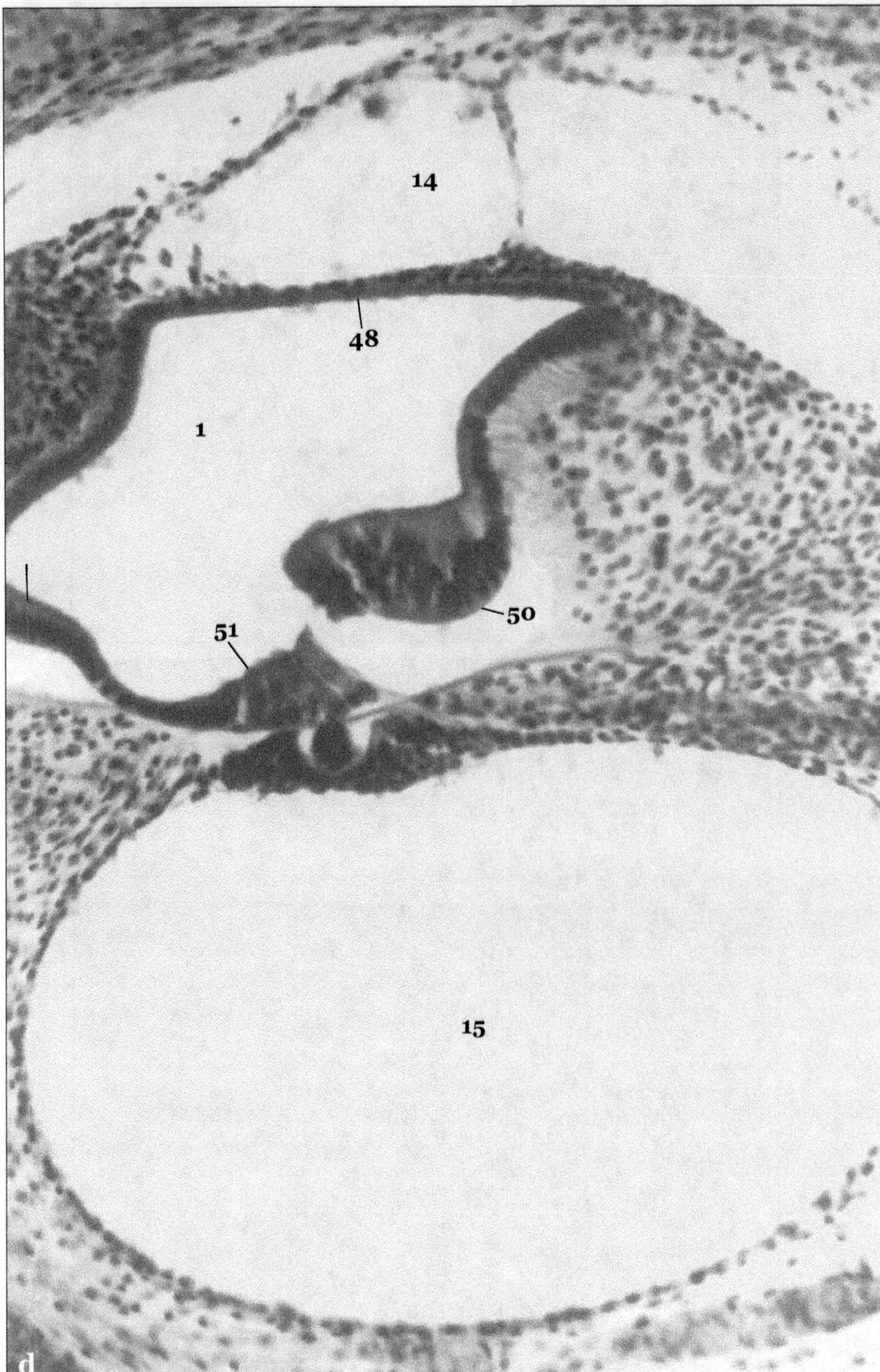

Fig. 31. Continued.

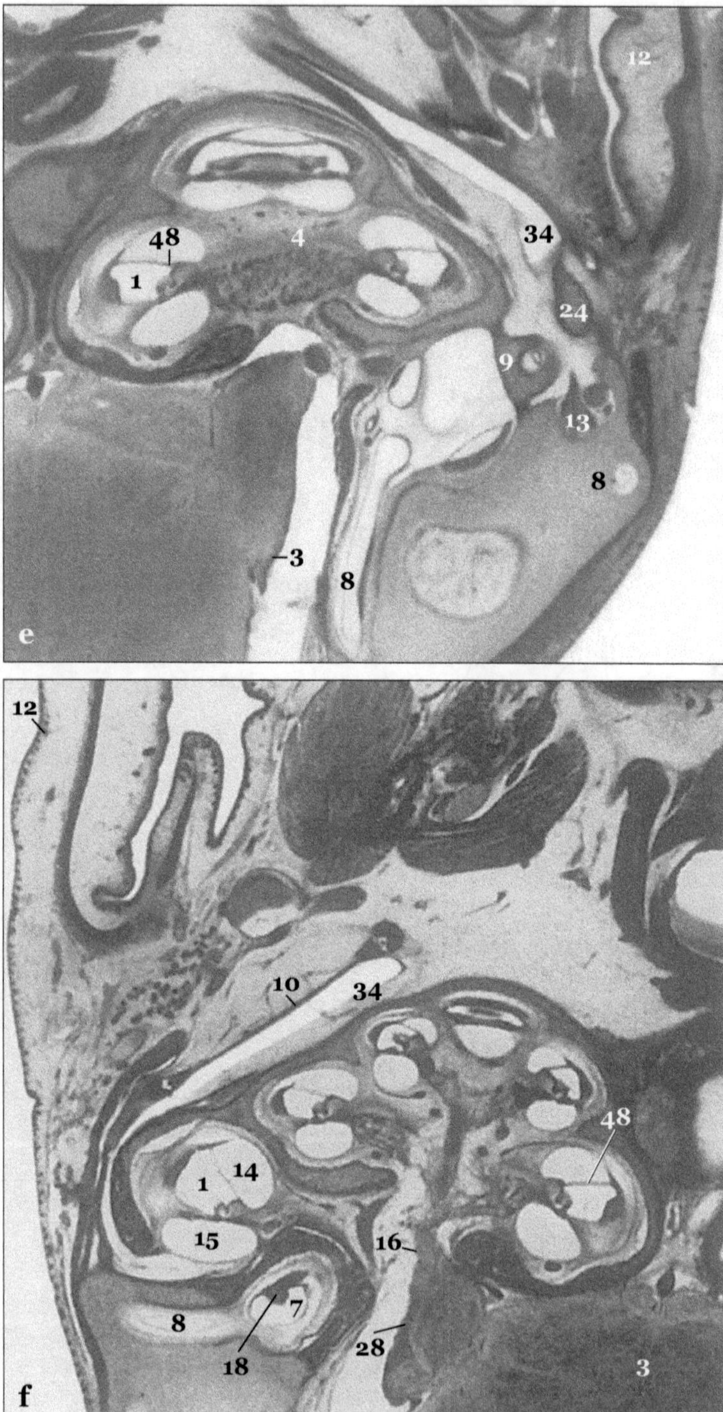

Fig. 31. Continued.

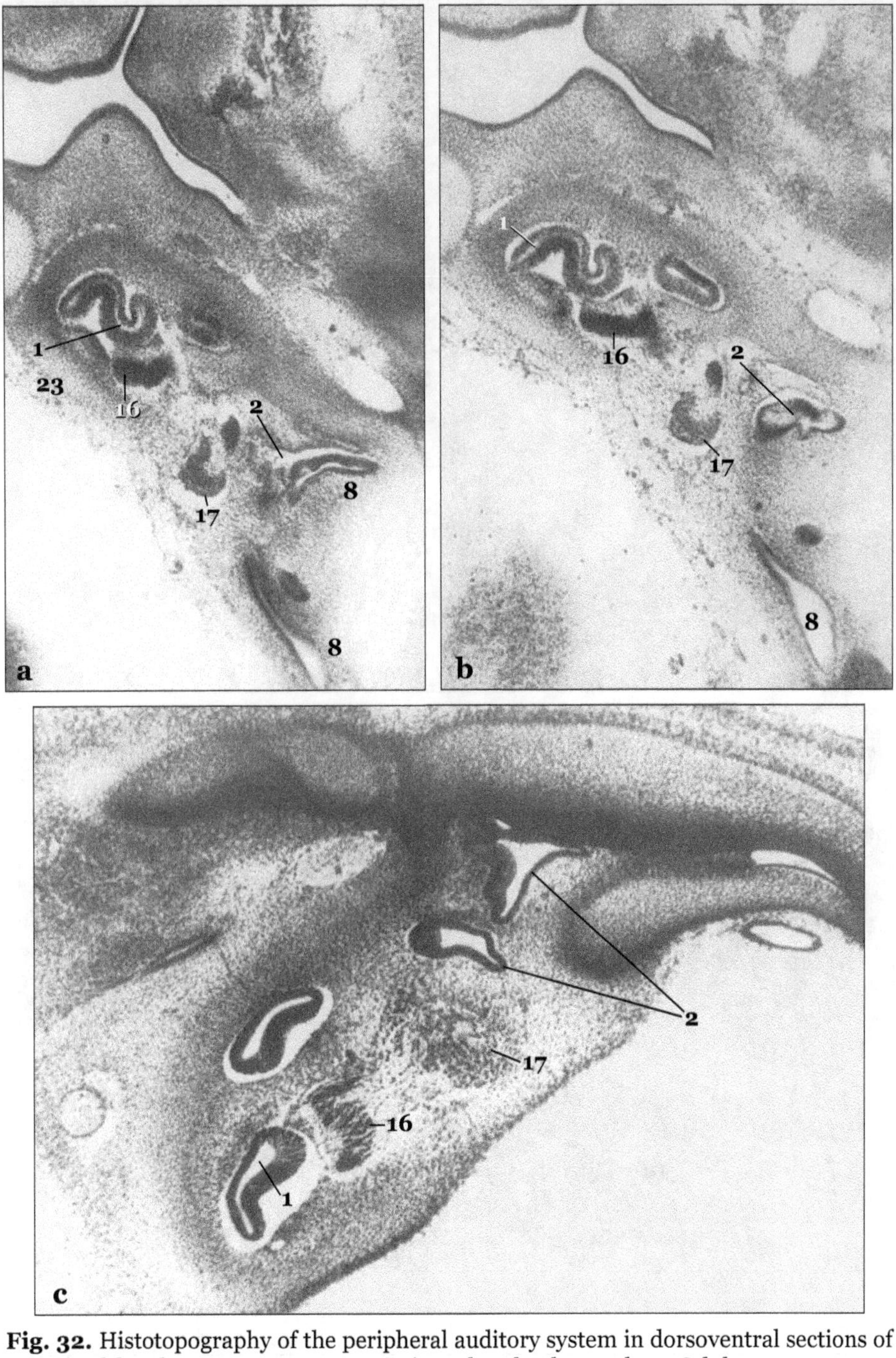

Fig. 32. Histotopography of the peripheral auditory system in dorsoventral sections of a prefetal head, stages 14, 15. a — *Erignathus barbatus*; b — *Odobenus rosmarus divergens*; c — *Delphinapterus leucas*; d — *Stenella attenuata*; e — *Balaenoptera acutorostrata*. Division into the cochlea and vestibular parts is shown.

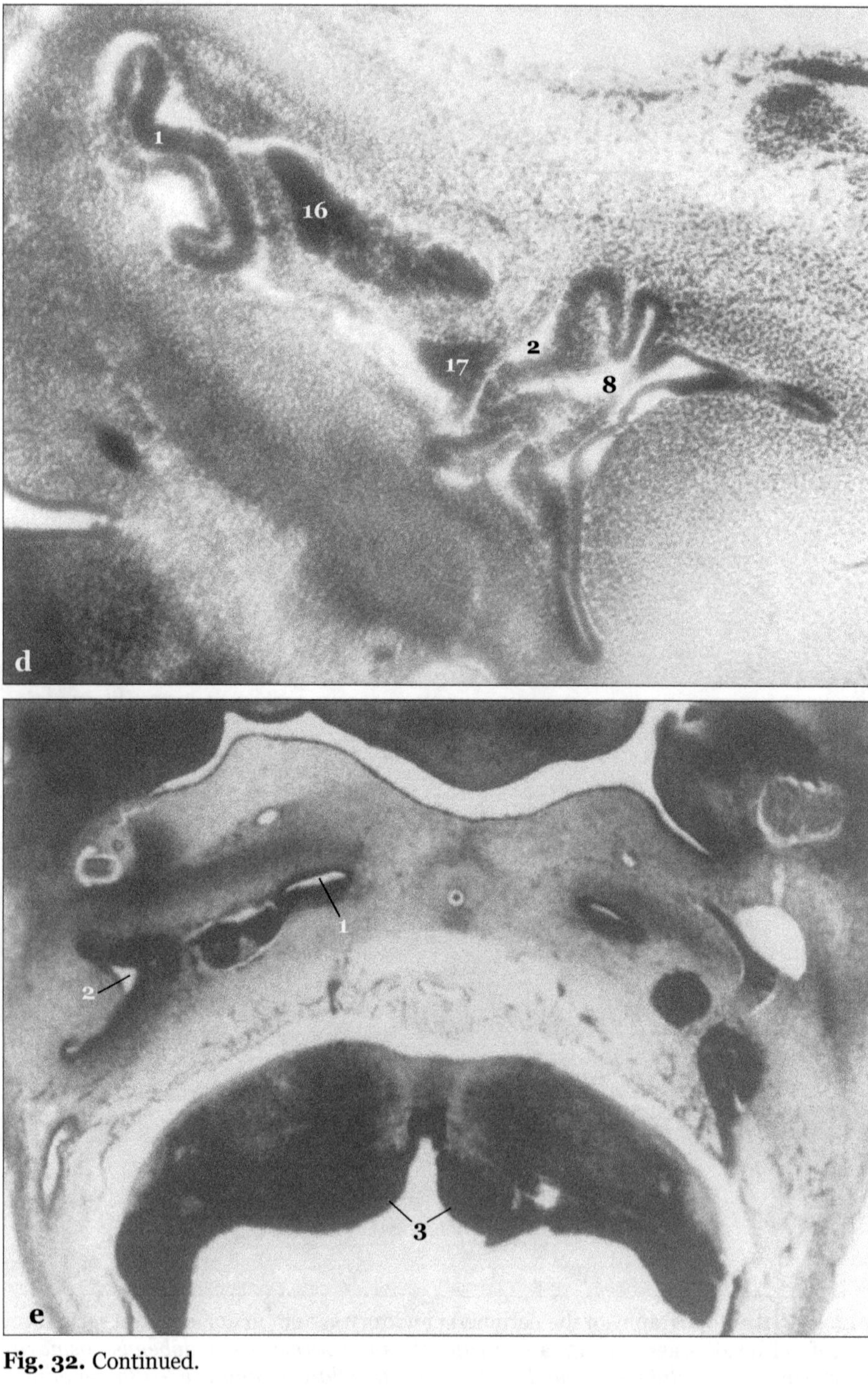

Fig. 32. Continued.

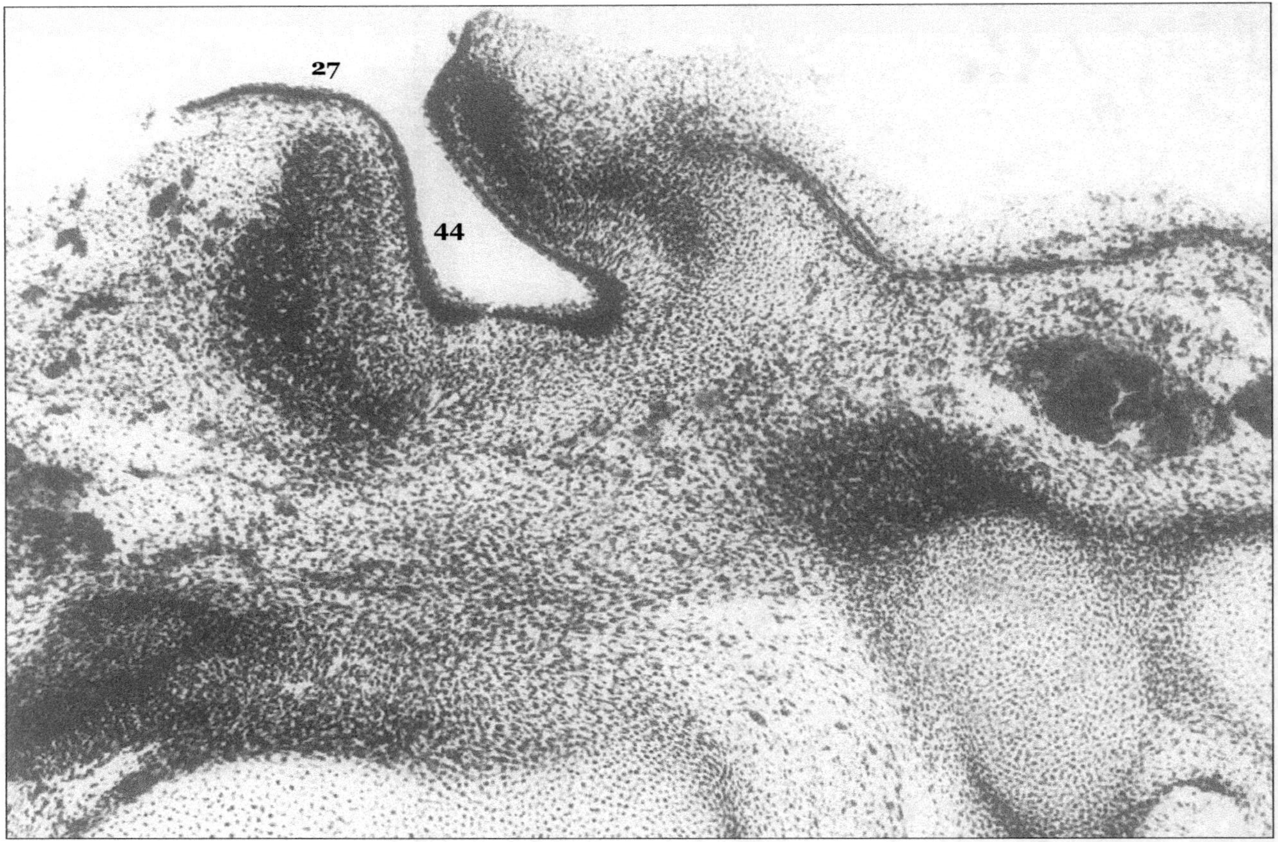

Fig. 33. Histotopography of the peripheral auditory system in dorsoventral sections of the head of a prefetal *Erignathus barbatus*, stages 14, 15; a cutaneous deepening is surrounded by prominences; an auricle is later to develop from the deepening.

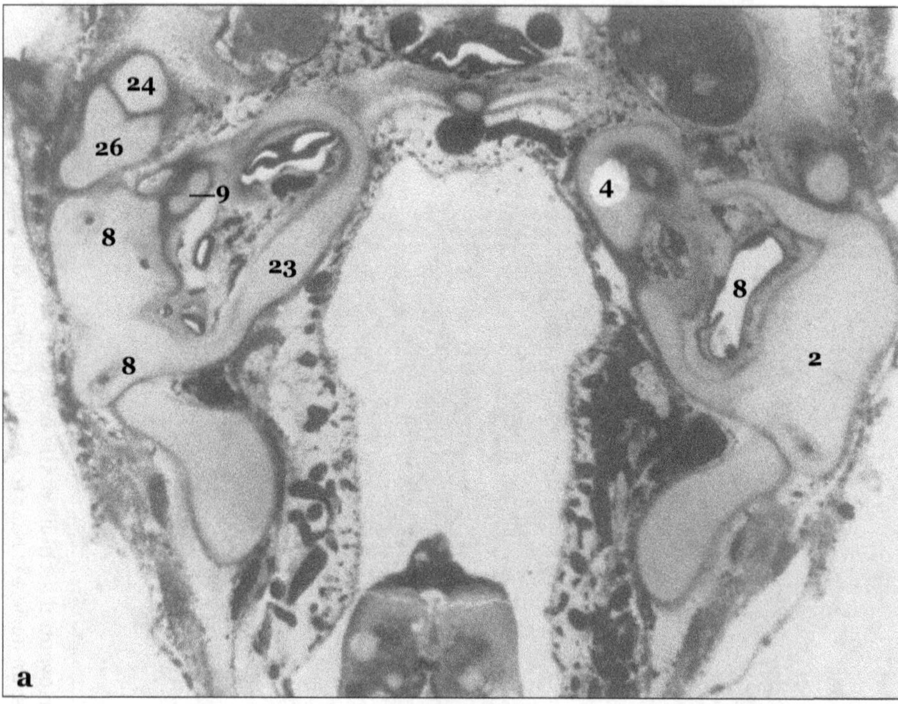

Fig. 34. Histotopography of the peripheral auditory system in a prefetal *Erignathus barbatus*. a — stage 16; a medial turn of the cochlea is formed; the cutis of the auditory ossicles is represented by independent formations, the malleus and incus are connected to each other rectangularly; b, c, d — stage 18; formation of the structures of the vestibular apparatus (sacculus, utriculus) is shown; differentiation of cochlear canal elements is marked; e — stage 19; the external auditory meatus is surrounded by embryonic cartilage; the auditory ossicles are massive; the malleus and incus are joined at a right angle; in the cochlea, differentiation of cochlear passage elements and of the cells of the organ of Corti occurs.

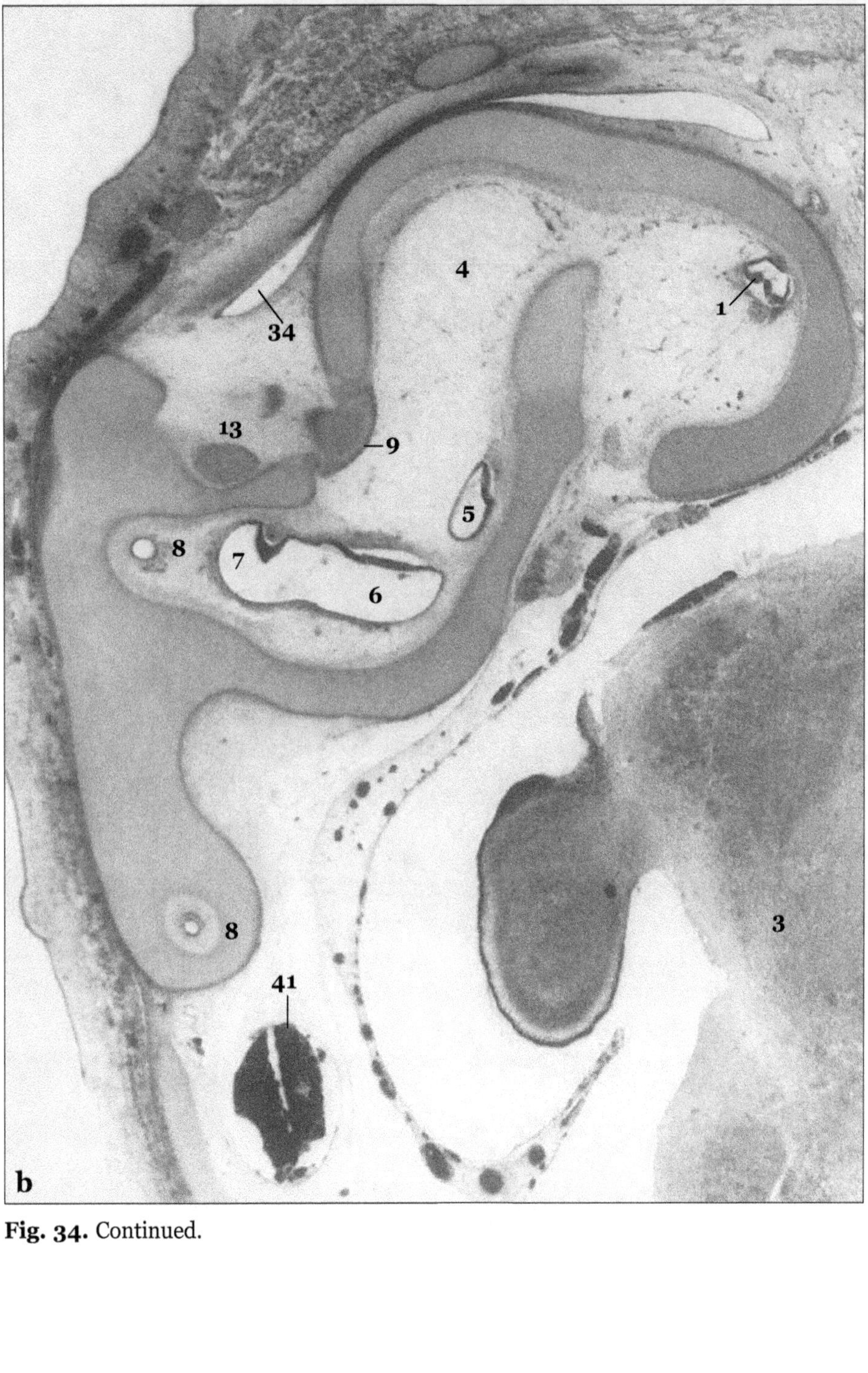

Fig. 34. Continued.

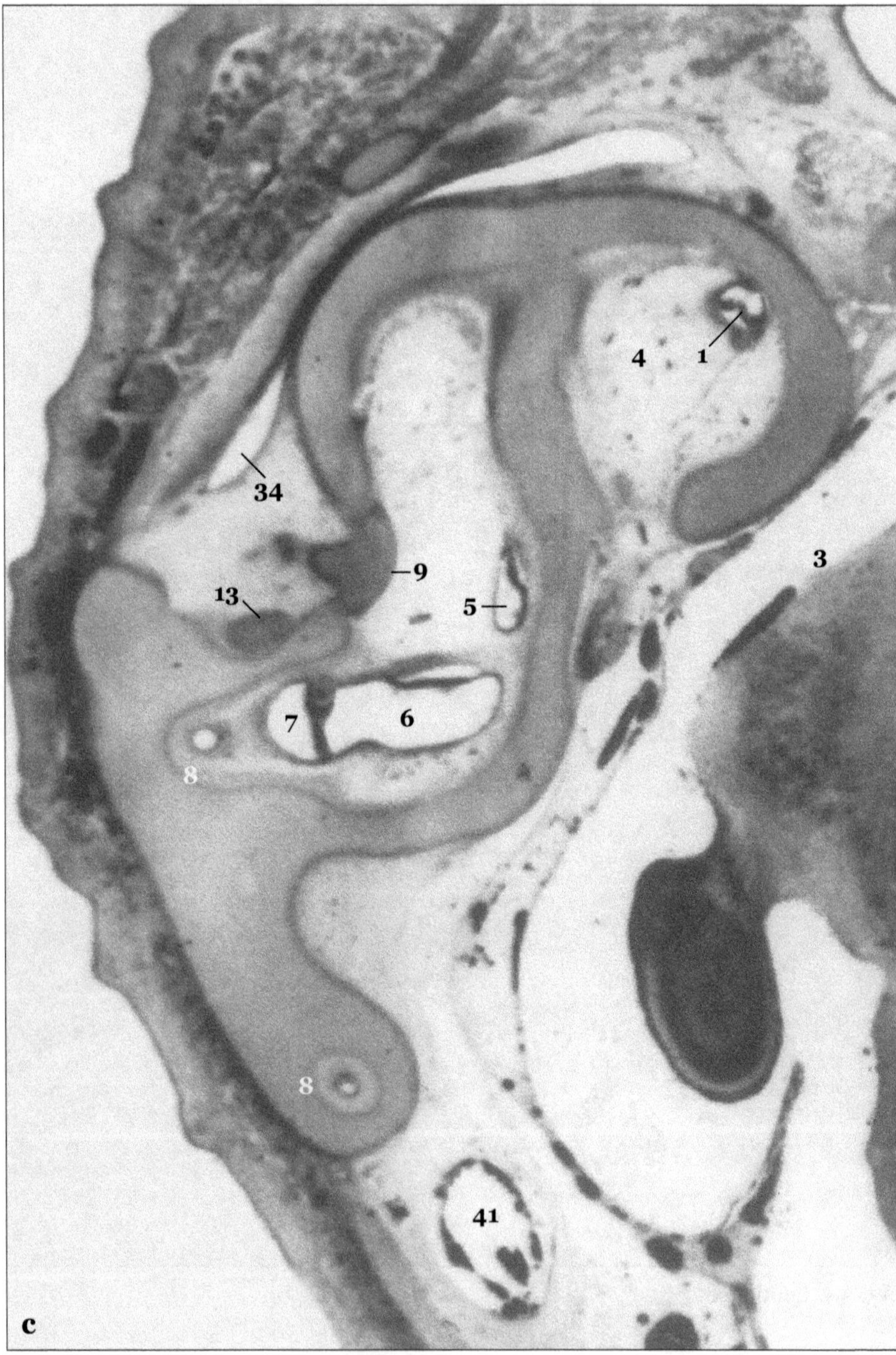

Fig. 34. Continued.

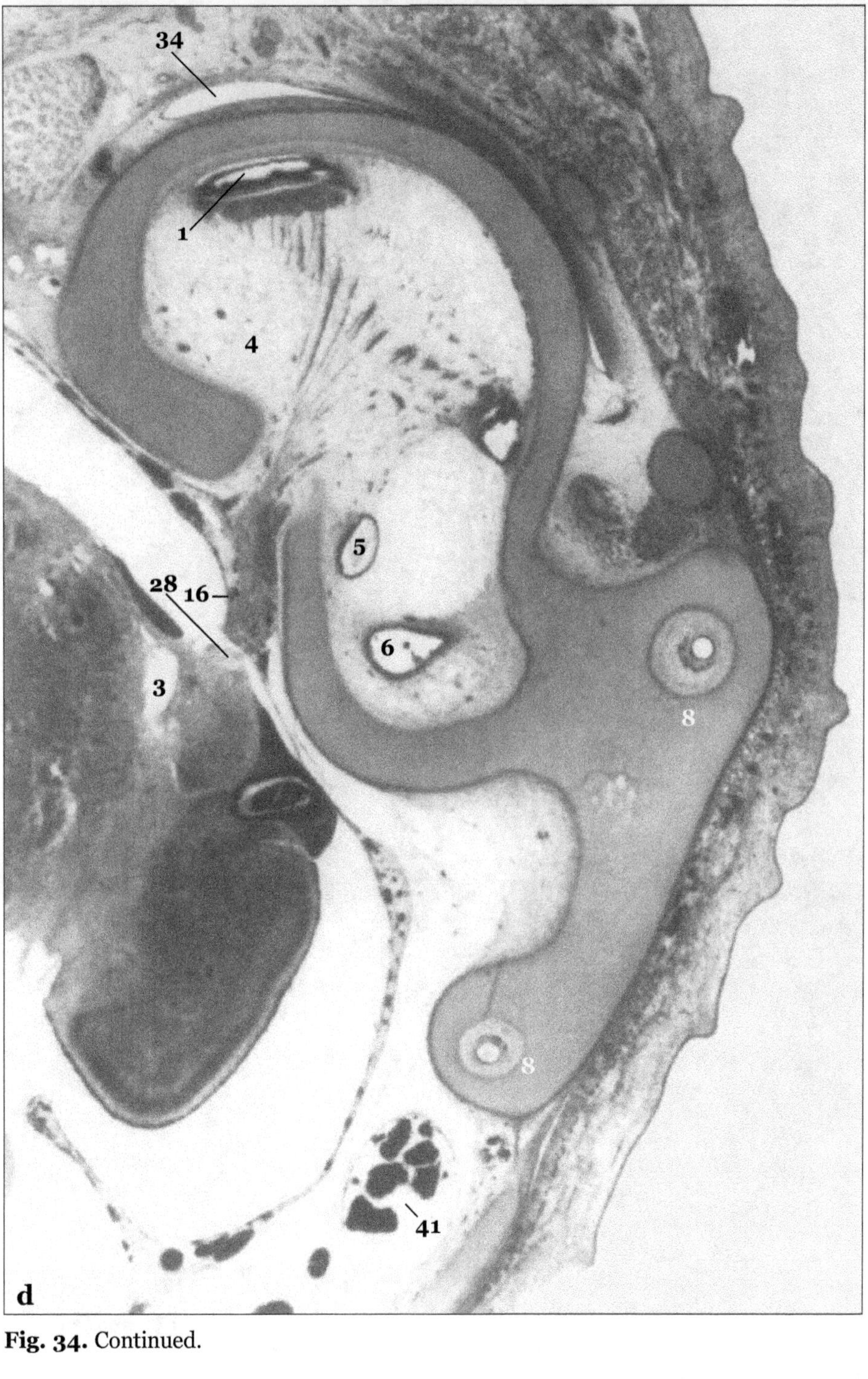

Fig. 34. Continued.

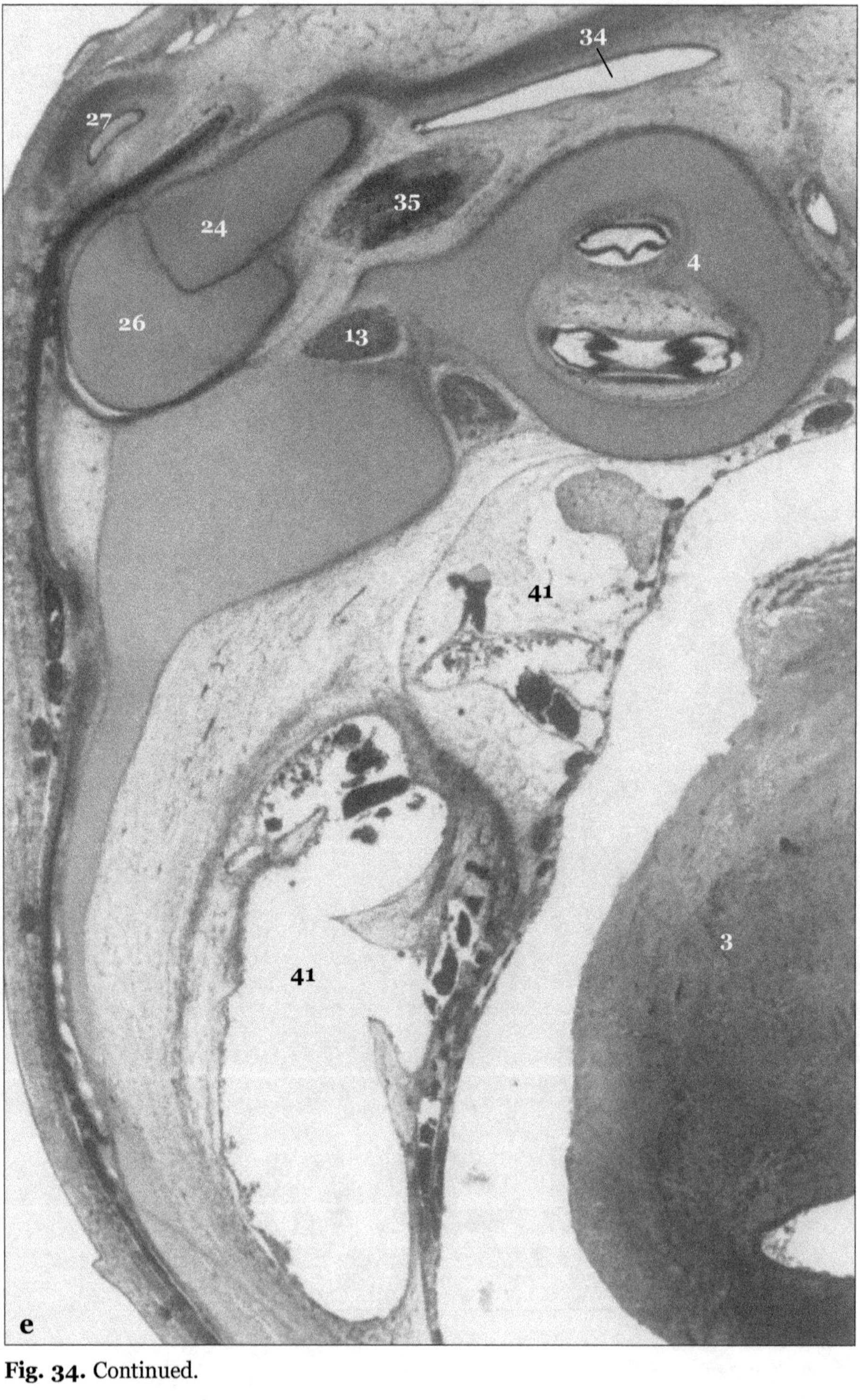

Fig. 34. Continued.

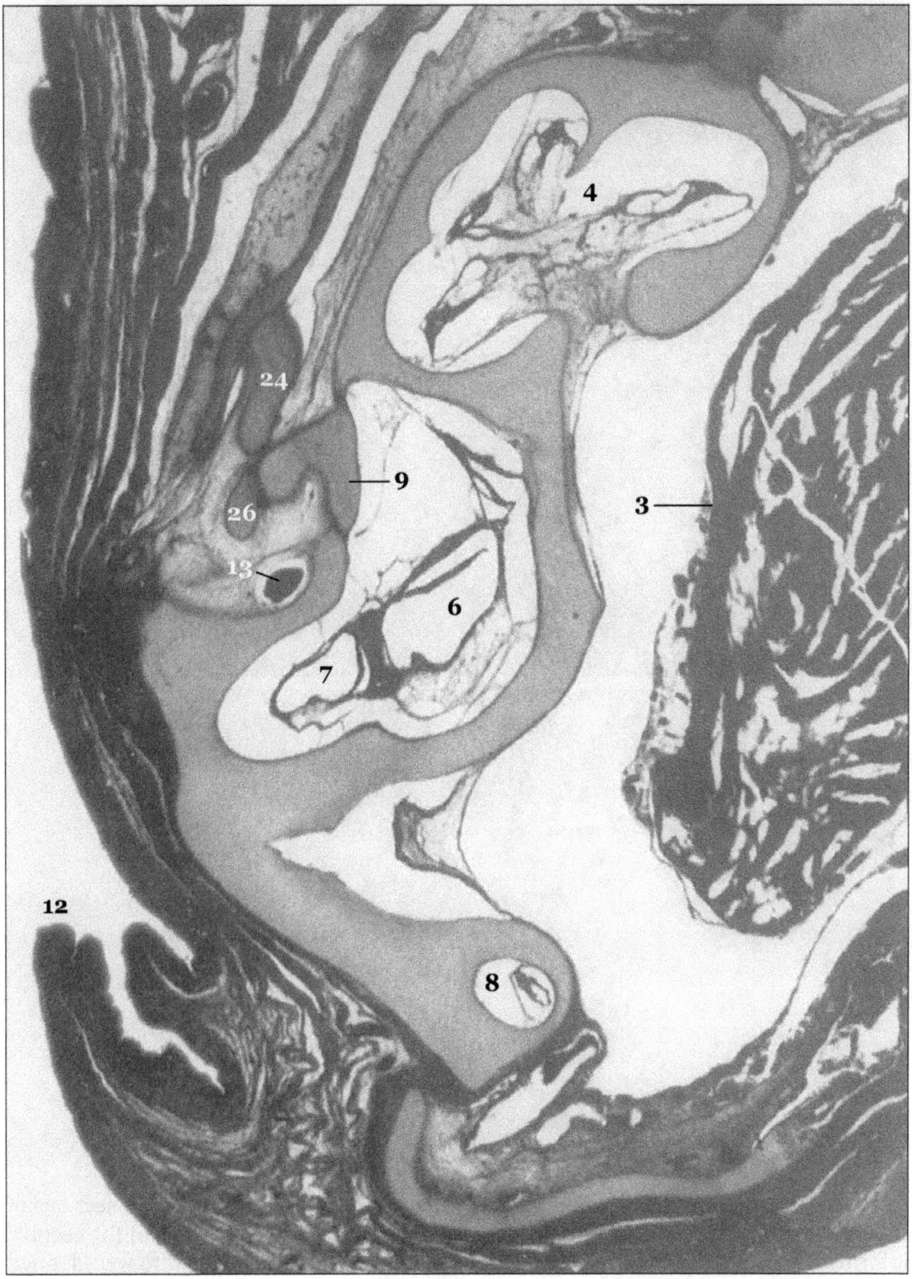

Fig. 35. Histotopography of the peripheral auditory system in dorsoventral sections of the head of a prefetal *Pusa hispida*, stages 18, 19. A rudimentary auricle is shown, which is not developed in the fetal period and is absent from adults, as in all representatives of phocids.

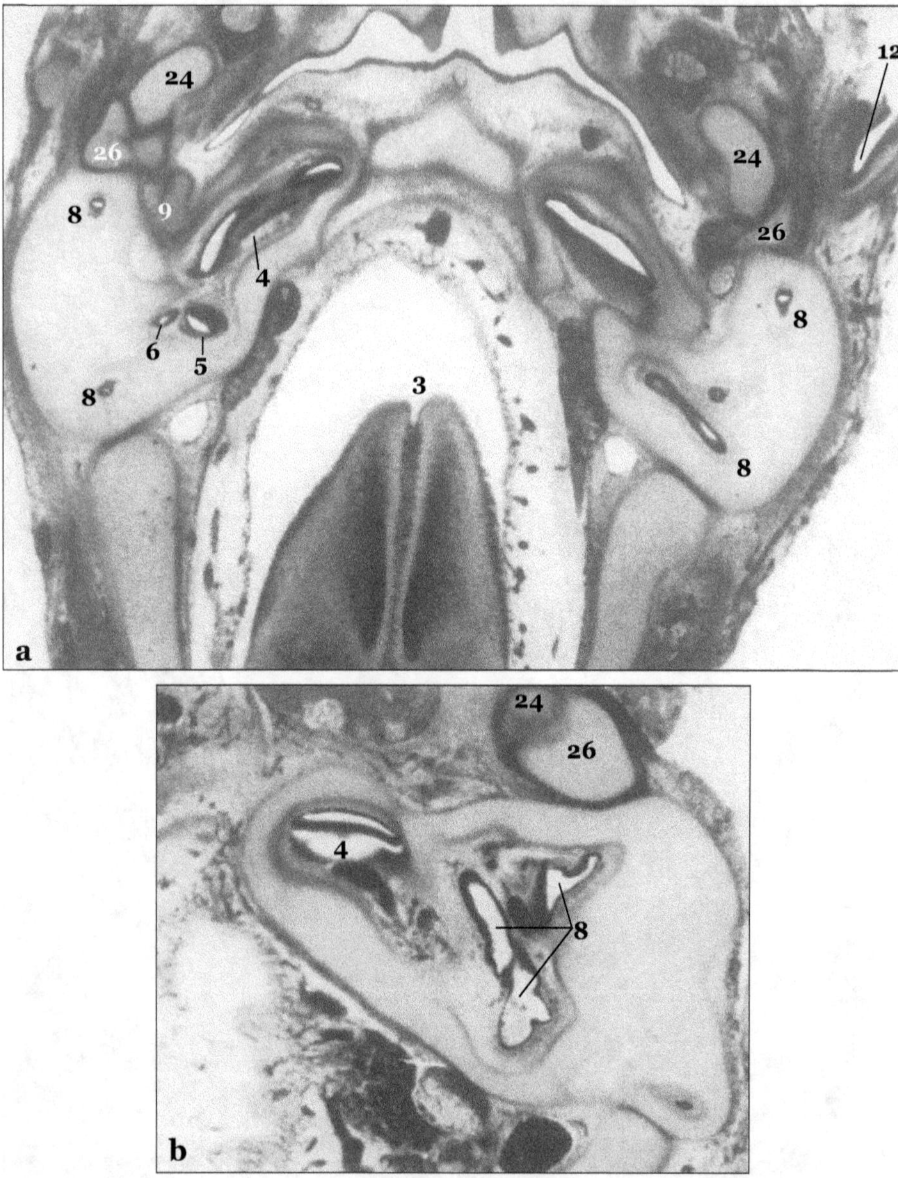

Fig. 36. Histotopography of the peripheral auditory system in dorsoventral sections of the head of a prefetal *Eumetopias jubatus*. a, b, c — stage 16; a basal turn of the cochlea is formed; auditory ossicles are represented by independent formations with well-defined borders; semicircular canals located at three mutually perpendicular planes are shown; the auditory meatus is elongated; d — stage 17; a medial turn of the cochlea is formed; differentiation of the elements of the cochlear canal and of those of the organ of Corti is absent; e — stage 18; the cochlea is formed by 2.5 turns; auditory ossicles are enlarged; f — stages 18, 19; an initial differentiation of the elements of the cochlear canal is marked in the cochlea; the cochlear branch of the acoustic nerve is shown.

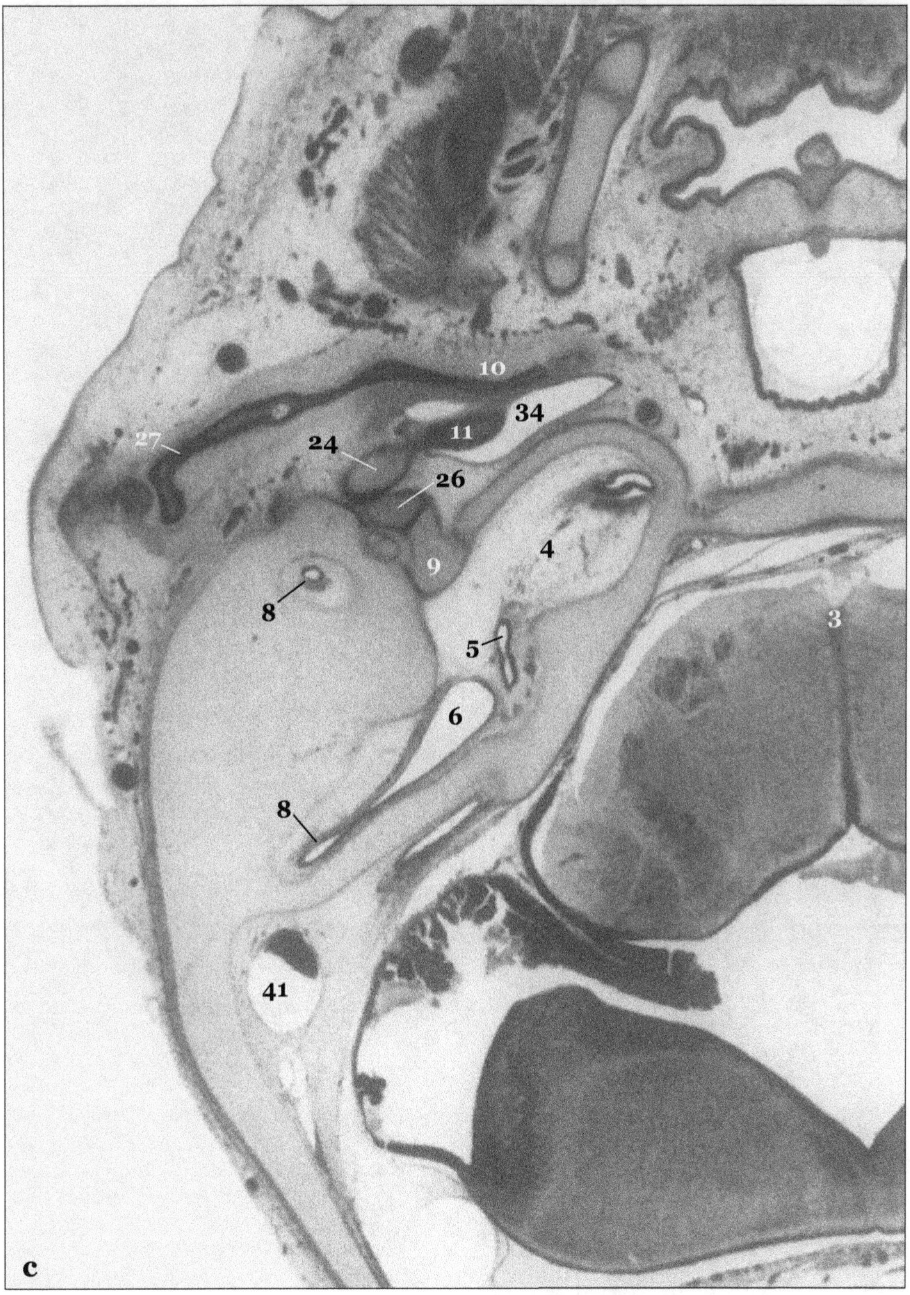

Fig. 36. Continued.

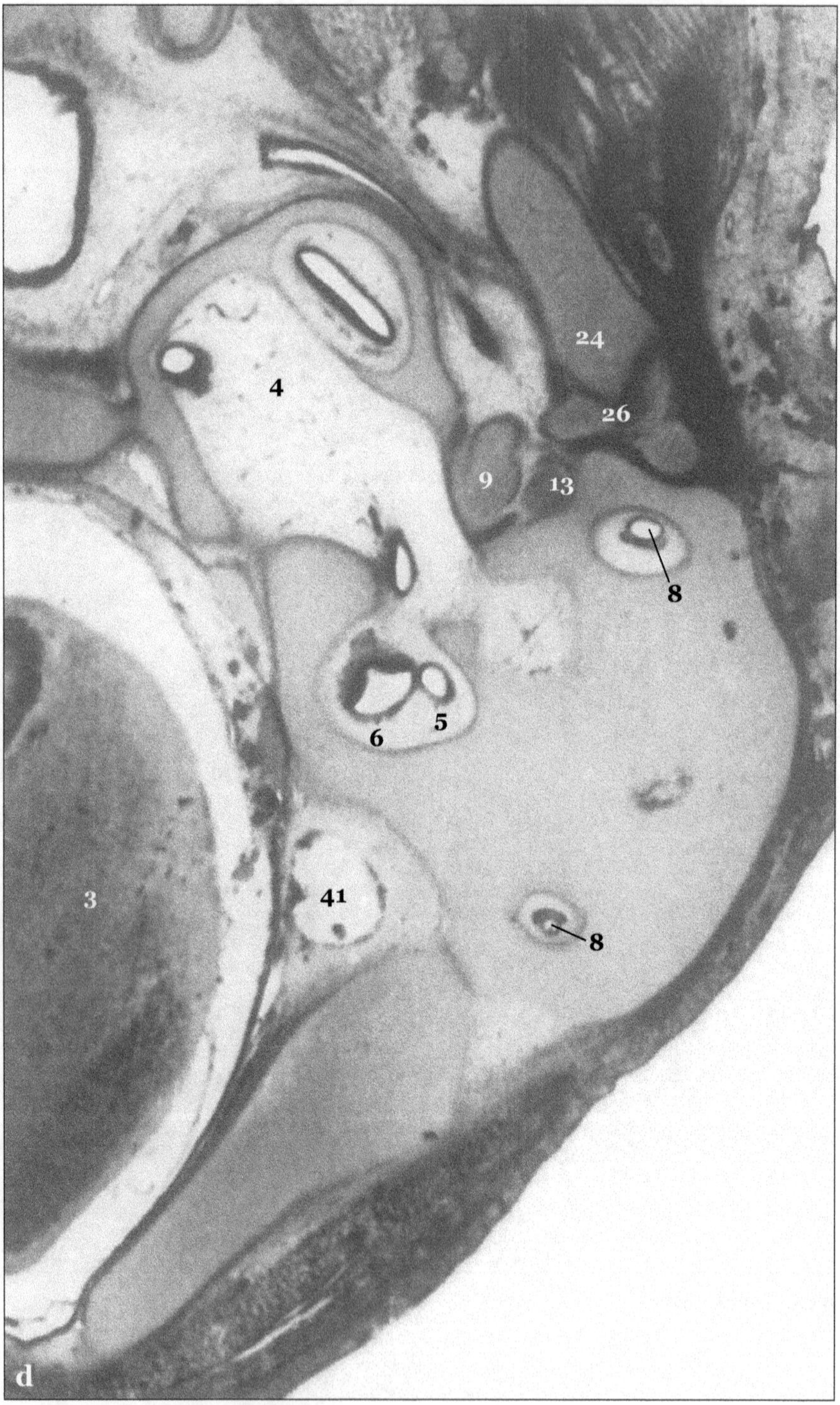

Fig. 36. Continued.

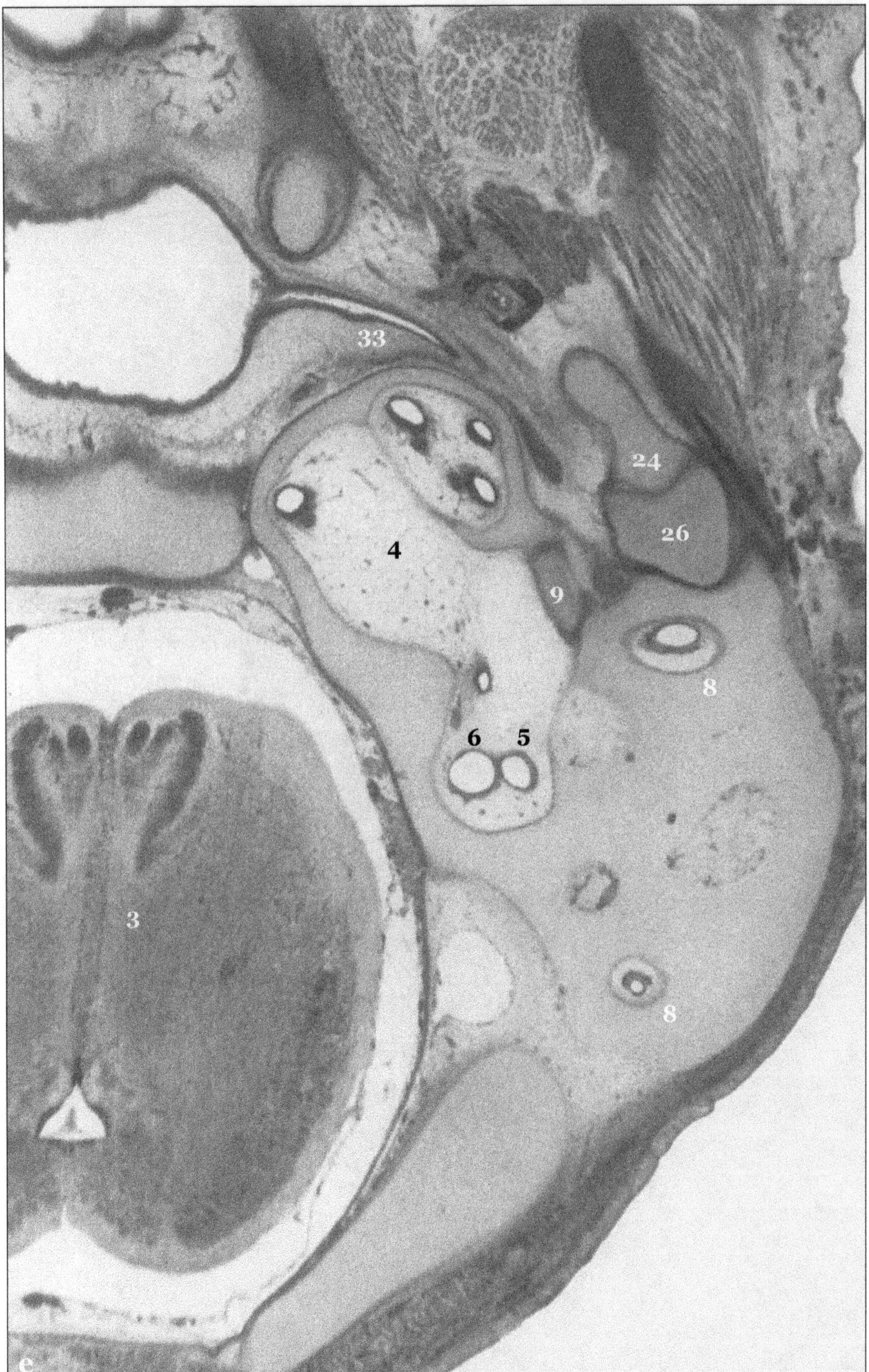

Fig. 36. Continued.

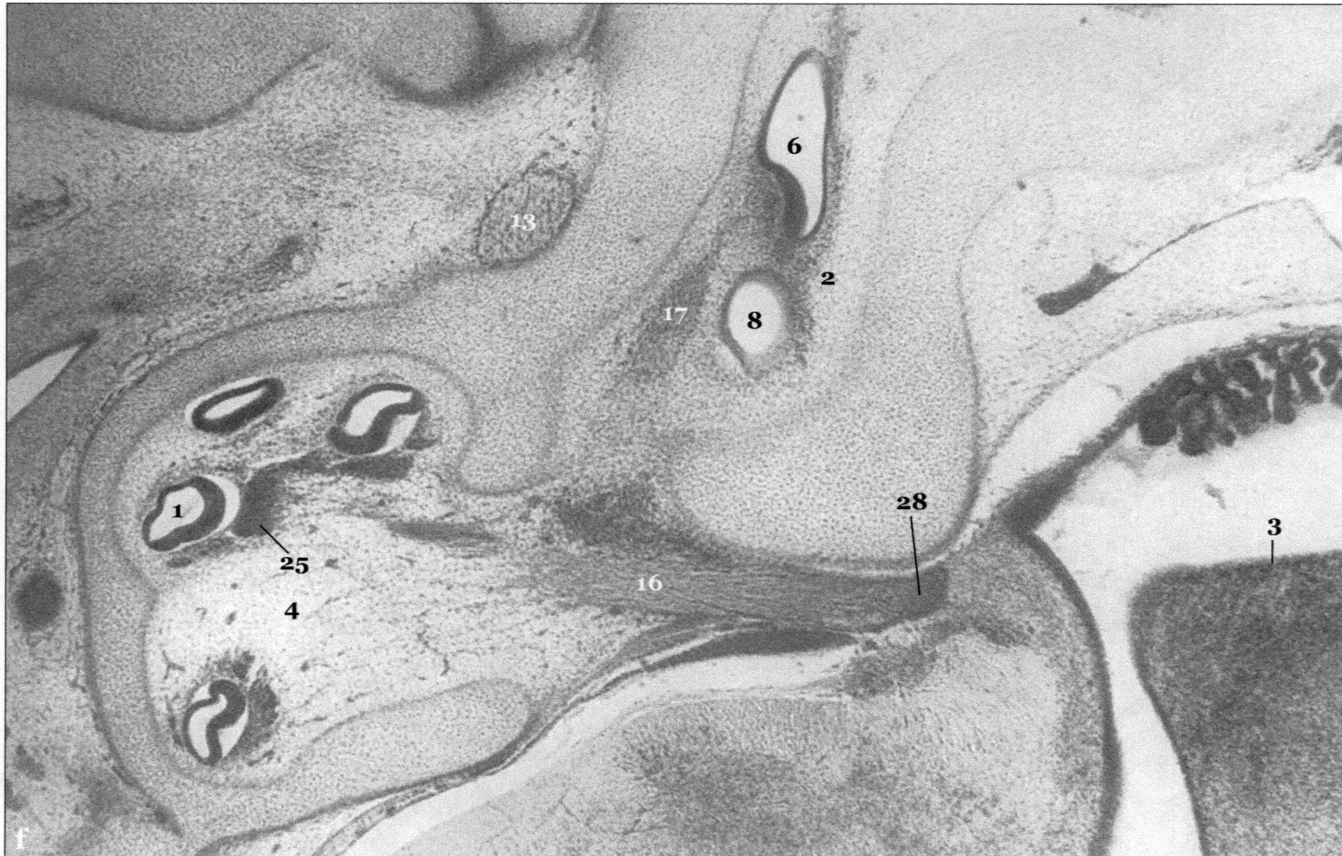

Fig. 36. Continued.

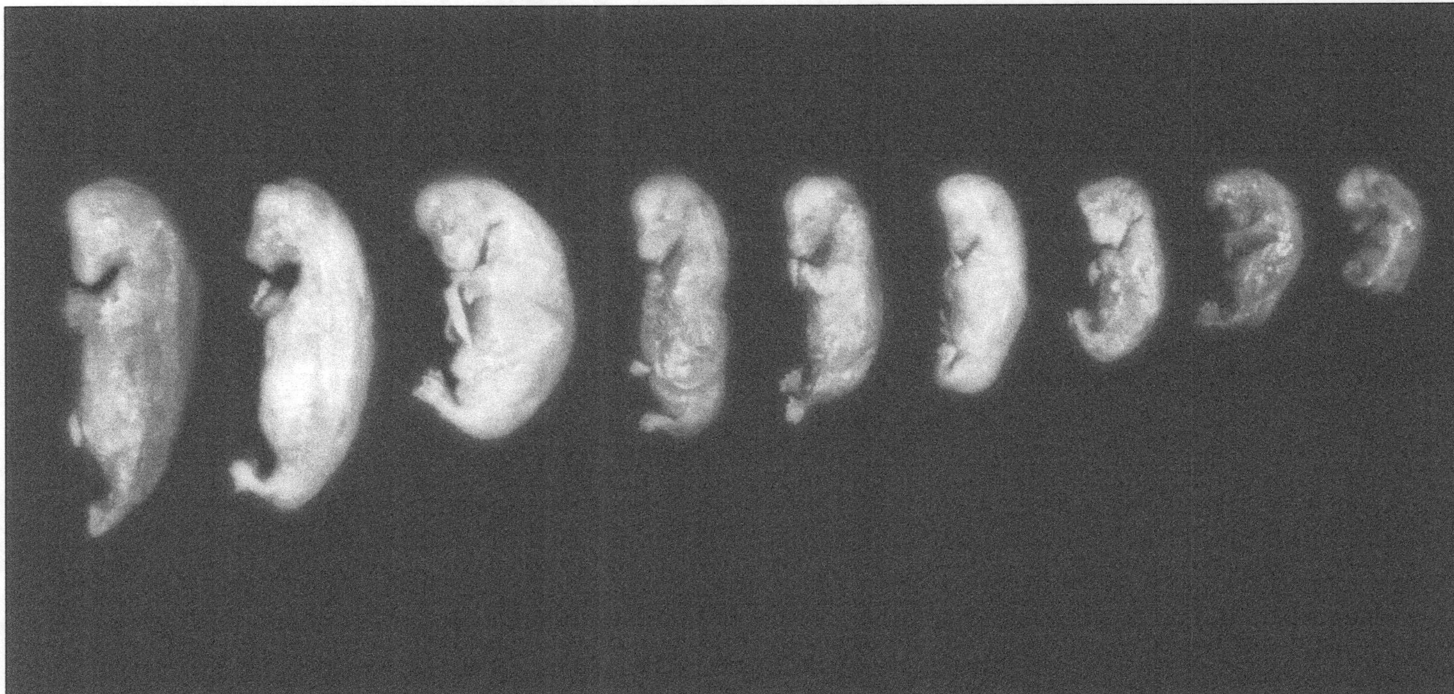

Fig. 37. General view of a prefetal *Odobenus rosmarus divergens*; 15–150 mm in sinciput-to-tail length.

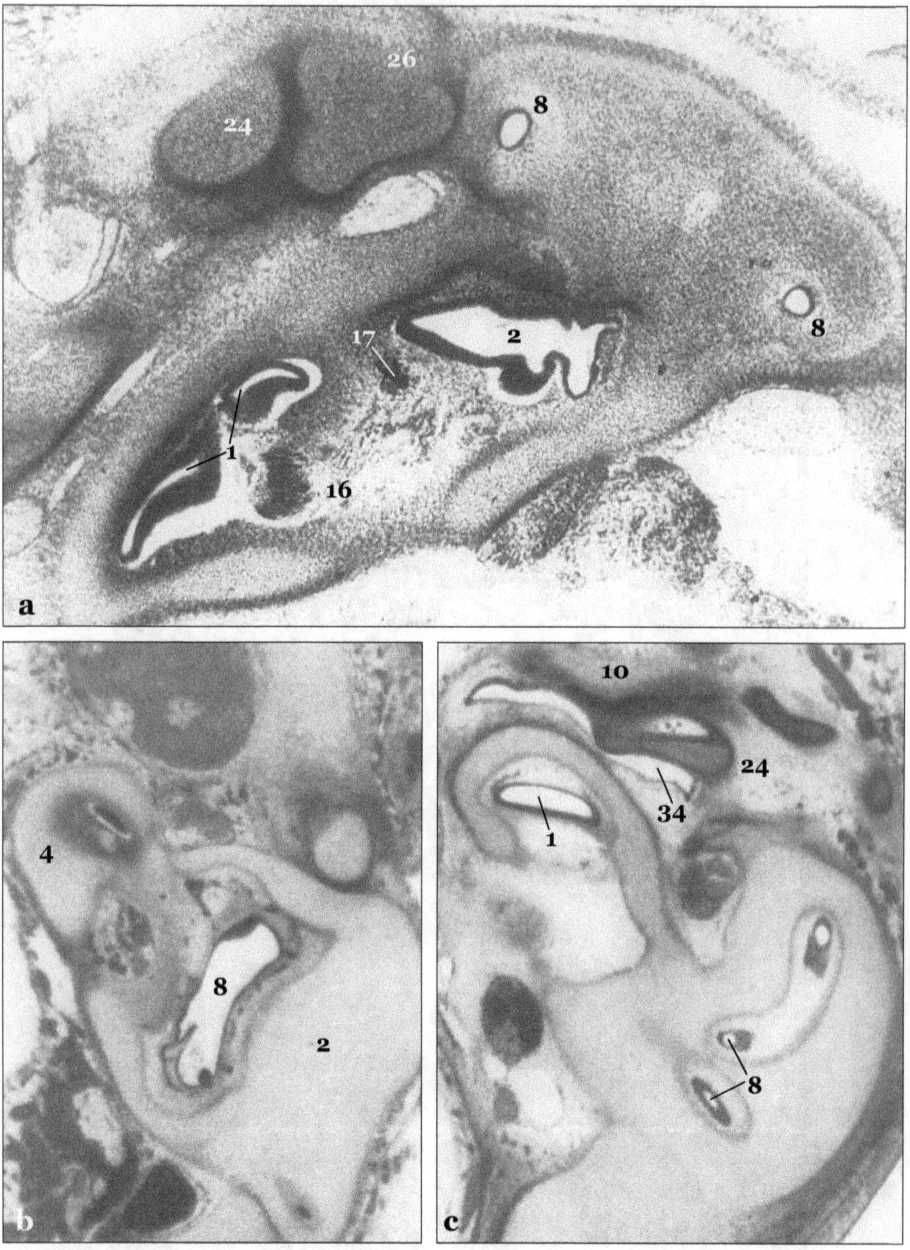

Fig. 38. Histotopography of the peripheral auditory system in dorsoventral head sections of a prefetal *Odobenus rosmarus divergens*. a — stage 15; subdivision into cochlear and vestibular parts; auditory ossicles with distinct borders; b — stage 16; a horizontal semicircular canal with a broad lumen; c — stages 16, 17; the manubrium mallei is joined to the tympanic membrane in the middle.

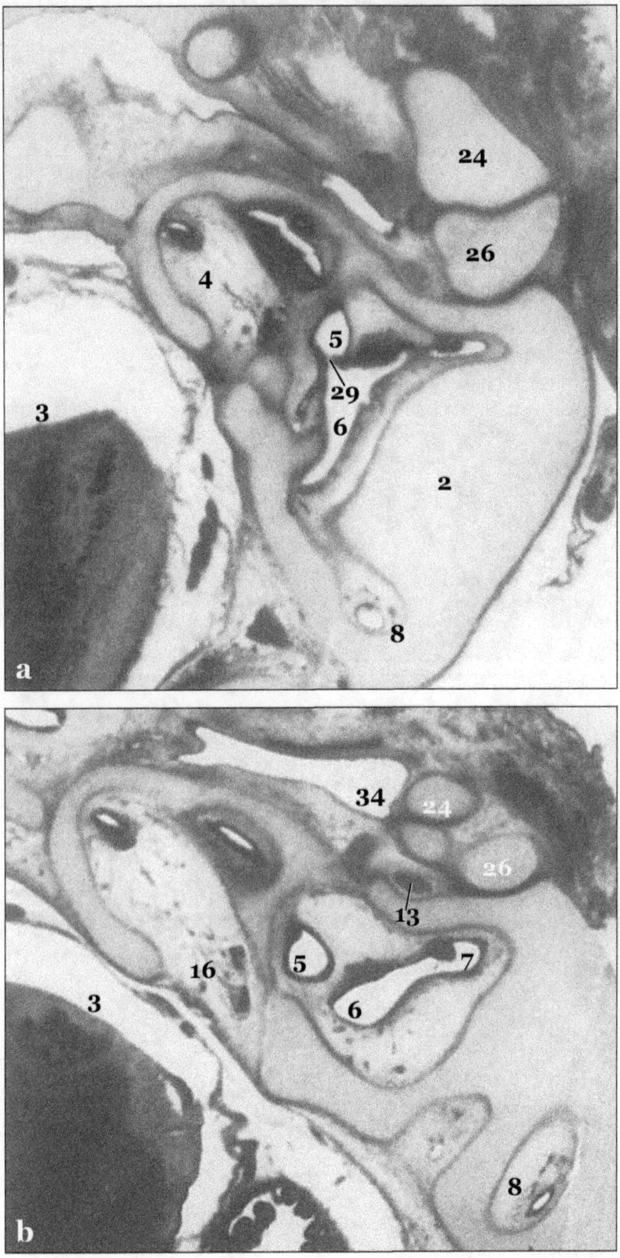

Fig. 39. Histotopography of the peripheral auditory system in dorsoventral head sections of a prefetal *Odobenus rismarus divergens*. a — stage 17; a sacculo-endolymphatic duct is shown; b — stages 17, 18; a medial turn of the cochlea is formed; c, d, e, f — stage 17; the auditory ossicles are massive; a broad lumen of the vertical semicircular canal is shown; the sacculus, utriculus, basal and medial turns are shown; an incudo-malleal articulation is the same as in terrestrial species; structures of the vestibular apparatus are formed; the formation of an auricle occurs.

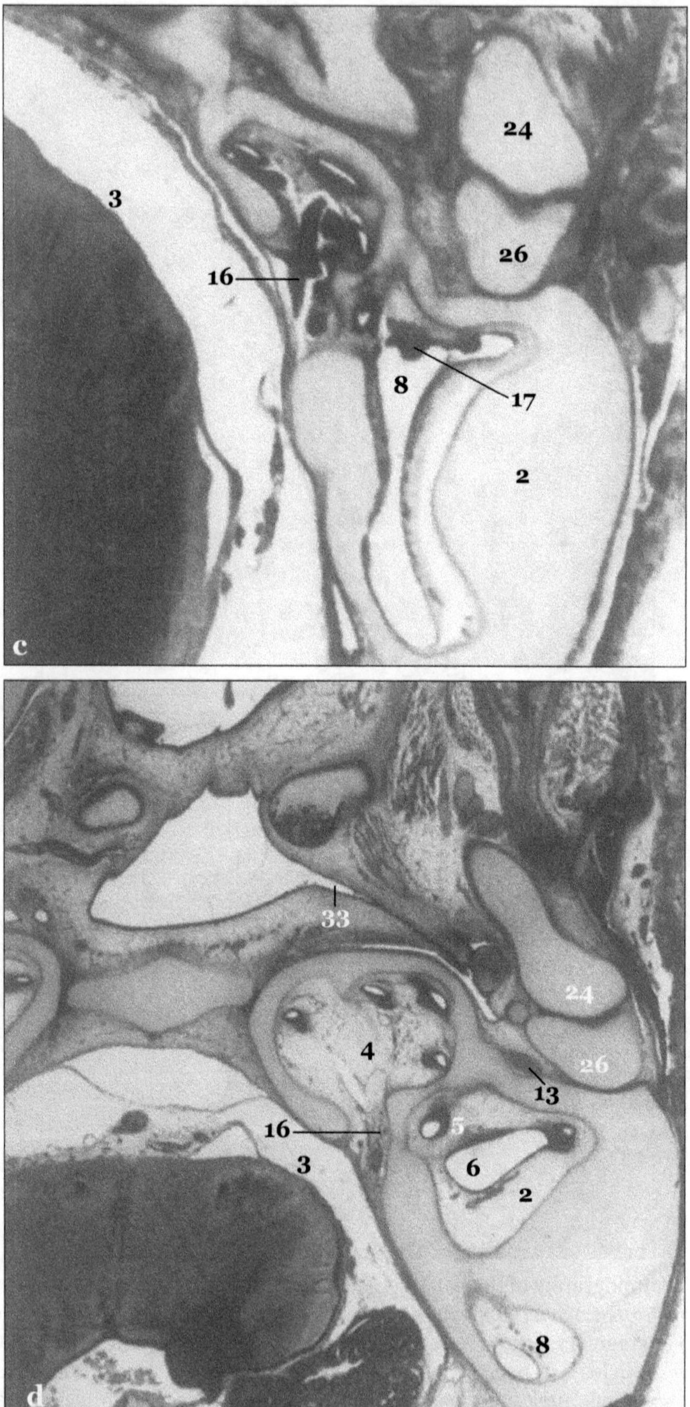

Fig. 39. Continued.

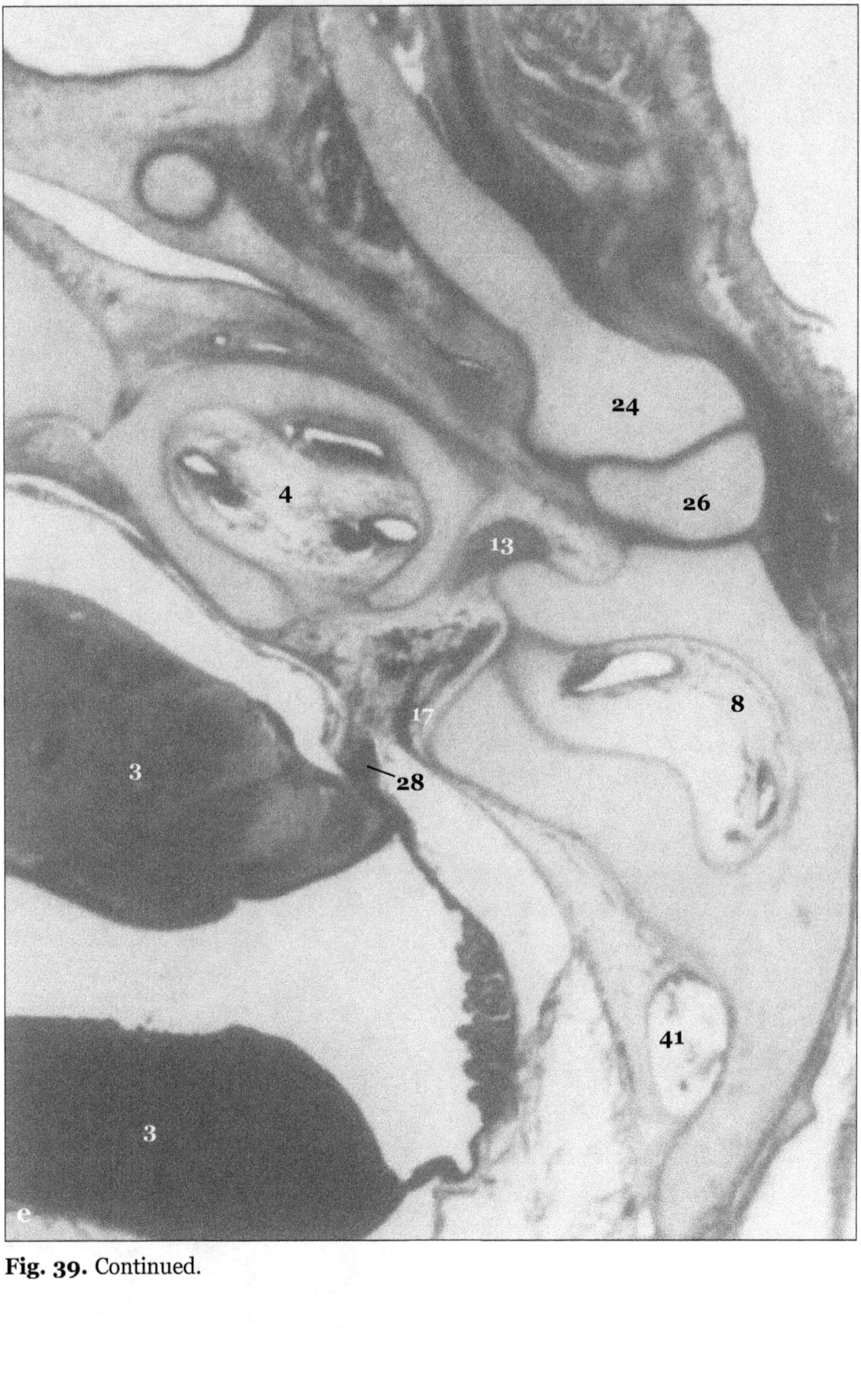

Fig. 39. Continued.

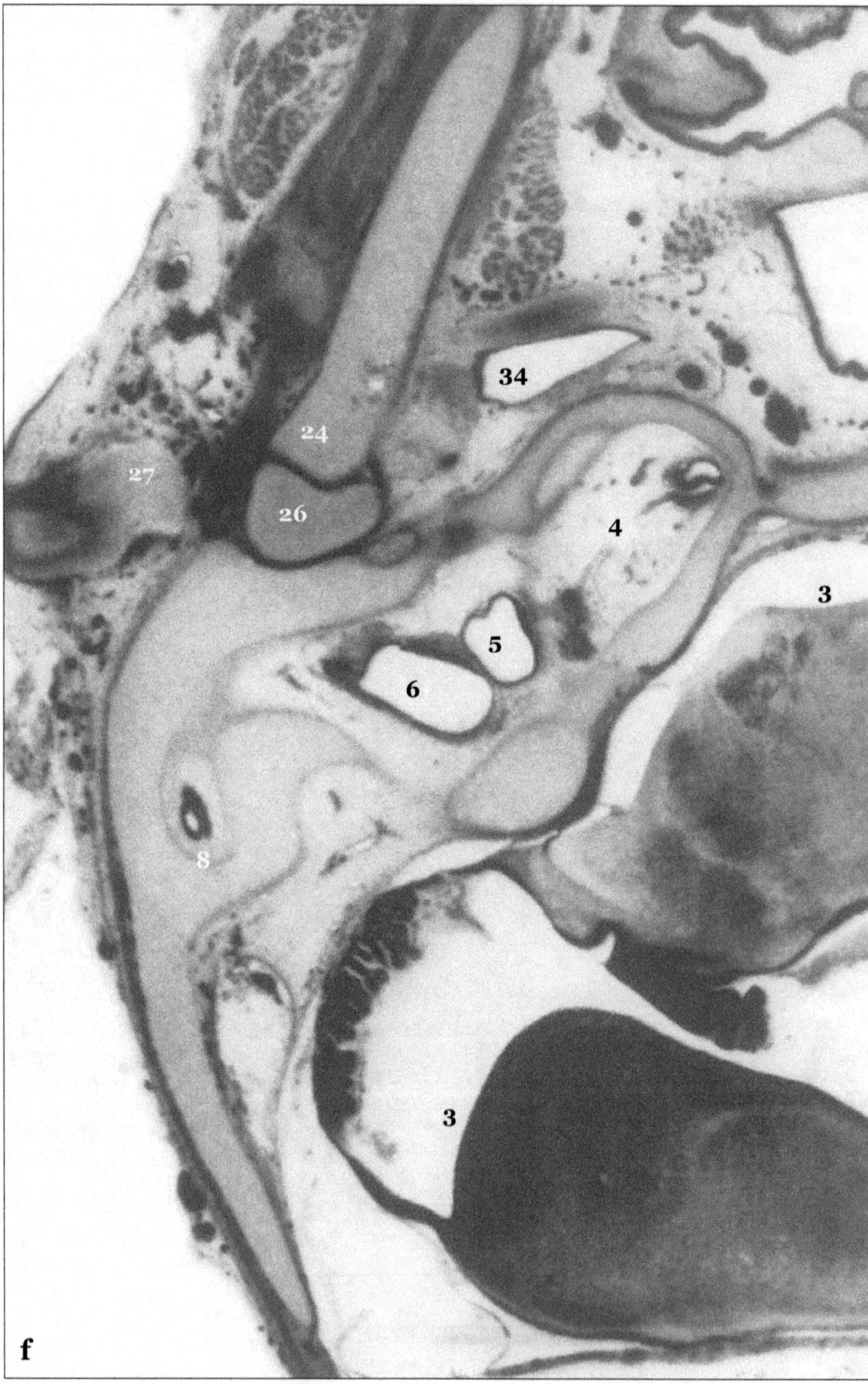

Fig. 39. Continued.

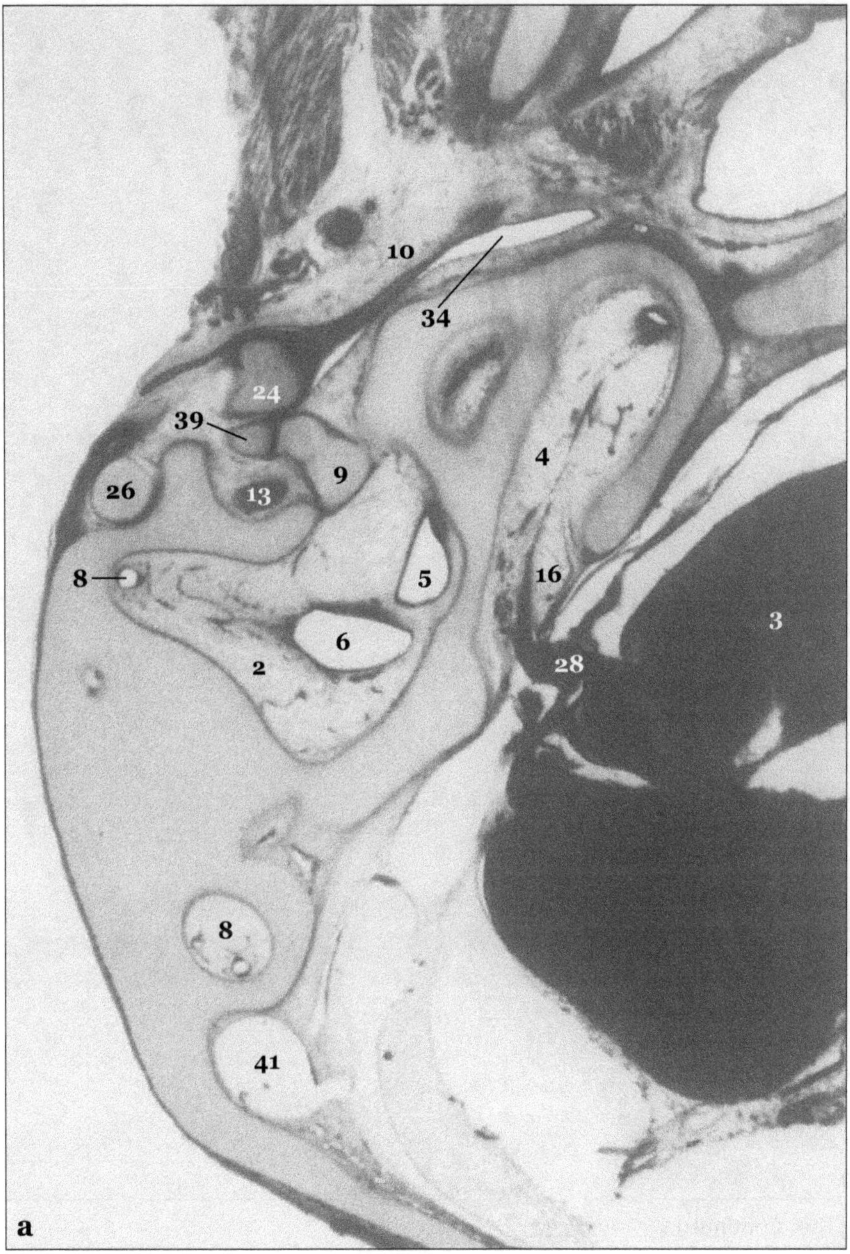

Fig. 40. Histotopography of the peripheral auditory system in dorsoventral head sections of a prefetal *Odobenus rosmarus divergens*. a — stage 18; the tympanic membrane is thickened; b — stages 18, 19; the auditory meatus is filled with epithelial cells; c — stages 18, 19; an anatomical formation of a cochlea, represented by 2.5 turns, is completed; d — stages 19, 20; both tympanic and vestibular scale formation is started in the basal turn of the cochlea; e — stage 19; the vestibular apparatus is twice as large as the cochlea, similar to the condition observed in terrestrial species.

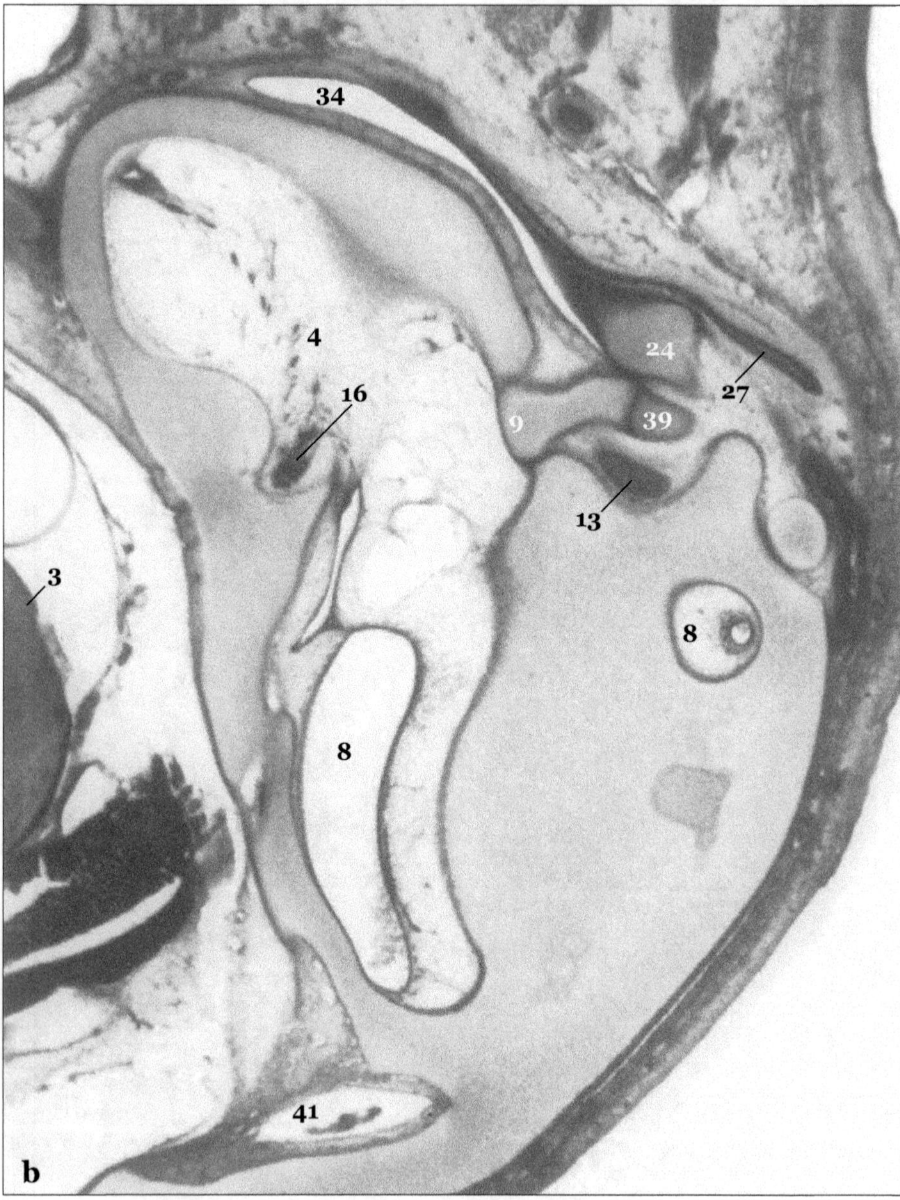

Fig. 40. Continued.

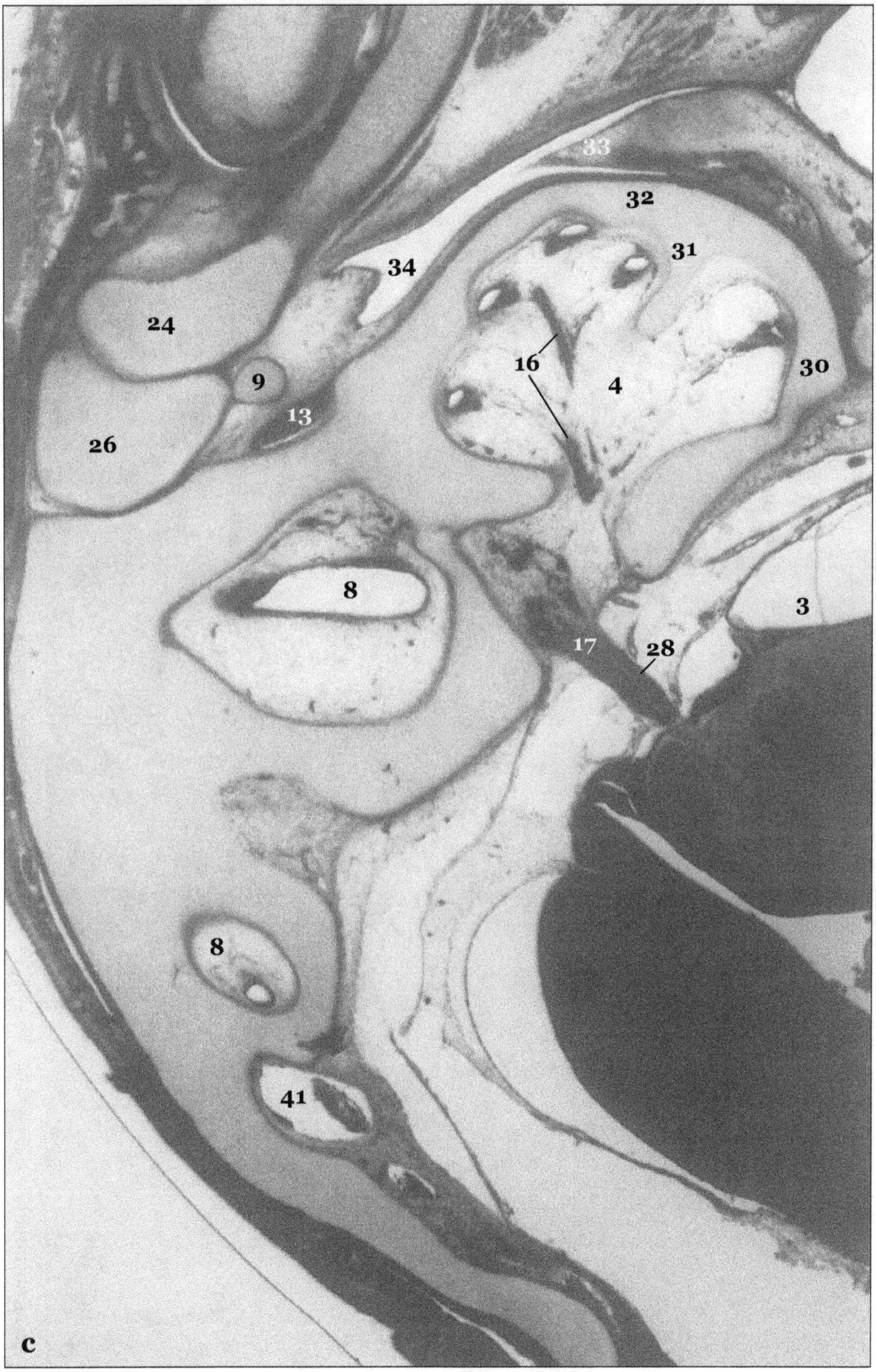

Fig. 40. Continued.

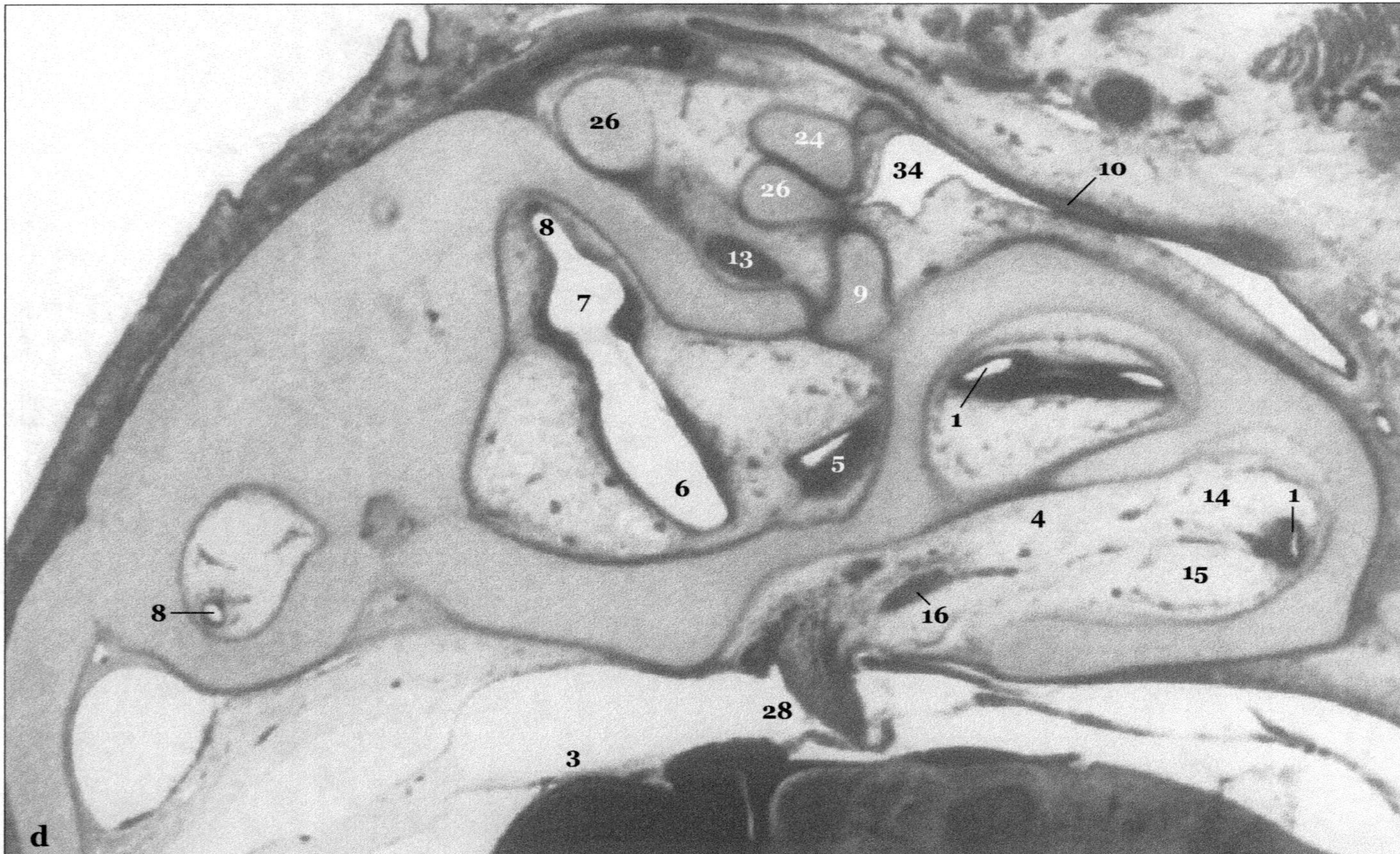

Fig. 40. Continued.

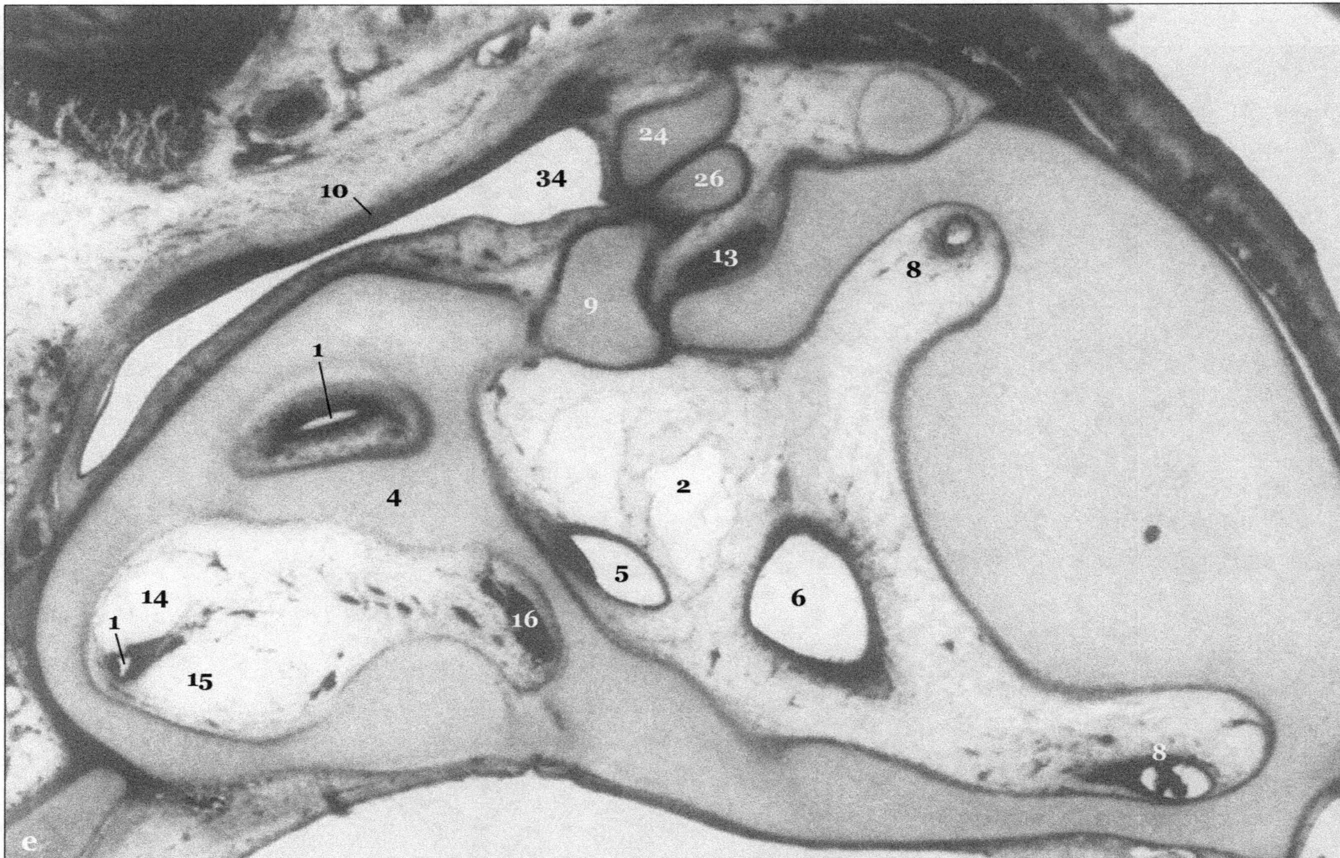

Fig. 40. Continued.

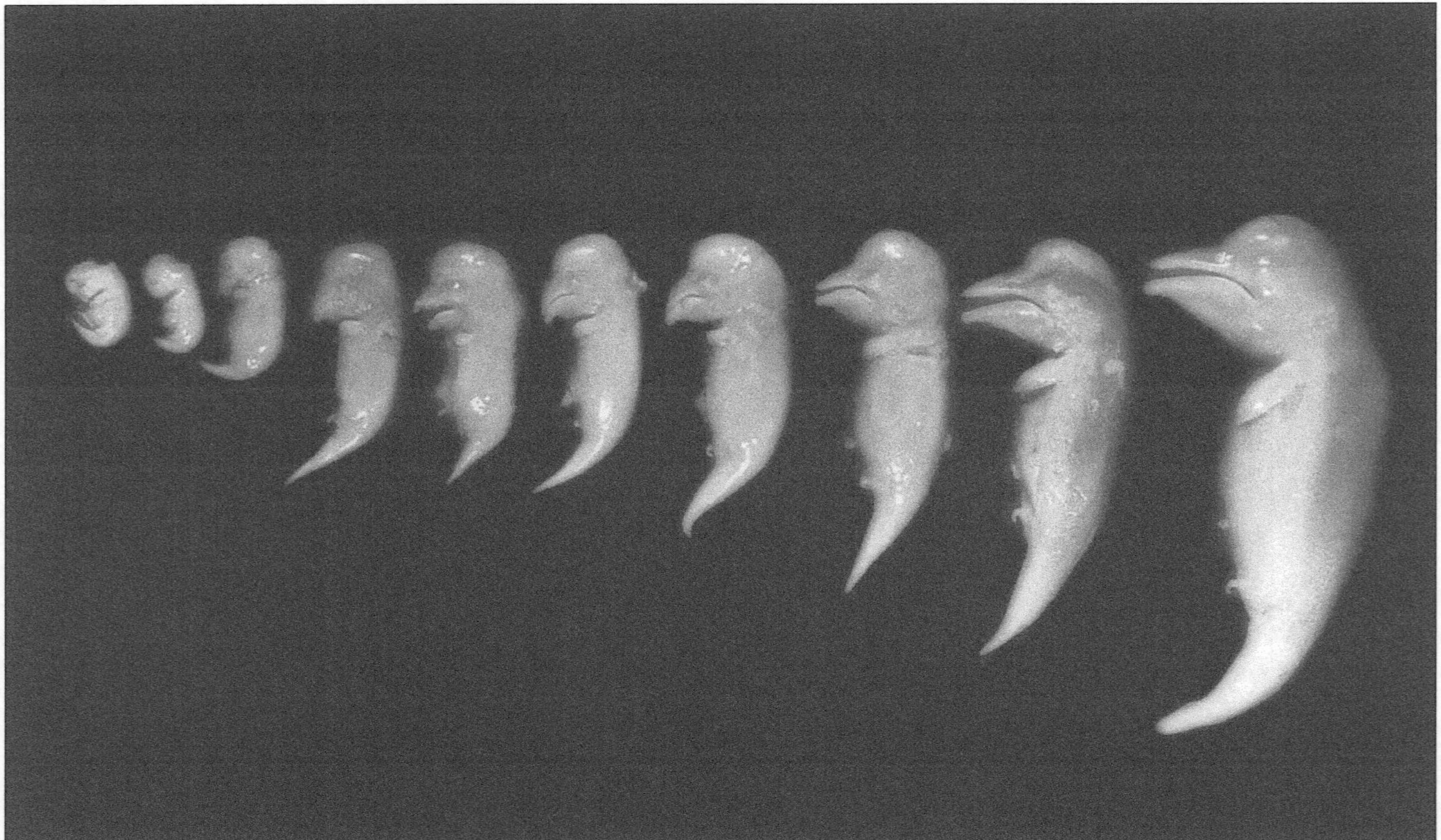

Fig. 41. General view of *Stenella attenuata* embryos; 15–150 mm in sinciput-to-tail length.

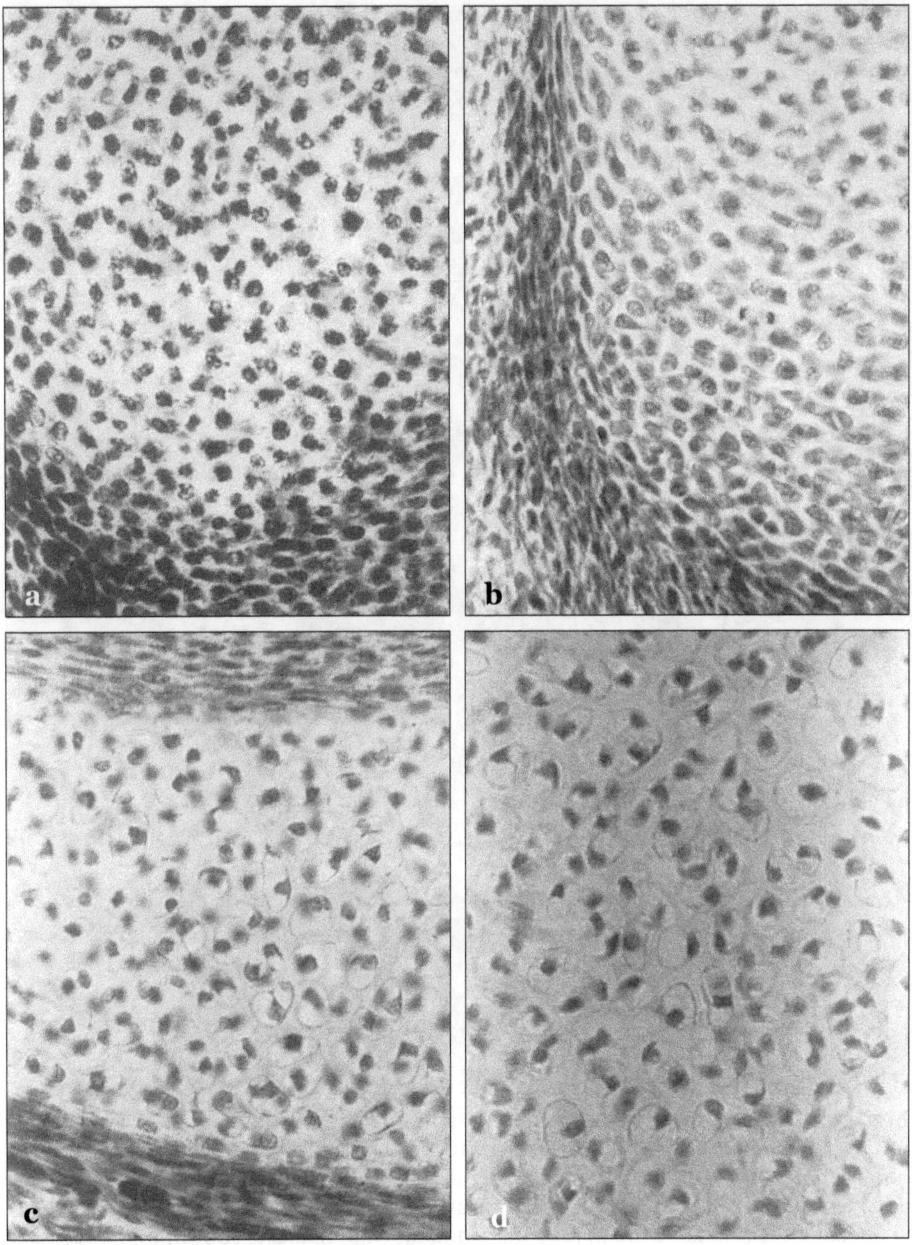

Fig. 42. Mesenchymal tissue transformation into embryonic hyaline cartilage in auditory ossicles of a prefetal *Stenella attenuata*: a — mesenchymal tissue; b — mature precartilaginous tissue; c — isogenic cell group formation, accumulation of the main substance; d — initial cartilaginous tissue.

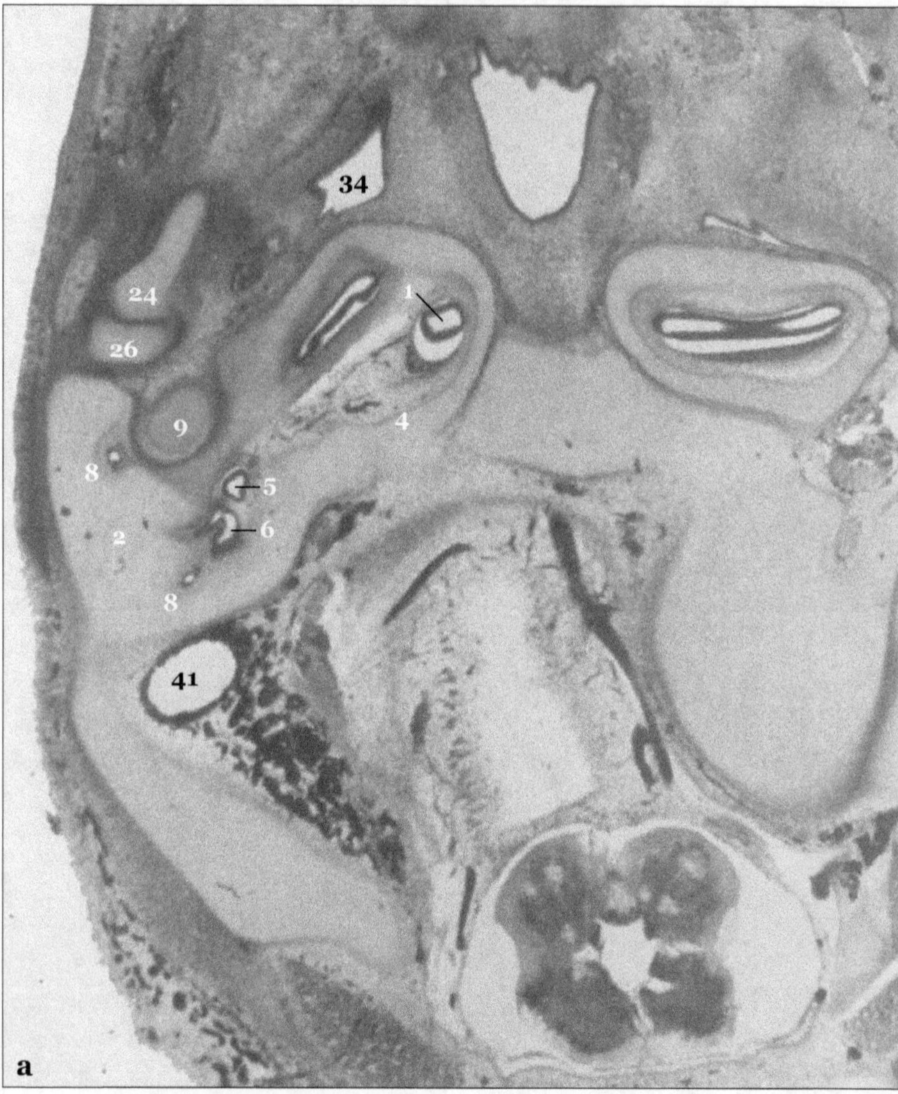

Fig. 43. Histotopography of the peripheral auditory system in dorsoventral head sections of a prefetal *Stenella attenuata*. a — stages 16, 17; the auditory ossicles are small and not differentiated into elements; b — stage 17; the cochlea is formed by 1.5 turns; the malleus and incus are joined to each other rectangularly, as in the definitive forms; c — a ligament connecting the tympanic membrane to the manubrium mallei is shown; in the cochlear canal the connective tissue is loosened, while the vestibular and tympanic scales are formed; d — stages 18, 19; in the tympanum, the formation of a cavernous plexus characteristic of only cetaceans is marked; both vestibular and cochlear branches of the acoustic nerve are shown; the cochlea is twice as large as the vestibular apparatus; the latter shows an ampulla of the semicircular canal with an auditory crest (crista ampullaris).

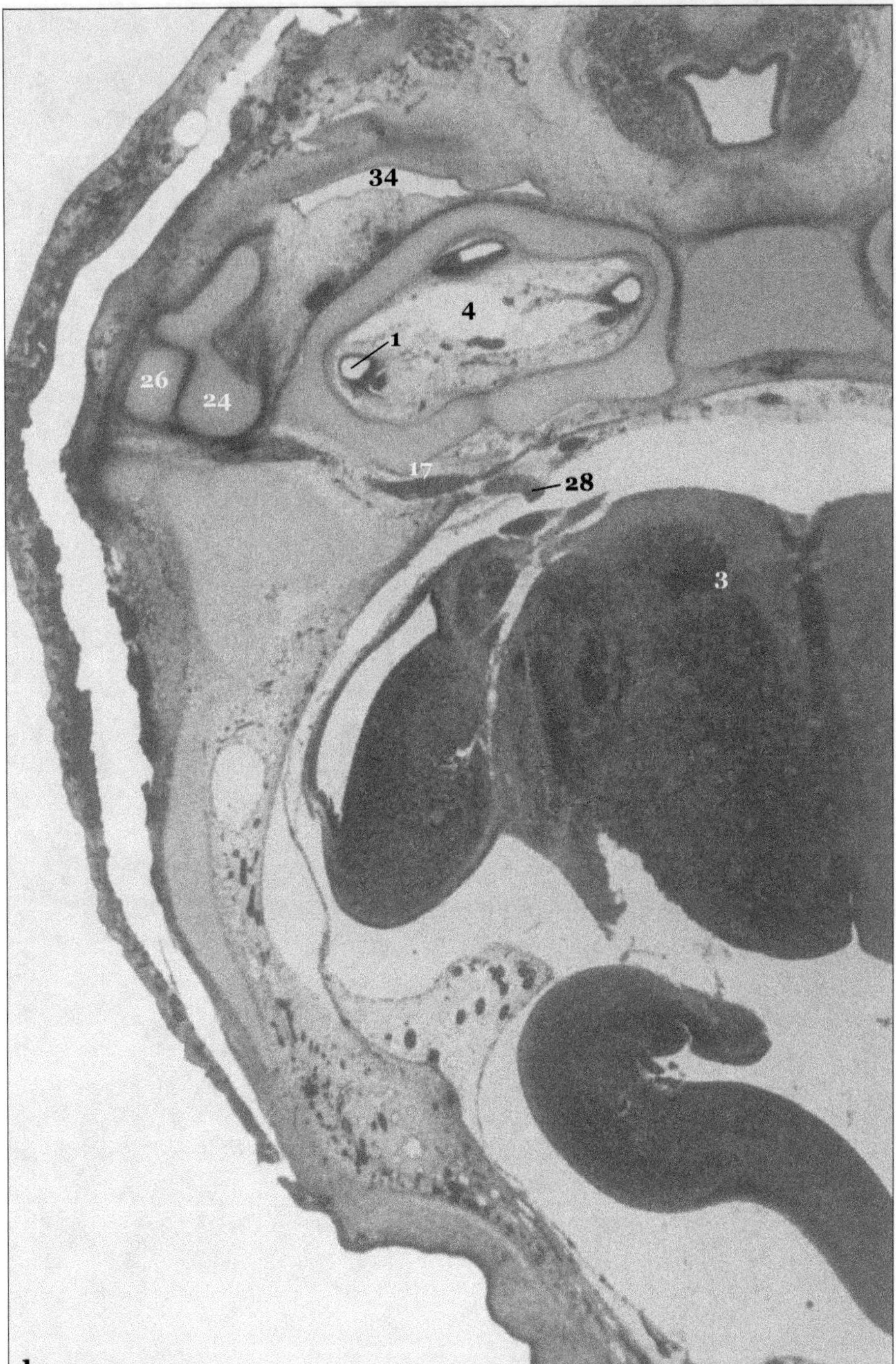

Fig. 43. Continued.

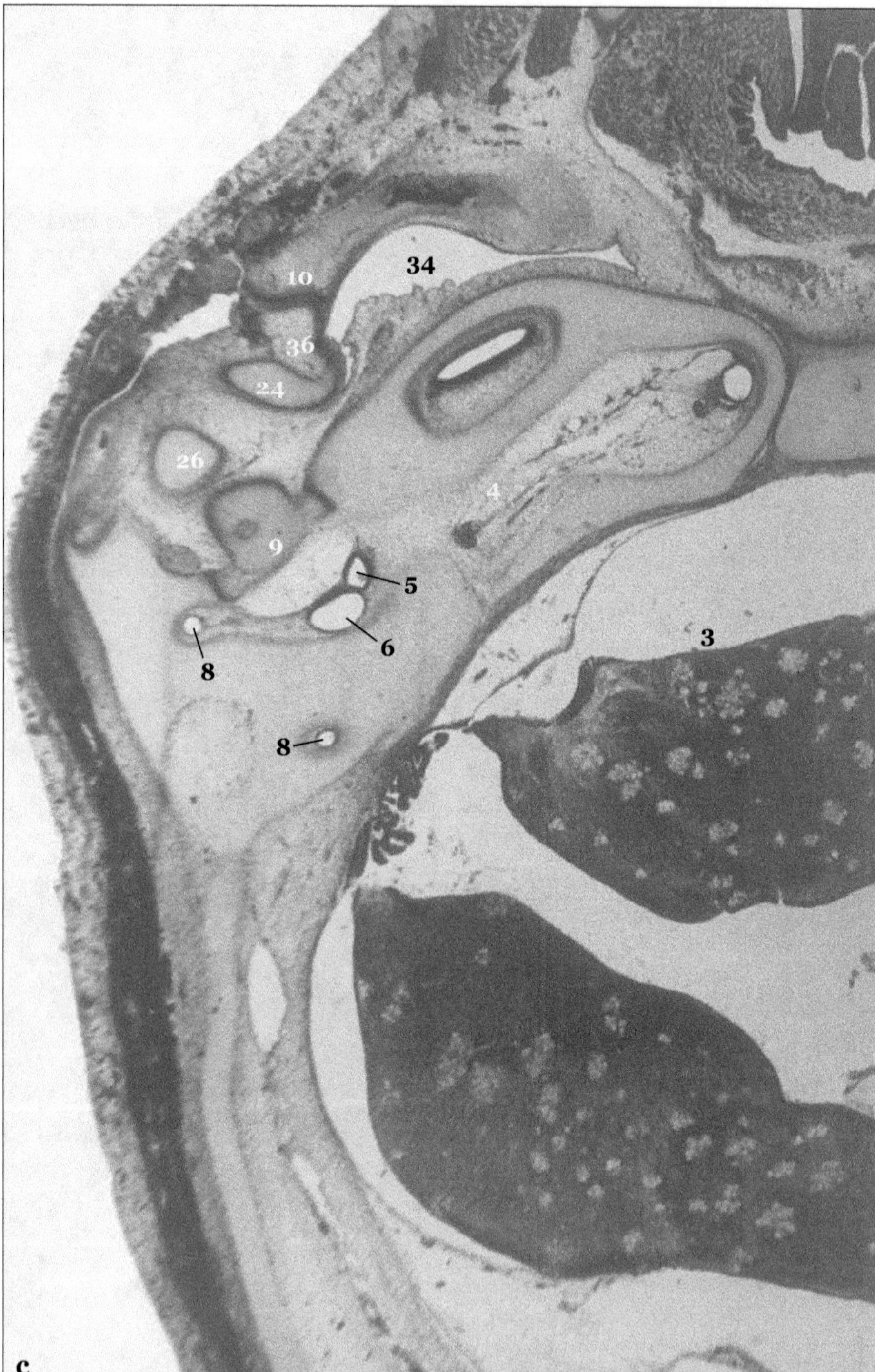

Fig. 43. Continued.

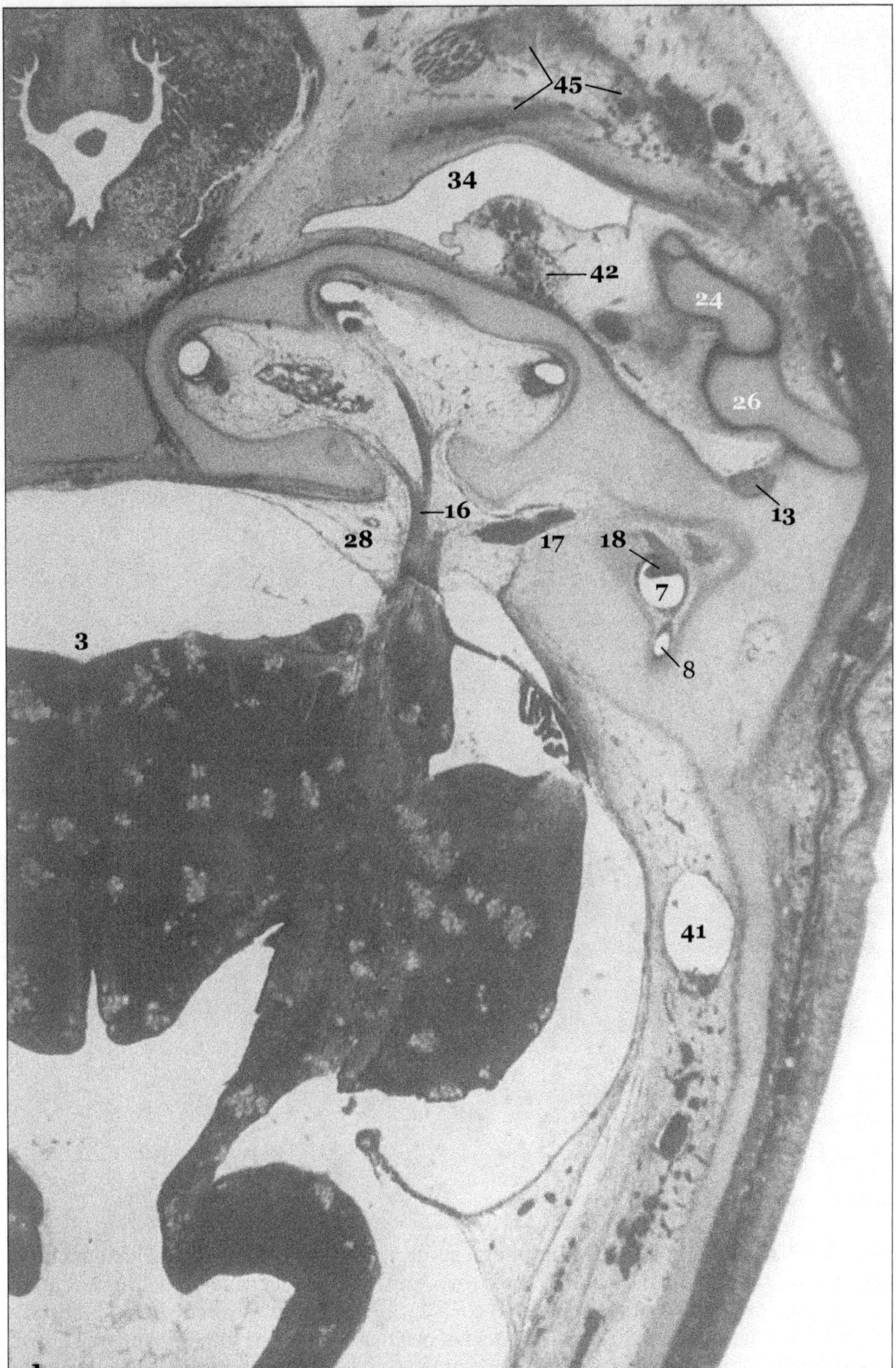

Fig. 43. Continued.

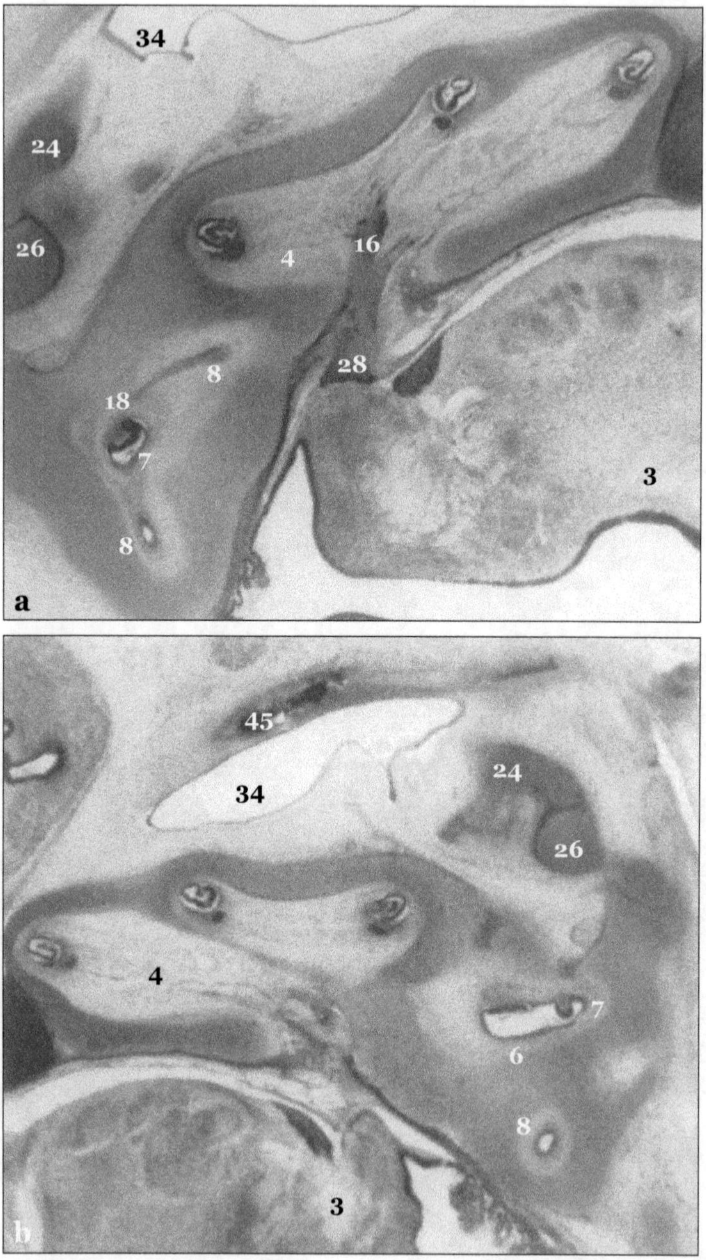

Fig. 44. Histotopography of the peripheral auditory system in dorsoventral head sections of a prefetal *Delphinapterus leucas*, stage 18. a — in the cochlear canal, cell differentiation in the organ of Corti occurs; a thickened cochlear branch of the acoustic nerve is shown; b — the auditory ossicles are small; the foci of ossification in the tympanicum are shown; c — a tympanic membrane ligament is shown; the cochlea is twice as large as the vestibular apparatus; d — both the sacculus and urticulus are small; the semicircular canals are narrow in diameter.

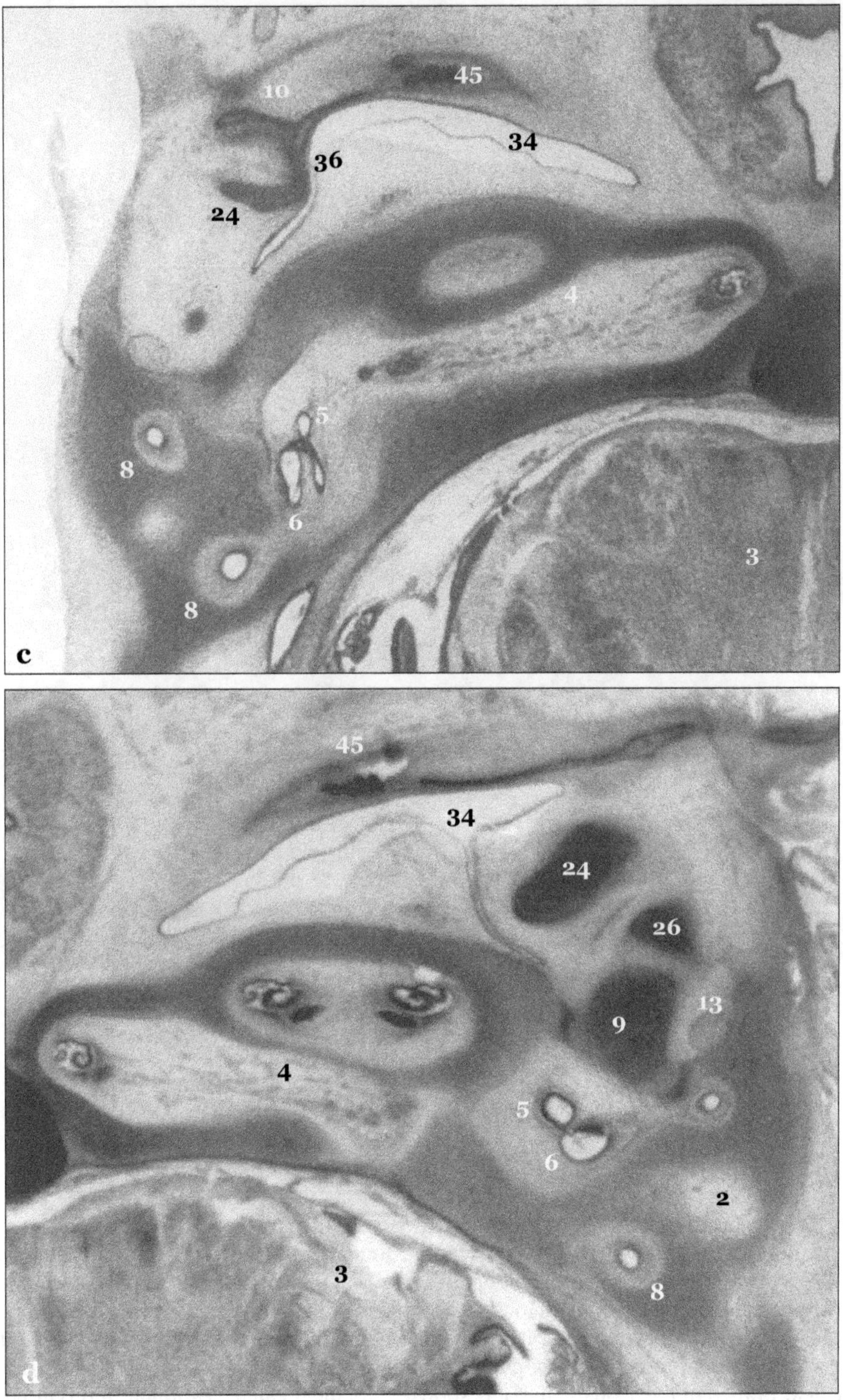

Fig. 44. Continued.

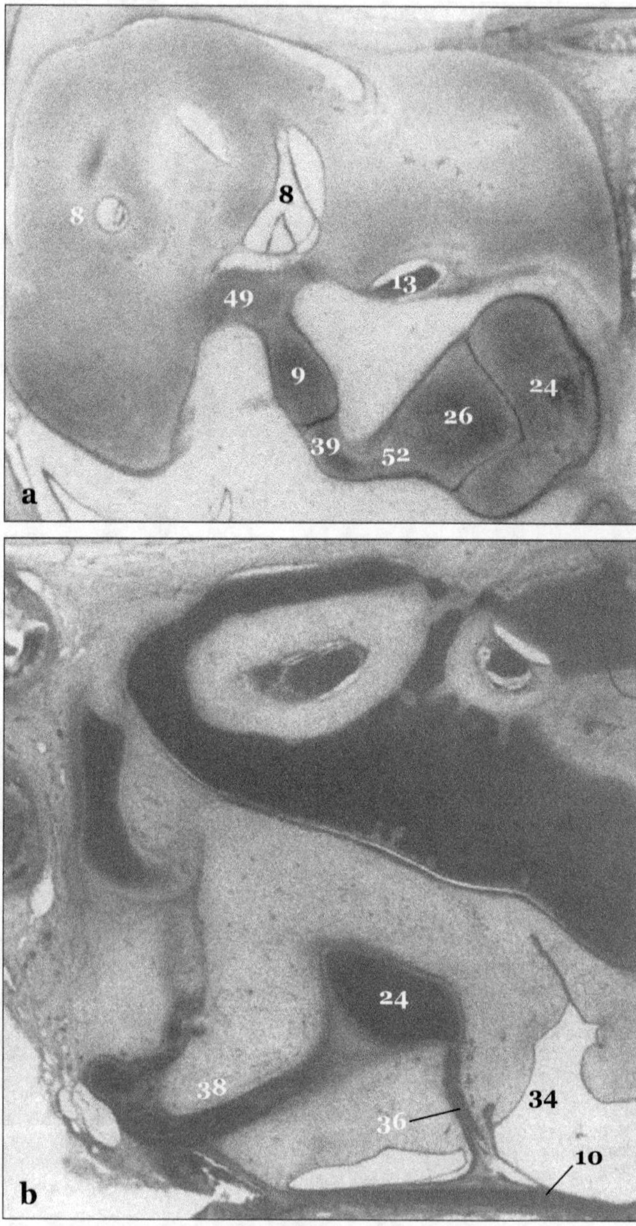

Fig. 45. Histotopography of the middle ear in dorsoventral head sections of a *Delphinapterus leucas* embryo (embryo length 250 mm). a — auditory ossicles arrangement in the cavity of the middle ear is shown; the malleus and incus are connected to each other at a right angle, thus making them incapable of independent movements; b — a long process of the malleus knitted with the tympanic bone; a tympanic membrane ligament connecting the tympanic membrane to the manubrium mallei; c — the cochlea is formed by 2 turns; a cochlear branch of the acoustic nerve is shown; a cochlear canal, vestibular and tympanic scales, Reissner's membrane, a spiral ganglion and a spiral ligament are formed.

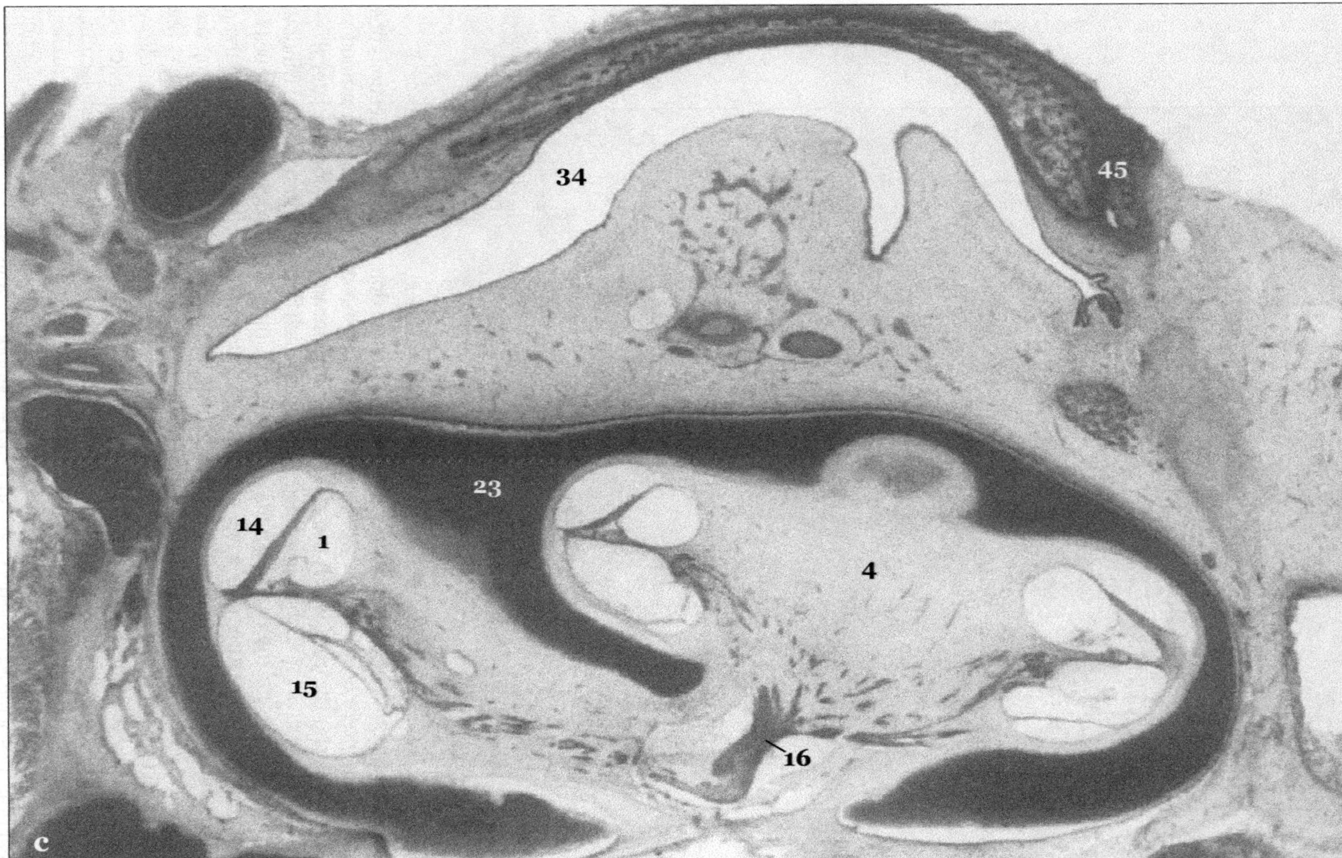

Fig. 45. Continued.

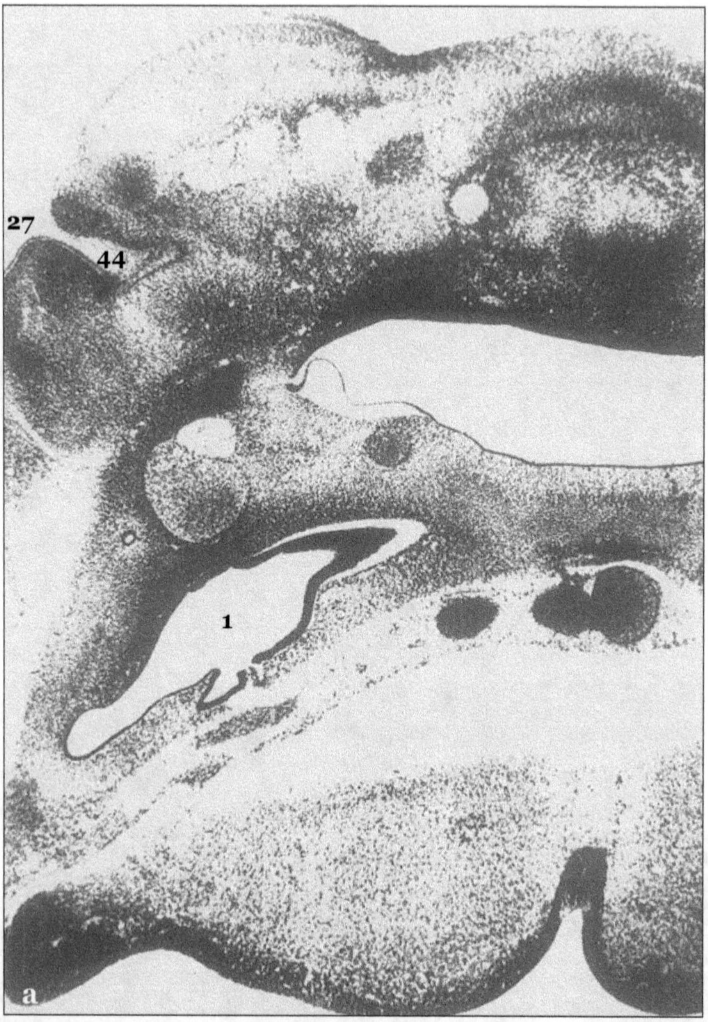

Fig. 46. Histotopography of the peripheral auditory system in dorsoventral head sections of a prefetal *Balaenoptera acutorostrata*. a — stages 14, 15; a cutaneous deepening surrounded by prominences is shown, from which the external auditory meatus will be formed later; b — stage 16; a basal turn of the cochlea is formed; the stapes which fits tightly into the oval window is shown; c — stage 17; the early formation of a medial turn of the cochlea; the stapes is differentiated into two crura; d — stage 17; auditory ossicles with well-defined contours; a tympanic bone is formed; a peribullar sinus is shown; e — stage 18; a ligament joining the tympanic membrane to the manubrium mallei is shown; formation of a vestibular scale in the cochlear canal is marked; the cochlea is twice as large as the vestibular apparatus; the auditory meatus is opened along its entire length; f — stages 18, 19; a long process of the malleus is knitted with the tympanic bone, this being characteristic of only aquatic and echolocating species; g — a thick ligament joins a thickened tympanic membrane to the manubrium mallei; h — stages 18, 19; the cochlea is formed by 2.5 turns, as in definitive forms; both vestibular and cochlear branches of the acoustic nerve are shown; i — peribullar sinuses are shown; the auditory ossicles are massive; foci of ossification in the os tympanic are formed.

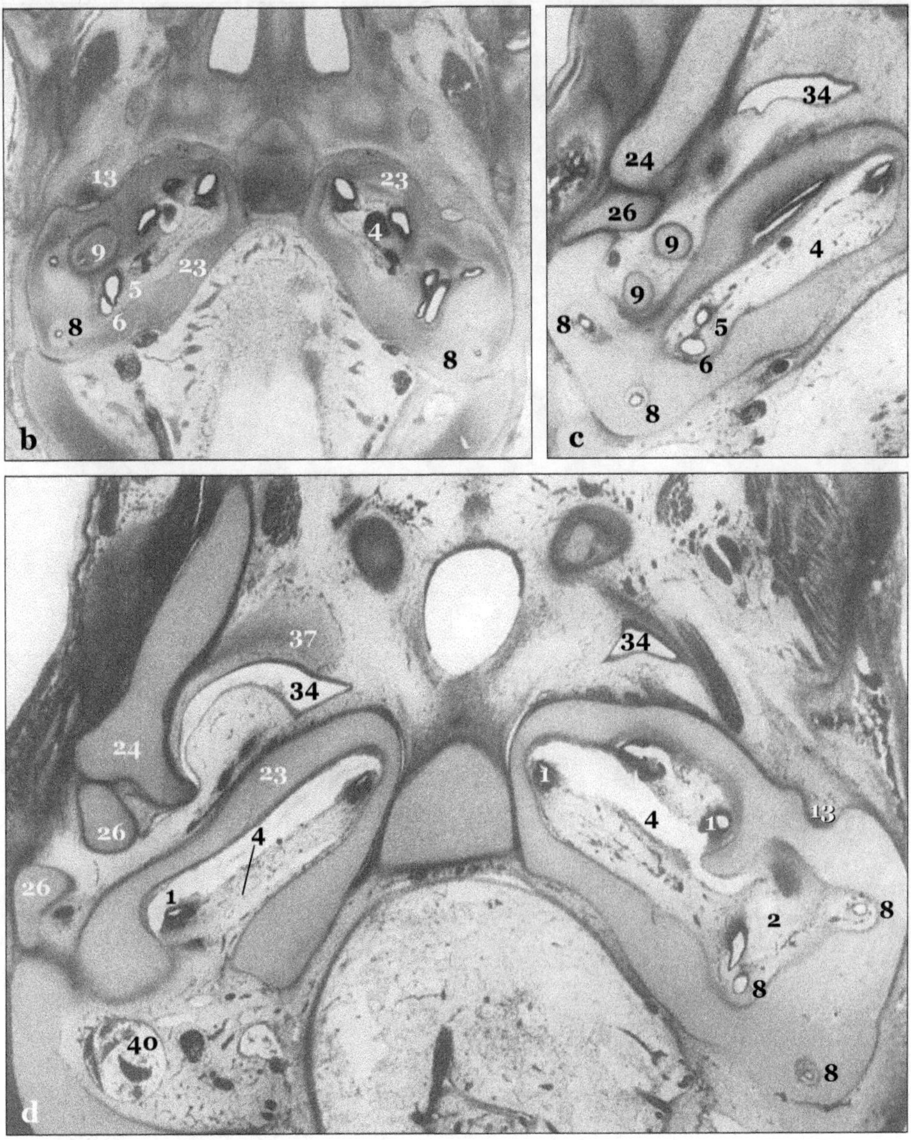

Fig. 46. Continued.

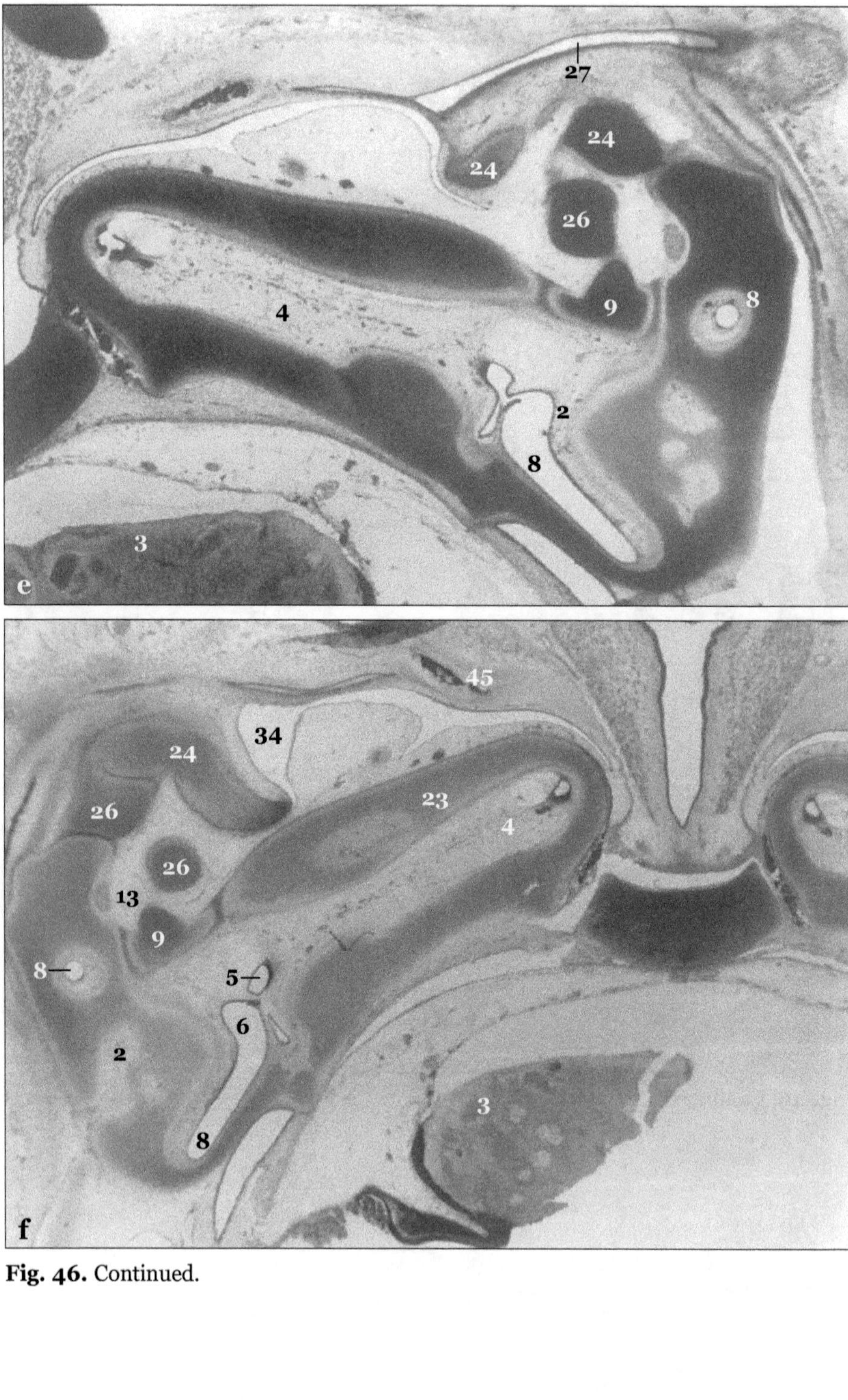

Fig. 46. Continued.

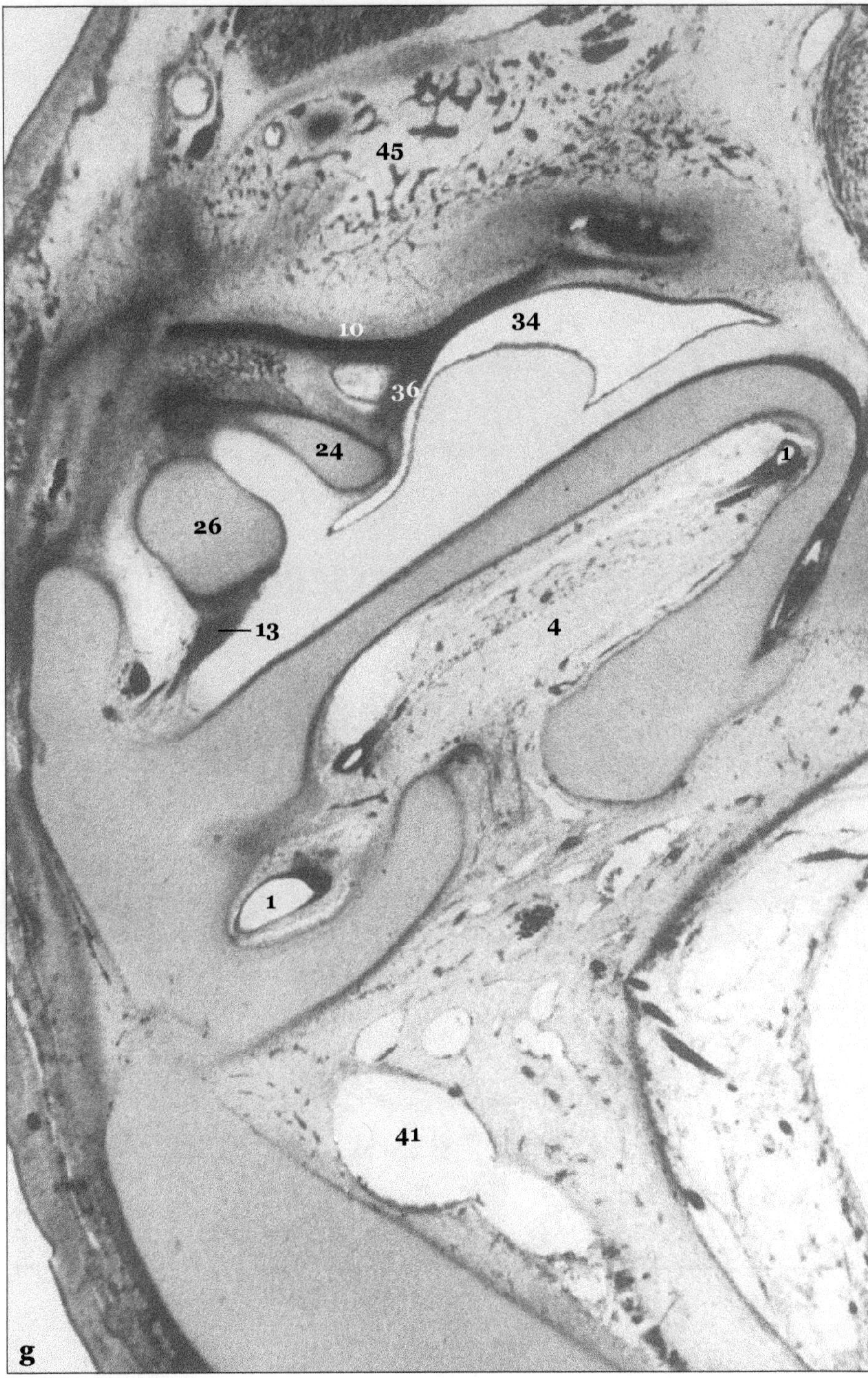

Fig. 46. Continued.

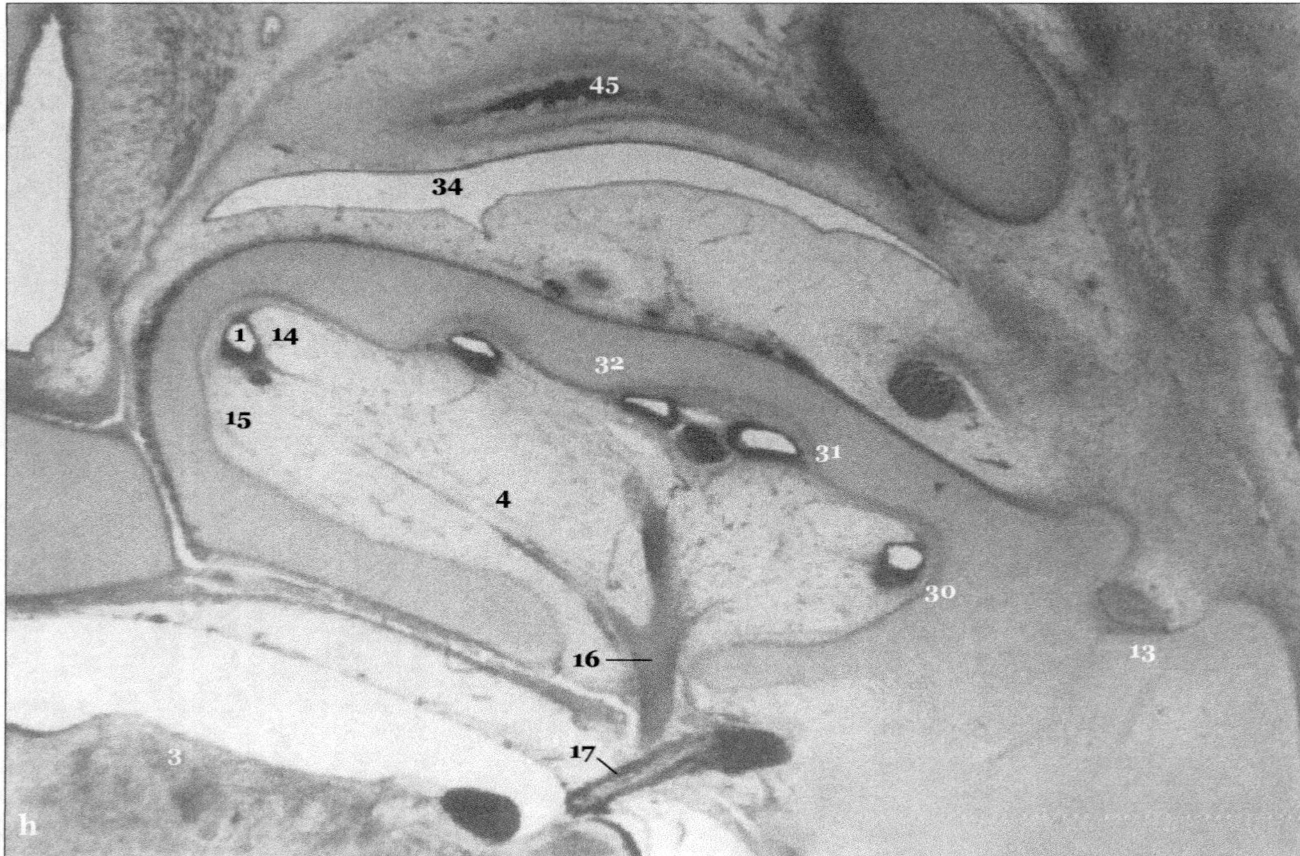

Fig. 46. Continued.

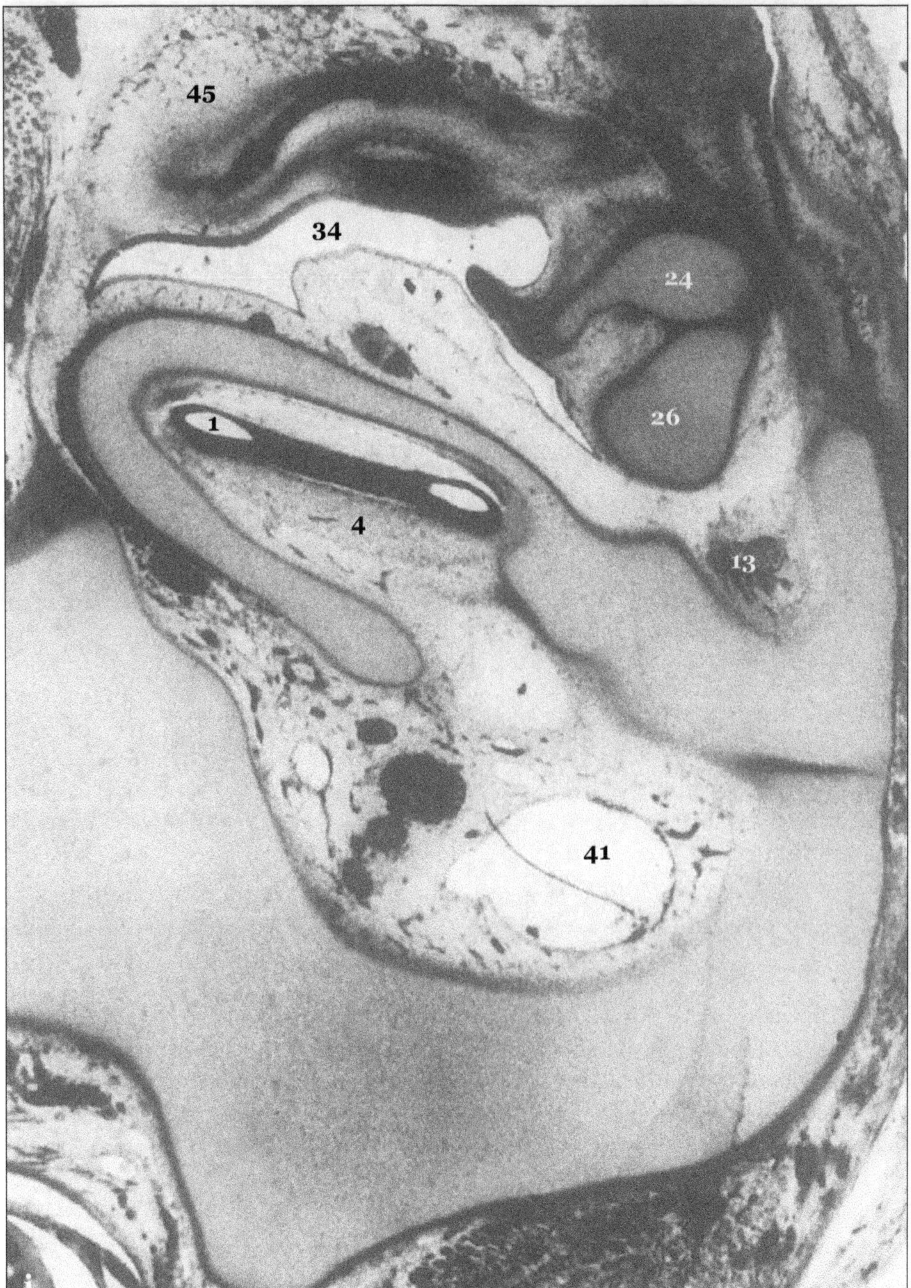

Fig. 46. Continued.

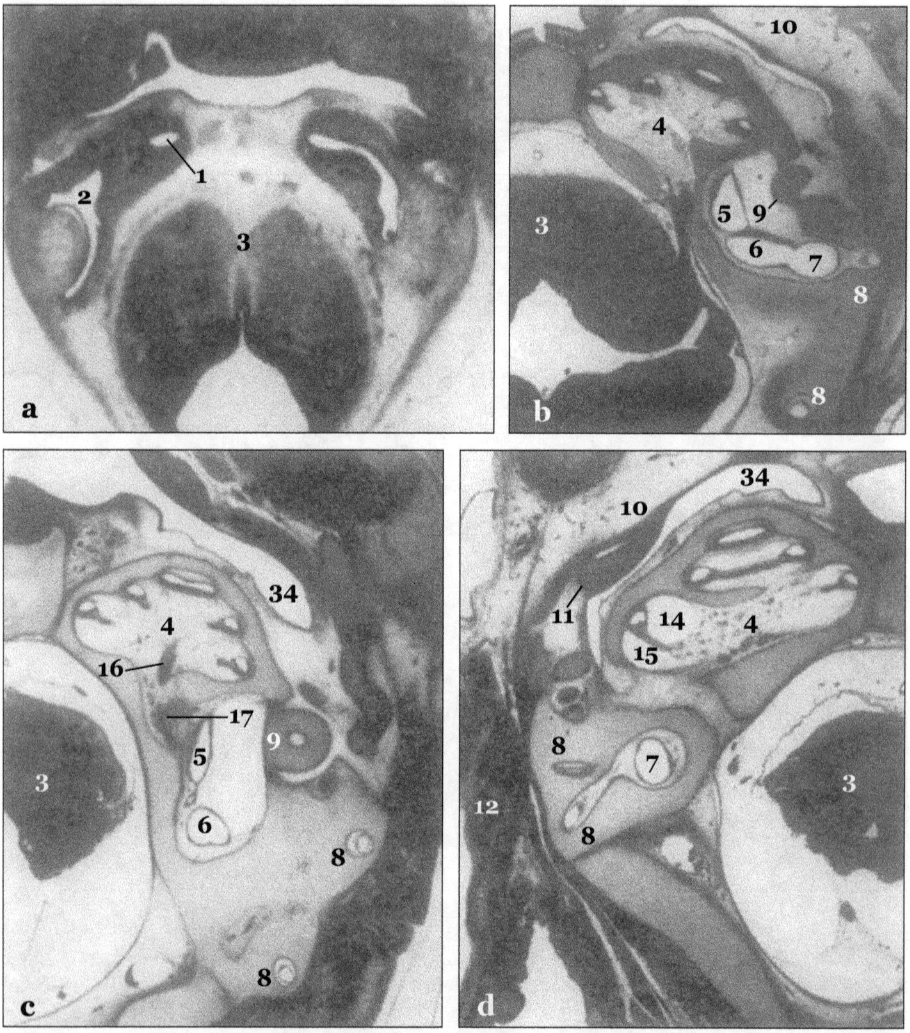

Fig. 47. Histotopography of the vestibular and cochlear structures in dorsoventral head sections of a prefetal *Rattus norvegicus* (a — stages 14, 15) and of *Sus scrofa domestica* (b — stages 17, 18; c, d — stages 19, 20).

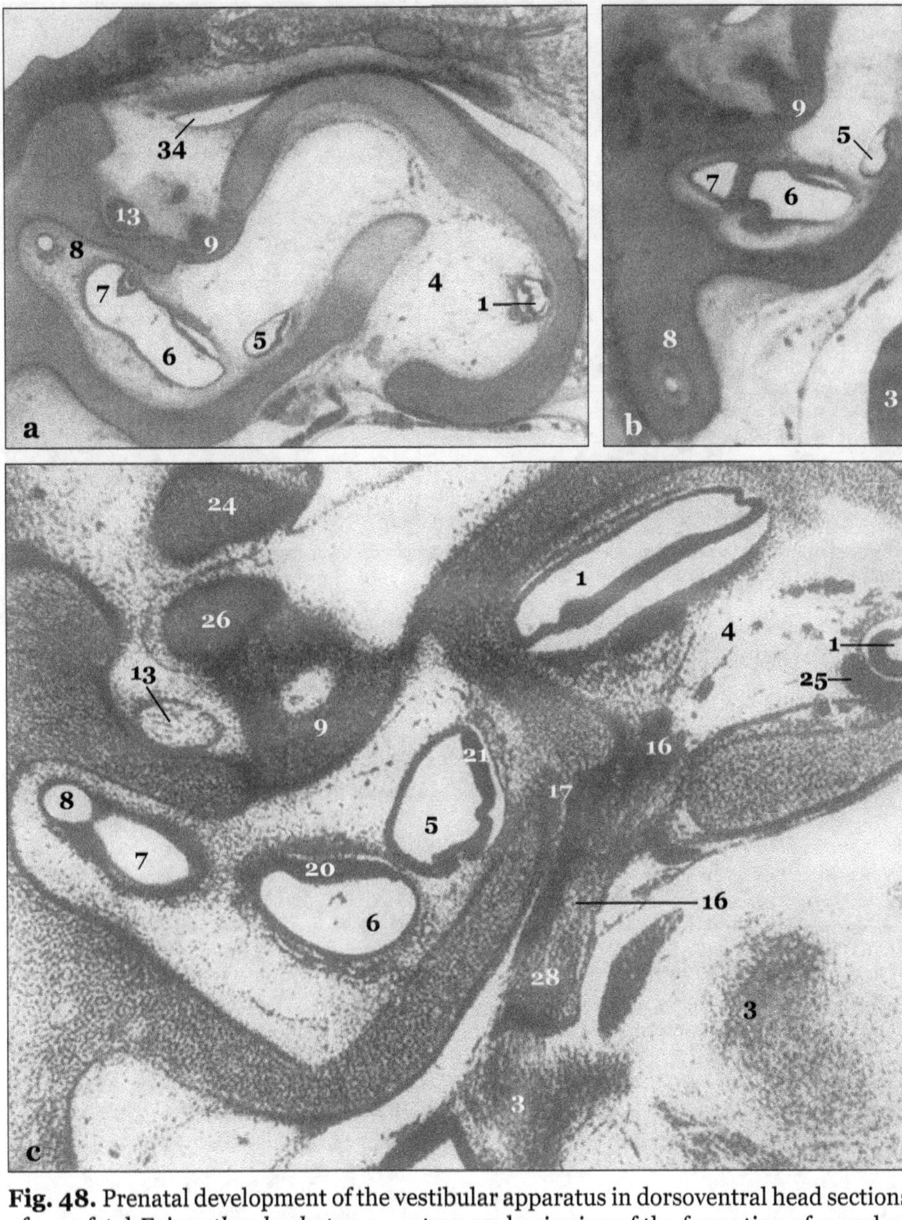

Fig. 48. Prenatal development of the vestibular apparatus in dorsoventral head sections of a prefetal *Erignathus barbatus*. a — stage 17; beginning of the formation of sacculus, utriculus and semicircular canal ampulla; b — stage 18; sacculus, utriculus and semicircular canal ampulla are formed; c — stage 19; saccular and utricular maculae are located at a right angle to each other.

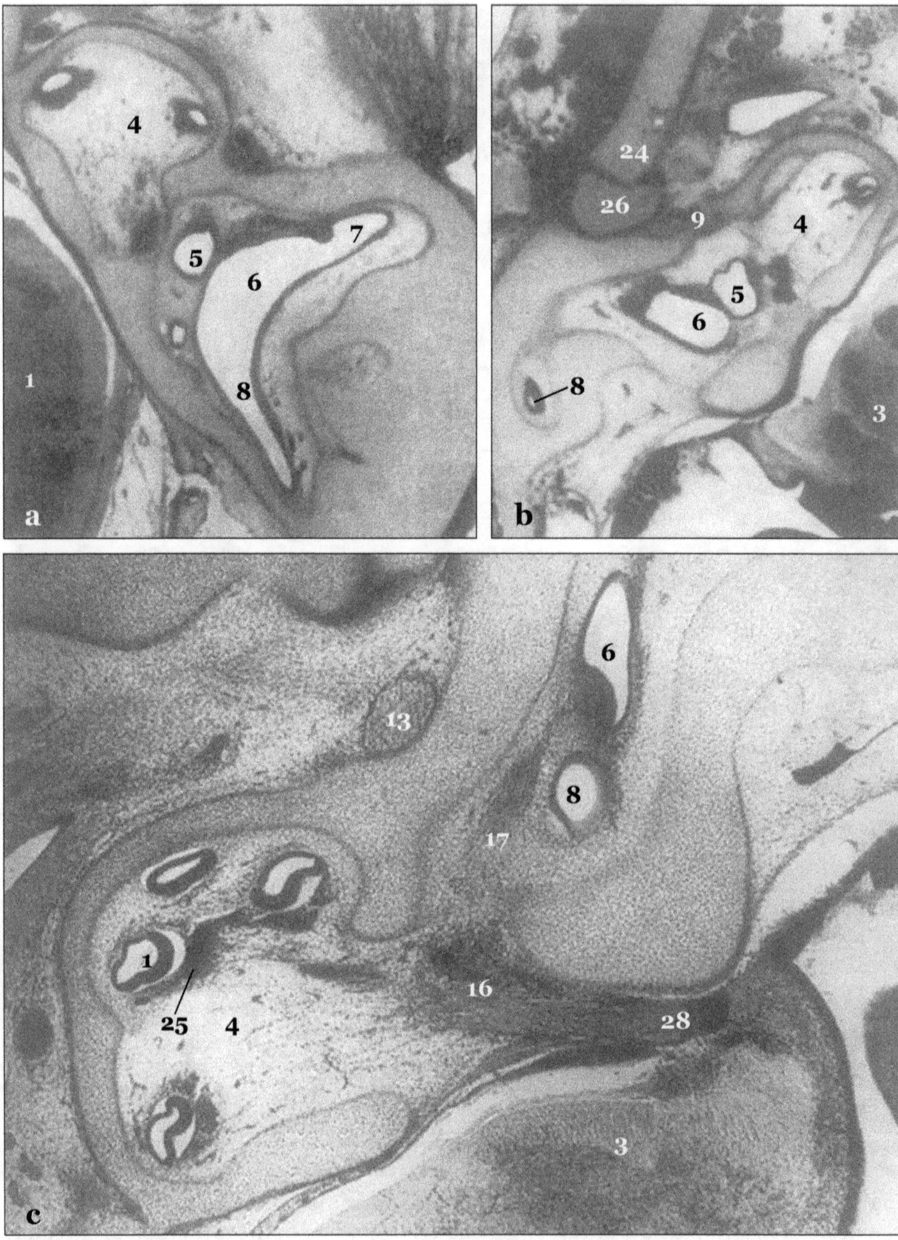

Fig. 49. Prenatal development of the auditory and vestibular organs in dorsoventral head sections of a prefetal *Eumetopias jubatus*. a — stage 16; formation of the semicircular canal ampulla; b — stage 17; sacculus and utriculus; c — stages 18, 19; initial differentiation of the structures of the cochlear canal.

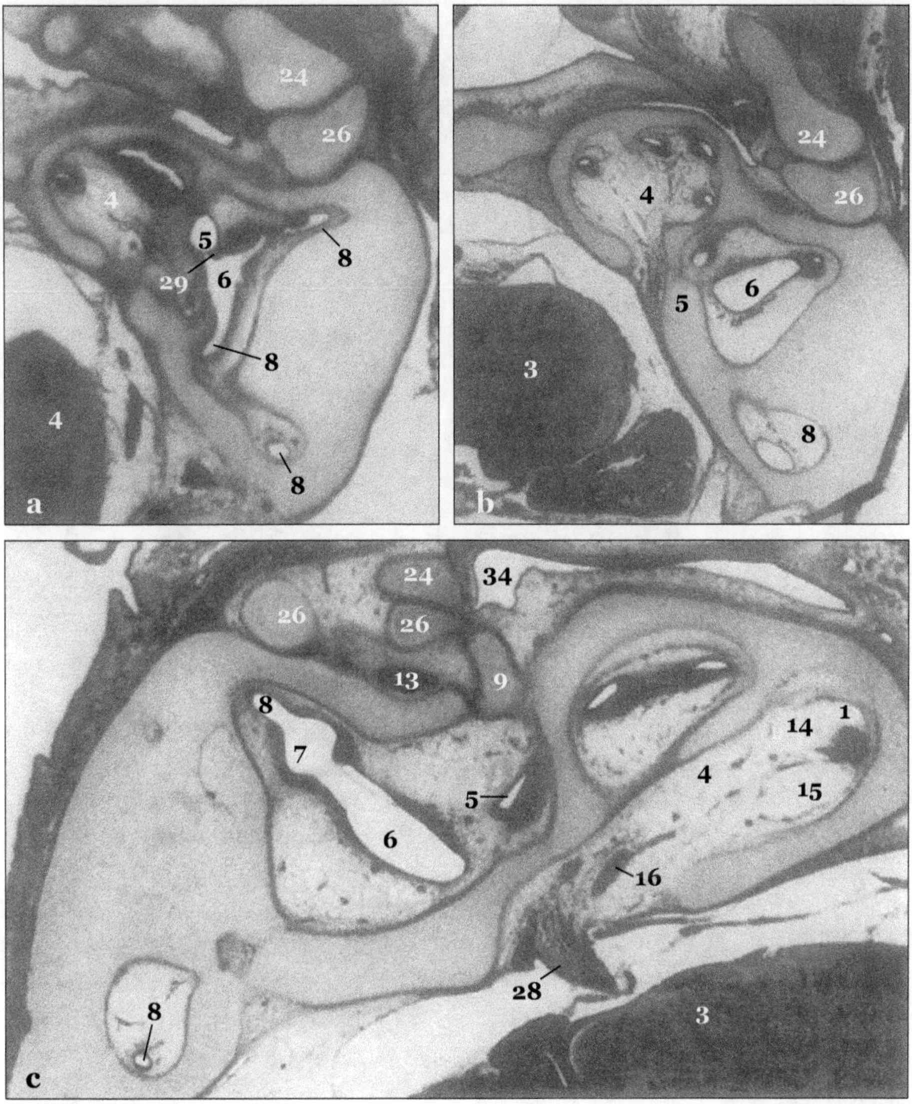

Fig. 50. Structures of the cochlear and vestibular apparatus in dorsoventral head sections of a prefetal *Odobenus rosmarus divergens*. a — stages 16, 17; formation of sacculus, utriculus and semicircular canals; b — stages 17, 18; sacculus and urticulus anlagen; the cochlea is twice as large as the vestibular apparatus; c — stages 19, 20; the sacculus, utriculus, semicircular canal ampulla and semicircular canals are fully formed.

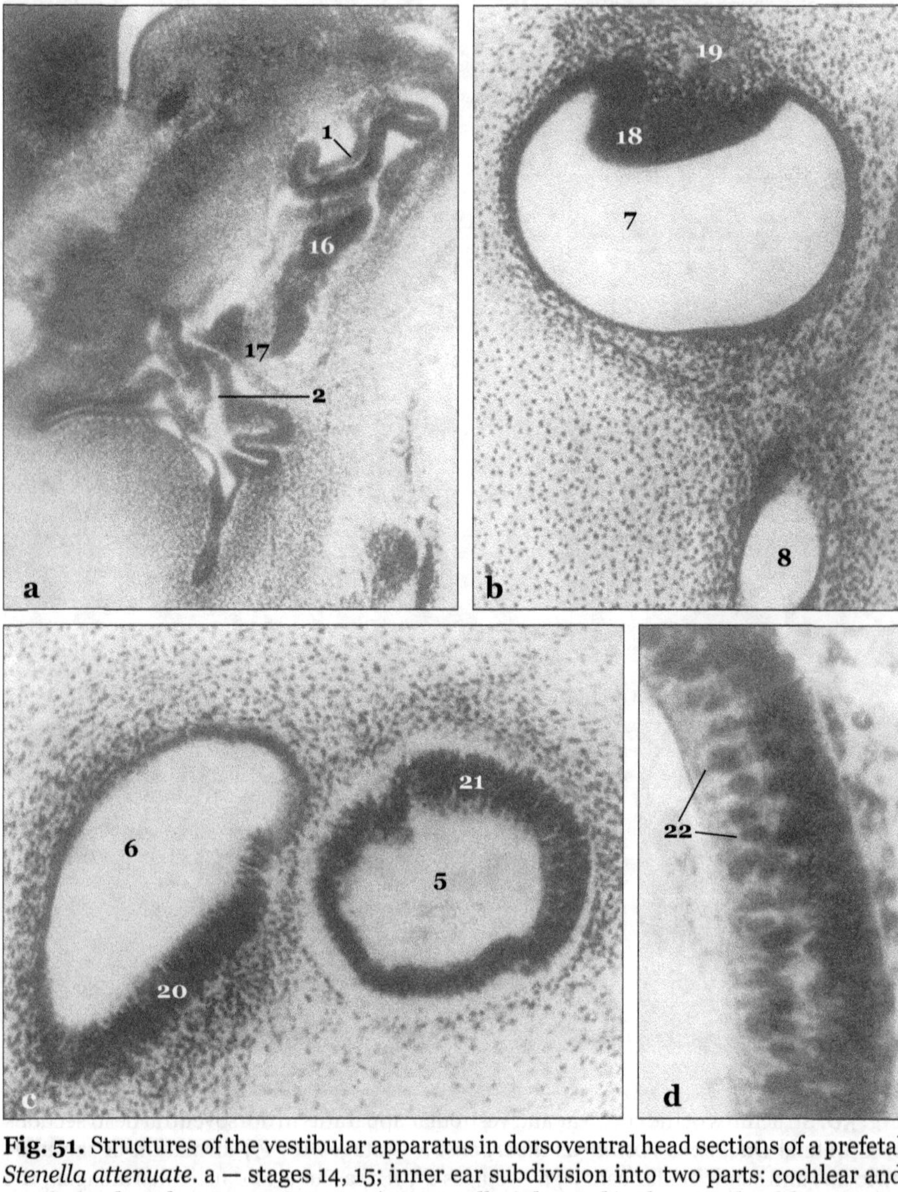

Fig. 51. Structures of the vestibular apparatus in dorsoventral head sections of a prefetal *Stenella attenuate*. a — stages 14, 15; inner ear subdivision into two parts: cochlear and vestibular; b, c, d — stages 18, 19; a crista ampullaris located in the ampula of the anterior vertical semicircular canal; a sacculus and utriculus; reception cells of the saccular macula.

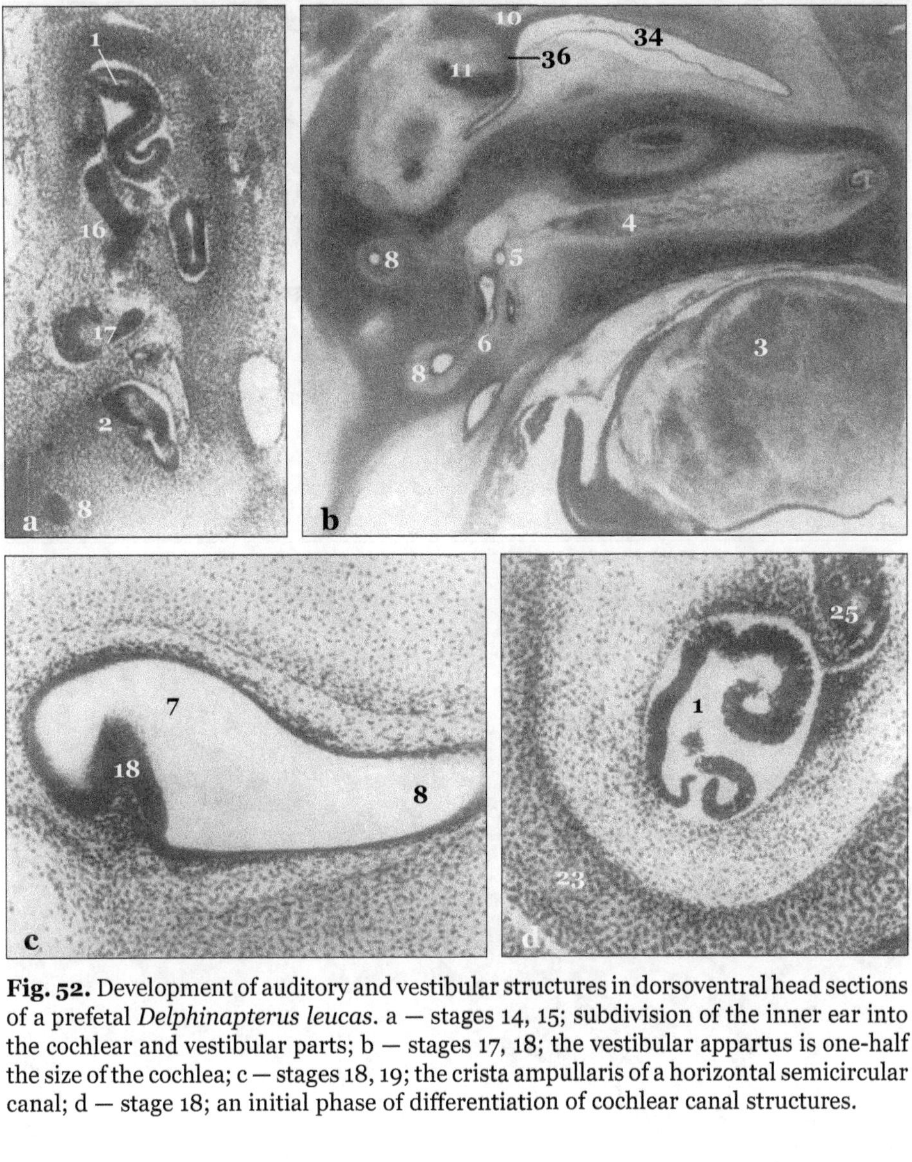

Fig. 52. Development of auditory and vestibular structures in dorsoventral head sections of a prefetal *Delphinapterus leucas*. a — stages 14, 15; subdivision of the inner ear into the cochlear and vestibular parts; b — stages 17, 18; the vestibular appartus is one-half the size of the cochlea; c — stages 18, 19; the crista ampullaris of a horizontal semicircular canal; d — stage 18; an initial phase of differentiation of cochlear canal structures.

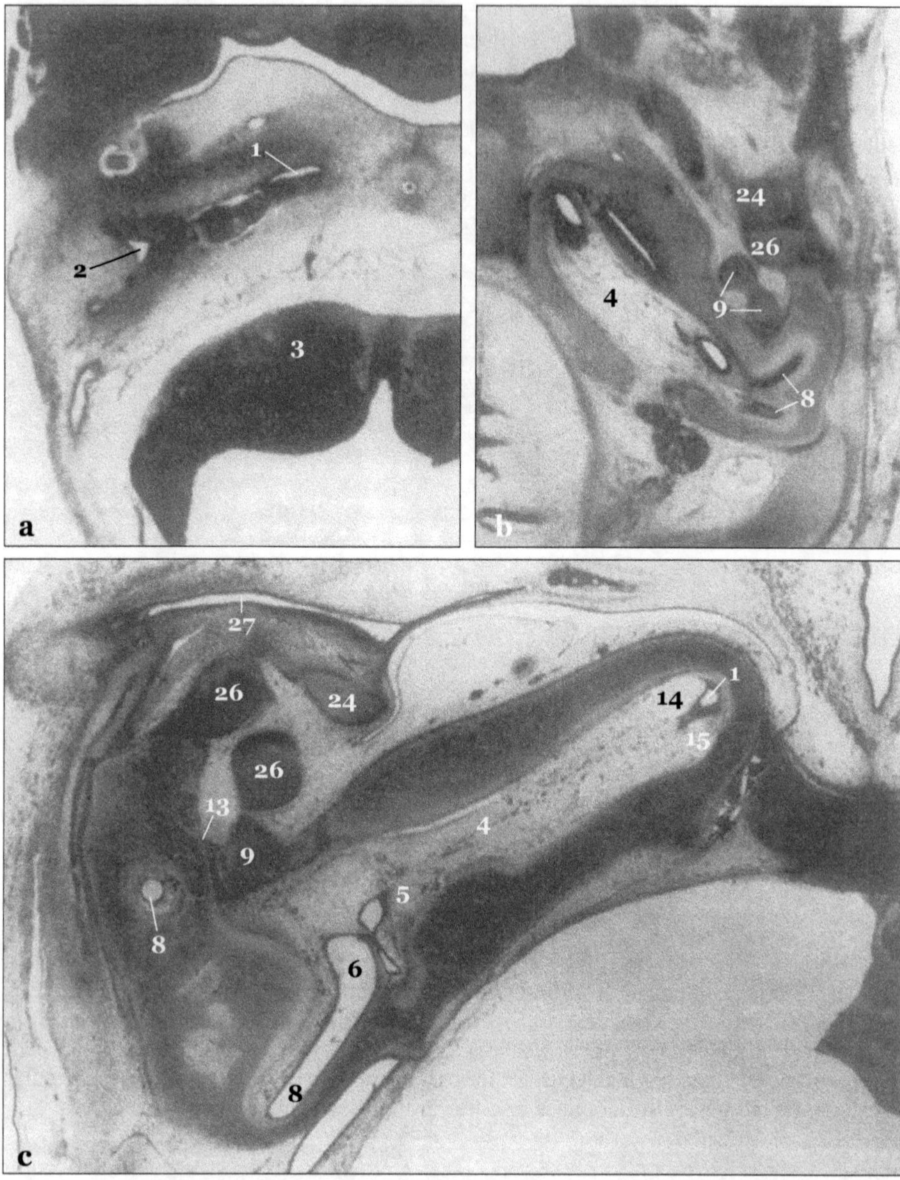

Fig. 53. Prenatal development of auditory and vestibular structures in dorsoventral sections of the head of a prefetal *Balaenoptera acutorostrata*. a — stages 14, 15; subdivision of the inner ear into the cochlear and vestibular parts; b — stage 17; the vestibular apparatus is one-half the size of the cochlea; c — stages 18, 19; formation of the sacculus and utriculus.

Conclusion

The aim of the present research was to reveal the specific and adaptive features in the structural organization of the peripheral part of the auditory system in representatives of many species, including marine mammals, comprising a wide ecological series from the terrestrial to the subterranean, aerial, semi-aquatic and aquatic mammals.

For the first time, the comparative and embryological analysis of the peripheral part of the auditory system was carried out using unique collections of cetaceans and pinnipeds, which has allowed for the study of its structural organization in more detail, as well as the determination of the stages of formation of structures of the outer, middle and inner ears in species that are phylogenetically close but different in ecological specialization, and in forms that are phylogenetically distant but similar in ecology.

As has been demonstrated, different parts of the peripheral part of the auditory system in terrestrial species of mammals are the least variable in evolutionary terms compared to the representatives of the class of mammals belonging to other ecological groups. Therefore, during the further statement of the material, the structure of the peripheral part of the auditory system in representatives of different ecological groups will be compared with its structural organization in terrestrial species. This is important to reveal those evolutionary transformations that have appeared in the peripheral part of the auditory system in connection with various physical properties of habitats as channels of acoustic communication.

In terrestrial species, the outer ear possesses common structural features: the presence of the auricle, the size, form and mobility of which vary in different species, depending on features of their biotopes; the presence of an external auditory meatus, which has the form of a tube opened along its whole extent; and the presence of a horseshoe-shaped cartilaginous lamina, which surrounds the auditory meatus. This structure of the outer ear is adapted for the optimal conduction of sound in air.

For the middle ear of terrestrial species, the following features are typical: a thin, medium-sized tympanic membrane, which does not possess elasticity; a chain of auditory ossicles, which is constructed as a pendulous system in which independent fluctuations of each of the auditory ossicles are possible, and the size of which is proportional to the size of the tympanic membrane; an elastic (by means of a ligament) connection of the long process of the malleus with the tympanic bone; in the center, the long handle of the malleus is connected with the tympanic membrane.

The basic features of the cochlea's structure in terrestrial species are the following: the sizes of the cochlea's basal turn differ insignificantly from the turn located above; the secondary osseous spiral lamina is not developed; the basilar membrane is wide and thick.

This type of peripheral auditory system structure is adapted to conduct air sound in an average frequency range from 16 up to 30,000 Hz.

In mammals, the change to a subterranean way of life essentially affected the structure of their auditory system, since underground passages represent filters whose band pass corresponds to the low-frequency range of sound [Konstantinov, Movchan, 1985].

In subterranean species, the greatest changes occurred in the outer ear as it came into direct contact with their habitat. As a result, the reduction of the auricle occurred. In the cavity of the auditory meatus, the semilunar valve appeared, due to the mobility in the connection of the cartilaginous laminae surrounding the auditory meatus, and with the help of the ear muscles it closed the lumen, which, most likely, is an adaptation protecting the auditory meatus from possible mechanical damage and from its filling with the soil during the animals' movement in underground burrows.

In addition to the outer ear, essential changes also occurred in the middle ear, in which a bigger and, compared to the terrestrial forms, more thickened tympanic membrane is typical. The functional value of the increased surface of the tympanic membrane lies in the improvement of the conditions for transferring low-frequency signals, and its thickening can be considered an adaptation to protect it from mechanical damages in the conditions of a compacted habitat. The tympanum is penetrated with osseous beams, which are presumed to be for the filtration of biologically important signals from noises, which the habitat is filled with. The auditory ossicles are big, which corresponds to the large surface of tympanic membrane. In the auditory ossicles, incisures have appeared, which is probably linked to the necessity for amplification of sound transmission in the conditions of a compacted habitat.

The structure of the inner ear of the subterranean species does not differ significantly from terrestrial species.

Aerial echolocating species acquired new adaptive features not only in the structure of the outer and middle ears, but also in the inner ear (cochlea). In the outer ear of bats, adaptations which protect the cochlea from superintensive signals appeared. One of the adaptations is the folding of the auricles, and another is the closing of the external auditory meatus with the help of a semilunar protuberance located near the tympanic membrane.

Being influenced by superloud sounds, the semilunar valve, by means of tension of the *m. antitragicus*, closes the auditory meatus like a plug in the area of the lumen.

Also, significant changes have occurred in the middle ear: the surface of the tympanic membrane has decreased; it is thin, but it is rigidly fixed on the tympanic ring, which has caused an increase in its elasticity. The reduction of the tympanic membrane's size has led to a corresponding reduction in the size of the auditory ossicles, which in echolocating species only are most rigidly connected with each other, especially in the area of the incudomalleal joint. The long process of the malleus knits with tympanic bone; therefore, the pendulous system of the auditory ossicles of terrestrial forms is replaced by an elastic vibrating system, which is capable of transferring signals of a wide frequency range, including ultrasound. The auditory ossicles have deep incisures, which is probably connected with the conduction of high-frequency sounds. The muscles of the middle ear in bats are more developed than in non-echolocating species.

In connection with the perception of high-frequency signals, radical changes in the inner ear (cochlea) have occurred: (1) the size of the basal cochlea's turn considerably surpasses the medial turn, located above it, which has a direct relation to the perception of high-frequency signals; (2) in the cochlear canal, the secondary osseous spiral lamina is well developed; (3) the thin, narrow basilar membrane is rigidly fixed between both laminae (primary and secondary); (4) in the organ of Corti, the Claudius cells are compactly located, increased in size and developed considerably [Pye, 1966a, b, 1967].

Thus, the echolocating ability of bats, caused by their way of life and based on their acquisition of structural features of the peripheral part of the auditory system, has led to the development of an auditory receiver of a new type, capable of providing sound conduction and sound perception of a wide range of frequencies, including ultrasound – from 30 Hz up to 98 kHz.

Semi-aquatic forms, having switched to an aquatic way of life, have not lost their connection with the land; therefore, in the peripheral part of their auditory system, some features directed to the improvement of underwater hearing appeared. On the whole, the auditory system of semi-aquatic mammals is adjusted to the perception of signals both in aerial and aquatic habitats.

In semi-aquatic insectivores, rodents and carnivores, whose way of life is to a grea extent connected with being on land, the morphological changes have affected only the outer ear by the way of formation of skin valves, folds, and hair bundles in the auricle, or simply its covering with thick fur.

The middle and inner ears of these mammals have kept the structural features typical of terrestrial species.

In semi-aquatic marine mammals belonging to different orders (sea otter, pinnipeds), there is a stronger connection with water as a habitat, which has caused the appearance of more advanced adaptive features for this habitat both in the outer and middle ears.

The outer ear of sea otters and otariids reveals similar structural features which become apparent in the following: (1) the presence of an auricle capable of closing the auditory meatus under water; (2) the auditory meatus represented by an open lengthened tube, the diameter and shape of its lumen change insignificantly along its whole extent; (3) the auditory meatus surrounded by horseshoe-shaped cartilage, as in terrestrial species; (4) the venous sinuses located in the osseous part of the auditory meatus – their functional value lies in the equalization of pressure in the tympanum during the animal's diving.

The outer ear of phocids and walruses reveal different structural features, which is connected to the fact that these animals spend significantly longer time under the water: (1) the auricle is reduced, the auditory meatus is lengthened, bent double, and surrounded by four cartilaginous laminae, which are flexibly connected with each other; (2) the diameter and the form of the auditory meatus's lumen sharply change throughout its extent.

The middle ear of sea otter and pinnipeds reveals structural adaptive features, which provide normal sound conduction both in air and under water. On the way of adaptation to an aquatic way of life, the following changes have occurred: (1) the thickening of the tympanic membrane; (2) in the auditory ossicles the size and form of separate elements change; (3) the thickening and shortening of the handles of the malleus; (4) the lengthening of the long process of the incus; (5) the reduction of the inter-ear aperture of the stapes; (6) in the area of the incudomalleal joint the auditory ossicles are connected with each other at right angles, which increases the rigidity in the auditory ossicles' chain; (7) the long process of the malleus is connected with the tympanic bone by means of a ligament; (8) the development of the venous sinuses in the walls forming tympanic bone; (9) the thickening of the mucous membrane covering the tympanum by means of the development of a dense network of blood vessels in its middle layer.

For the inner ear, the following features in the structure are characteristic: (1) the basal turn of the cochlea is not increased in comparison with the turn located above it; (2) the secondary osseous spiral lamina is not developed; (3) the basilar membrane is wide and thick.

As a result, the inner ear remains similar to the inner ear of animals possessing low- and middle-frequency hearing.

As has been shown by different authors, the upper range of auditory perception in different representatives of pinnipeds varies in the low- and mid-frequency ranges: up to 7.5 kHz in a walrus; 10 kHz in a Steller sea lion; 25 kHz in the air and 40 kHz in the water in a fur seal; 12 kHz in the air and 32 kHz in the water in a Caspian seal [Mohl, 1968; Terhune, 1974; Terhune, Ronald, 1975 Babushina et al., 1986; Kupin, 1986].

Thus, the sound conduction in two acoustically different environments and the necessity for deep-water submersion in semi-aquatic animals have caused the complication of the structural organization of the outer and middle ears [Harrison, Tomlinson, 1963].

In typically aquatic mammals (cetaceans), a more advanced adaptation of the peripheral system of underwater perception of sounds is evident. Different types of auditory receivers are found not only in two suborders of cetaceans, but also in the suborder of odontocetes, in which two types of reception of acoustic signals have developed: with the help of the tympanic membrane-ligament (Delphinidae) and without it (*Kogia, Physeter catodon, Berardius*) [Fleischer, 1975]. In connection with the absence of the tympanic membrane, the lateral wall of the tympanic bone serves as a specialized surface for sound perception.

Constant dwelling in aquatic habitats has led to significant transformations of the outer and middle ear, including the appearance of new additional structures in them. For the outer ear of the aquatic forms, the following structural features are typical: (1) in all cetaceans, the auditory meatus is constantly closed; this is provided by the presence of an ear cerumen in its cavity in mysticetes and the full epithelial obliteration of the auditory meatus in its distal part in odontocetes; (2) in mysticetes, as well as in terrestrial forms, the auditory meatus has the form of a cone-shaped tube; (3) in odontocetes, the auditory meatus is lengthened, bent double, the diameter and the shape of its lumen vary throughout its extent; (4) the auditory meatus is surrounded by horseshoe-shaped cartilage in both mysticetes and odontocetes.

The structure of the middle ear differs in mysticetes and odontocetes, which is connected with the different frequency tuning of their auditory system.

Experimentally, by means of conditioned reflex, behavioral and electrophysiological methods of research, it was established that the upper range of auditory perception in dolphins reaches 120 to 140 kHz with the frequency of best sensitivity equal to 60 to 70 kHz [Schewill, Lawrence, 1953; Bullock et al., 1968; Supin, Sukhoruchenko, 1970; 1971; Sukhoruchenko, 1971; Belkovitch, Dubrovsky, 1976]. Mysticetes perceive sound signals up to frequencies of about 10 to 14 kHz.

In dolphins, two sound generators are present; in many cases, they work simultaneously and in coordination with each other [Lilly, 1967]. According to the data of V.I. Markov and V.A. Tarchevskaya [1978], up to 40% of the bottlenose dolphin's signals are formed by two generators, which are switched on in parallel and in series.

For the middle ear of mysticetes, the following structural features are characteristic: the tympanic membrane is strongly thickened and has the form of a "glove finger;" the handle of the malleus is reduced; the long process of the malleus knits with the tympanic bone; the connection in the area of the incudomalleal joint is rigid (i.e., the malleus and the incus are connected with each other at right angles); the connection of the tympanic membrane with the reduced handle of the malleus is carried out with the help of a ligament, as opposed to those species of mammals in which the handle of the malleus is connected directly to the tympanic membrane; the venous sinuses are developed; new structures have appeared: the cavernous plexus is located in the tympanum and the peribullar sinuses; the reduction of the short process of the incus that deprives the incus of its support against the wall of the tympanic bone, therefore the middle ear does not form an elastic vibrating system in the auditory ossicles' chain; the tympanic bulla of mysticetes knits with the cranium at two points only and does not participate in the formation of a cranial wall; in the elements connecting the periotic bone with the cranium there are sesamoid ossicles, which provide some mobility of the tympanic bulla relative to the cranium [Yamada, 1953 a, b]. The presence of such mobility in the tympanic bulla-cranium connection makes the auditory receivers independent from each other to some extent and capable of providing directed reception of acoustic signals.

In odontocetes (dolphins), the most important structural features of the middle ear are as follows: (1) the tympanic membrane is rounded, reduced in size, thick and elastic; (2) the triangular ligament is asymmetrically attached to the tympanic membrane and connects it to the reduced handle of the malleus; (3) the auditory ossicles are reduced in size; (4) in the incus both arms are developed, its short arm is attached to the wall of the tympanic bone; (5) the malleus and the incus are connected with each other at right angles, the long arm of the malleus knits with the tympanic ring, which, as in bats, transforms the sound conduction apparatus into an elastic vibrating system, which is capable of transferring signals of a wide frequency range, including ultrasound; (6) the presence of a cavernous plexus in the tympanum, the venous sinuses in the walls of the tympanic bone, and the thickened mucous membrane of the middle ear, which provide pressure equalization in the tympanum during the animal's diving and remerging; (7) the development of peribullar sinuses around the tympanic bulla, which are filled with foam; (8) the tympanic bone does not knit with the cranial bones; their connection is carried out by means of the short tendinous ligament that makes the auditory receivers independent of each other, which is necessary for directed hearing under water.

The inner ear (cochlea) possesses structural features that are characteristic of echolocating forms: (1) the basal turn of the cochlea is widened; (2) in the cochlear canal the secondary osseous spiral lamina, which, according to its development, surpasses that in echolocating bats, is well developed; (3) the basilar membrane is very narrow, thin and rigidly fixed between the spiral laminae; (4) in the organ of Corti, the Claudius cells are increased in size; (5) the quantity of the cells of the spiral ganglion is increased 3 times in comparison with non-echolocating species. The vestibular apparatus has the same components typical of the majority of mammals, independent of their ecological status. However, only in the absolute hydrobionts are the sizes of the vestibular apparatus smaller than the sizes of the cochlea, which indicates the paramount value of the auditory function in the vital activity of these mammals.

The comparative morphological analysis of the peripheral part of the auditory system in mysticetes possessing low-frequency hearing, and in odontocetes capable of echolocation, has shown that their acquiring of different methods of spatial orientation and communication has caused a divergence in the evolution of the development of the auditory receivers, each of which is adjusted to the perception of a certain spectrum of frequencies.

On the whole, similar features in the structure, connection and attachment of structures of the sound-conduction apparatus in the tympanum of cetaceans are viewed as adaptations, which increase the coefficient of sound pressure transmission by the middle ear that provides the efficiency of the sound conduction. Unlike echolocating dolphins, in the middle and inner ears of mysticetes, the structures on which the perception of high-frequency signals depend have not developed fully.

In echolocating mammals belonging to different ecological and taxonomical groups, the development of the peripheral auditory system has acquired common properties. This has occurred due to its parallel evolution, during which the conditions for their intraspecific acoustic communication in situations adverse for vision, and in connection with specific properties of the environment as the channel of acoustic communication, have been created.

In spite of the fact that a general plan of the structural organization of the peripheral auditory system is revealed in the majority of species of mammals, in each ecological group this has evolved in its own direction, showing independently functional development depending on the acoustic properties of the habitat.

It is not only the structure and development of the mammalian peripheral auditory system functioning in the specific frequency spectrum for each group which is connected with the ecology of the species, but also the structure and development of the system of sound signals [Airapetiantz, Konstantinov, 1974], which is directed to the optimization of the intraspecific communication of animals in their specific habitat.

Evolutionary changes to the peripheral auditory system in various ecological groups became apparent due to polymorphism and the appearance of new structures. This has been shown by examples in the variety of the outer, middle and inner ear structures.

Thus, the peripheral part of the auditory system in different species of mammals possesses common basic features in its structure and development. Nevertheless, the use of only certain acoustic properties of the habitat by separate groups of animals has caused strongly pronounced polymorphism in all of its parts, beginning with the outer ear. The occupation of various habitats by mammals, as well as the development of different methods of spatial orientation and communication, were accompanied by significant morphological transformations of all parts of the peripheral auditory system, in particular the phylogenetically young parts, which are typical only for the mammalian class.

During the adaptive specialization in the representatives of different ecological groups, the outer ear, being in direct contact with the habitat, is mainly exposed to the greatest morphological reorganizations connected with the adaptation to terrestrial, subterranean, aerial, semi-aquatic and aquatic ways of life.

Using bioacoustic analysis and taking as an example the closed auditory meatus, it has been shown in dolphins that the tendency of shutting the auditory meatus in semi-aquatic species and its total obliteration in absolute hydrobionts are progressive adaptations for the provision of underwater hearing.

The middle ear keeps the general basic principle of structure in most mammals. In species that are phylogenetically distant but close by ecological specialization or the frequency tuning of their auditory system, there are parallel features in the development of the separate elements of the auditory ossicles and also in the way of their junction with each other and fastening in the tympanum. During the adaptation to an aquatic way of life, semiaquatic and aquatic species have developed new additional structures that are unusual for initial terrestrial forms.

The biomechanical features of the sound conduction apparatus are directed to the increase of the coefficient of sound pressure transmission by the middle ear to the cochlea, which determines the effectiveness of the auditory organ's functioning under water and considerably expands the range of perceived frequencies.

The inner ear adopts some specific structural features. However, at the similar set of the inner ear's components, the size, form and arrangement of the structures of the cochlear and vestibular apparatuses vary. The structure of receptor formations in the organ of Corti and the auditory maculae is similar in most species. The exceptions are the echolocating species in which the peculiarities in size and arrangement of the cells of the organ of Corti are marked.

As is apparent from the description, all parts of the peripheral auditory system are multicomponent formations. As opposed to the outer and middle ears, which are characterized by different structural variations and a wide spectrum of adaptable transformations connected with the peculiarities of species ecology, in the representatives of different ecological groups, the inner ear possesses a variety of functions and therefore keeps a similar structural organization. Usually, both in the cochlear and vestibular analyzers, the topography, form and size of separate components vary. In echolocating mammals, the substantial growth of the cochlea's sizes in comparison with the sizes of the vestibular apparatus, as well as other features in the structure of the cochlear canal and the cells of the organ of Corti serve as the cochlea's adaptations to the perception of frequencies of a wide range, including ultrasound (dolphins, bats). At the same time, the large cochlea and extraordinarily small size of the vestibular apparatus in absolute hydrobionts, which posses varied orientation of hearing, can be considered to be the adaptation of the inner ear to aquatic life, as the hearing of aquatic mammals dominates among distant analyzers, thus providing the survival rate of these animals in conditions of constant dwelling in an aquatic environment.

In the representatives of different ecological groups, in the early embryogenesis, the auditory and vestibular structures are simultaneously separated from each other and reveal similar structural features. In the first half of the early prefetal period (stages 13 to 15), both the auditory and vestibular parts possess common structural features in most mammals. Specific features in the structural organization of the hearing and equilibrium organs are formed in the second half of the early prefetal period (stages 16 to 20) at similar stages of development and in a certain sequence. These structural features are caused by peculiarities of habitat and develop parallel to each other from the homologous anlages of the peripheral auditory system in phylogenetically distant and close forms; the morphological features of the mammalian peripheral auditory system, which were formed in the early prefetal period, continue to develop in the fetal period and during the whole period of postnatal ontogenesis.

The anatomic formation of the structures of the inner ear finishes in the early prefetal period, while the cellular differentiation of the sensory epithelium of the cochlea, maculae and cristae in immature-born species continues up to the early stages of postnatal ontogenesis. In mature-born species (cetaceans, ungulates), the differentiation of the inner ear structures is complete by the moment of birth.

In studied groups of mammals, the features connected with the stages of differentiation of the sensory epithelium of maculae and cristae into receptor and supporting cells are revealed. In terrestrial and semi-aquatic mammals (otariids, walruses) whose way of life to a great extent is connected with staying on a firm substratum, the initial cellular differentiation of the sensory epithelium occurs in the utricular macula, which indicates the important role of organ gravitation in the vital activity of mammals. The simultaneous cellular differentiation of the sensory epithelium in the saccular and utricular maculae, and also the similarity of the sizes of the cochlear and vestibular parts of the inner ear in phocids, allowed for the assumption that in these species the organs of gravitation and vibration are equally vital. Each of these organs is adapted to function in a habitat determined by physical properties. In absolute hydrobionts (cetaceans), the initial cellular differentiation in the saccular macula indicates that in aquatic mammals the organ of vibration carries out a more important function than the organ of gravitation.

One of the aspects of comparative embryological study of the peripheral auditory system's formation in cetaceans and pinnipeds is the determining of their phylogeny.

From the point of view of the researchers, who accept the monophyletic origin of cetaceans, the present-day groups of cetaceans and pinnipeds are descended from a common ancestor [Weber, 1886; Kellogg, 1922; Slijper, 1936; Romer, 1939, Tomilin, 1957, Van Valen, 1968; Borisov, 1969 a, b].

According to the opinion of adherents to the diphyletic origin of cetaceans, the evolution of two suborders was preceded by a convergence rather than a divergence; therefore, in their opinion, it would be more correct to consider these suborders as two independent orders [Ragovitza, 1903; Gregori, 1910; Kükenthal, 1922; Kleinenberg, 1958; Andersen, 1967; Rice, Scheffer, 1968].

The monophyly of pinniped origins is advanced based on ecological and climatological research [Kellogg, 1922; Weber, 1928, Mattew, 1939; Davies, 1958; Scheffer, 1958], on biochemical studies [Borisov, 1969a; Sarich, 1969; Seal et al., 1970], on a comparative analysis of the karyotypes [Fay et al., 1967; Duffield Kulu, 1972 Anbinder, 1980] and a comparative embryological study of the formation of the external adaptive traits in phocids and otariids in the prenatal period [Kuzin, 1970].

The hypothesis of polyphyletic origins was advanced after the discovery of serious morphological differences among Otariidae, Phocidae and Odobenidae, and also in view of the peculiarities of the geography of their natural habitats [Howell, 1929; Mc Laren, 1960; Chapsky, 1970; Ray, 1976; Repenning, 1976].

Our comparative morphological research and also the comparative study of the peripheral auditory system's development have revealed the basic distinctions of its structure between the representatives of two suborders of cetaceans, as well as inside the order of pinnipeds between Otariidae on one hand, and Phocidae, Odobenidae on the other hand. Our results might seem to testify to a greater extent in favor of the diphyletic hypothesis, as opposed to the monophyletic origin of mammals.

However, the study of the peripheral auditory system's development has shown that in Mysticeti and Odontoceti at similar stages of development in the early prefetal period, the tympanic membranes reveal similarity in their structure, whereas in the fetal period they acquire species-specific traits.

The structure of the peripheral auditory system in adult forms allows for the assumption that Mysticeti are the most ancient group among cetaceans which have kept structural features peculiar to terrestrial forms. Odontoceti should be regarded as a more advanced group in evolutionary terms, whose peripheral auditory system has acquired qualitatively new, progressive adaptations related to the development both of a special method of spatial orientation and communication (echolocation). On this basis, my original data favor the earlier advanced hypothesis of monophyletic origins of cetaceans [Kellogg, 1922; Romer, 1939; Tomilin, 1957; Van Valen, 1968] and confirmed by immunological evidence [Borisov, 1969; Sarich, 1975], molecular DNA hybridization [Mednikov, Shubina, 1977], and a comparative karyotype analysis [Anbinder, 1980].

Among pinnipeds, the greatest similarity with terrestrial forms in the structure of the peripheral auditory system is revealed in Otariidae, while Phocidae and Odobenidae represent evolutionarily more advanced groups of mammals, whose auditory organ is to a great extent adapted for functioning in an aquatic habitat.

During the investigation of the development of the peripheral auditory system, the similarity between Otariidae and some representatives of Phocidae (*Pusa hispida*) was revealed, becoming apparent in the formation of the auricle, which in *Pusa hispida* does not develop in the fetal period and in adult forms is absent.

On the basis of the data obtained, it is possible to assume that the most ancient group among pinnipeds is Otariidae, from which Odobenidae and Phocidae have descended. This can also serve as additional evidence to support the hypothesis of monophyletic origins of pinnipeds, which has been confirmed by immunological data [Borisov, 1969a] and a comparative karyotype analysis [Anbinder, 1980].

BILIOGRAPHY

Agarkov G.B., Solukha B.V., Khomenko B.G. *1971*. About the sonar ability of dolphins. In: "Bionics". Kiev, 5: 32-57 (In Russian).

Airapetiantz E.Sh., Konstantinov A.I. 1965. To the question about the role of echolocation in spatial analysis by bats. In: "Bionics". M.: Nauka, 334-341 (In Russian).

Airapetiantz E.Sh., Konstantinov A.I. 1970. Echolocation in Nature. L.: Nauka (In Russian).

Airapetiantz E.Sh., Konstantinov A.I. 1974. Echolocation in Nature. M., Nauka: 205-208 (In Russian).

Akayevsky A.I. 1962. Anatomy of domestic animals. M.: Selkhozgiz, 1-82 (In Russian).

Akimov V.N. 1976. About morphological and functional heterogeneity of the rows of outer hair cells of the cochlear spiral ganglion in guinea pig. Vestnik otorinolaryngologii, 3: 16-19 (In Russian).

Alexander C.G. 1900. Beitrag zur Morphologie des Ohrlabyrinthes. Zbl. Physiol., 14: 604.

Alexander G. 1901. Uber Entwicklung und Bau der Pars inferior Labyrinth der höheren Säugetiere. Denkschr. Österr. Akad. Wiss., Math.-naturwiss. Kl., 70: 429-482.

Alexander G. 1904. Entwicklung und Bau des inneren Gehörorganes von *Echidna aculeata*. Denkschr. med.-naturwiss. Ges. Jena, 6 (2): 1-118.

Alexander G. 1905. Zur Frage der phylogenetischen, vikariierenden Ausbildung der Sinnesorgane. Über das Statische und das Gehörorgan von Tieren mit kongenital defektem Sehapparat: Maulwurf (*Talpa europea*) und Blindmaus (*Spalax typhlus*). Z. Psychol. Physiol. Sinnesorgane, Leipzig, 38: 24-33.

Altmann F. 1950. Normal development of the ear and its machanics. Arch. Otolaryngol., 52: 725.

Anbinder E.M. 1980 Karyology and Evolution of Pinnipedia. M.: Nauka, 152 p. (In Russian).

Andersen H.T. 1967. Cardiovascular adaptations in diving mammals. Amer. Heart J., 74 (3): 295–298.

Anson B.J., Black T.J. 1934. Early relation of auditory vesicle to ectoderm in human embryos. Anat. Rev., 58: 127.

Babushina E.S., Zaslavsky G.L., Krasnitsky B.Yu., Yurkevitch L.I. 1986. Sensitivity of the northern fur seal to frequency change in aquatic and air habitats. Abstr. of the IXth All-Union Conference on the Study, Conservation and Rational Use of Marine Mammals. Arkhangelsk: 18-19 (In Russian).

Beauregard U. 1894. Recherches sur l'appareil auditif chez les mammifères. J. Anat. Physiol., Paris, 366-413.

Belkovitch V.M., Solntseva G.N. 1970. Morpho-functional features of the auditory organ of some dolphins. Zool. Zhurn., 49 (2): 273-280 (In Russian).

Belkovitch V.M., Dubrovsky N.A. 1976. Sensory Foundations of the Orientation in Cetaceans. Leningrad: Nauka (In Russian).

Blair S.F. 1964. Perceptual theories of middle ear muscle function. Ann. Otolar., Rhinol. Laryngol., 73 (3): 724.

Böas J. E. 1912. Ohrknorpel und äusseres Ohr der Säugetiere: eine vergleichend-anatomische Untersuchung. Copenhagen, Nielsen and lydiche.

Boenninghaus G. 1903. Das Ohr der Zahnwales, zugleich ein Beitrag zur Theorie der Schalleitung. Zool. Jahrb., Anat., 19: 189–360.

Boettcher A. 1869. Über Entwicklung und Bau des Gehörlabyrinthes nach Untersuchungen an Säugetieren. Dresden, 203 S.

Bogoslovskaya L.S. 1975. Ultra-structure of the spiral ganglion and auditory nerve of the bottlenose dolphin. Abstr. of the VIth All-Union Conf. for the Study of Marine Mammals. Kiev, 1: 41-43 (In Russian).

Bogoslovskaya L.S., Lubinsky I.A., Posin N.V., Putsillo E.M., Shmelev L.A., Shura-Bura T.M. 1973. Excitation spread along a fiber with local heterogeneities (results of modelling). Biofizika, 18 (5): 886-890 (In Russian).

Bogoslovskaya L.S., Solntseva G.N. 1979. The Auditory System of Mammals. Moscow: Nauka, 238 p. (In Russian).

Bondy G. 1907. Beiträge zur vergleichenden Anatomie des Gehörorgans der Säuger (Tympanicum, Membrana Shrapnelli und Chordaverlauf.).Anat. Hefte, I. Abt., Wiesbaden, 35 (H. 106): 293-408.

Borisov V.I. 1969a. On the use of precipitation reaction for the study of the phylogeny and systematics of Pinnipedia and Carnivora. Zool. Zhurn., 48 (2): 248-255 (In Russian).

Borisov V.I. 1969b. Comparative analysis of the serum albumins of cetaceans and some terrestrial mammals. In: "Marine Mammals". Moscow: Nauka, 308-312 (In Russian).

Brant Cl. 1970. The vestibular system. Annu. Rev. Psychol., 21: 273-306.

Bredberg G. 1968. Cellular patterns aud nerve supply of the human organ of Corti. Acta Otolaryngol., Suppl., 236 p.

Brown A, Pye J. 1975. Auditory sensitivity at high frequencies in mammals. Adv. Exp. Physiol. a. Biochem., 6: 1-73.

Brunner H. 1934. Die Entwicklung der Vorhofsäckchen im menschlichen Innenohre. Monatsschr. f. Ohrenh., 68: 185.

Buchanan T. 1828. Physiological illustrations of the organ of hearing. London, Consitt and Goodwill, 102-135.

Bullock T.H., Grinnell A.D., Ikezono E. 1968. Electrophysiological studies of central auditory mechanisms in cetaceans. Zeitschr. Vergl. Physiol., 59: 117.

Burlet H.M. de 1934. Vergleichende Anatomie des statoakustischen Organs. Handb. Vergl. Anat. Wirb., 2 (2): 1293.

Cajal S. Ramon 1909. Histologie du système nerveux de I'homme et des vertébrés, Paris, 1 (2).

Carte E., Macalister A. 1868. On the anatomy of *Balaenoptera acutorostrata*. Trans. Roy. Phil. Soc. London, 158: 210-261.

Caspersson T. 1947. The relation between nucleic acid and protein synthesis. In: Symposia of the Soc. for Exptl Biol., Cambridge.

Chapsky K.K. 1970. The concept of the arctic origin of Pinnipedia and other ways of solving the problem. In: Tolmachev A.I. (Ed.), The Arctic Ocean and its Coast in the Cenozoic. Leningrad, Gidrometeoizdat, 166-173 (In Russian).

Chow K.L. 1951. Numeracial estimates of the auditory central nervous system of the rhesus monkey. J. Comp. Neurol., 95 (1): 159-175.

Clarke R. 1948. Hearing in *Cetacea*. Nature, 161: 979-980.

Claudius M. 1858. Physiologische Bemerkungen über das Gehörorgan der Cetacean und das Labyrinth der Säugetiere. Kiel, 1-35.

Claudius M. 1865. Das Gehörlabyrinth von *Dinotherium giganteum* nebst Bemerkungen über den Werth der Labyrinthformen für die Systematik der Säugetiere. Kassel, 13 (2): 65-74.

Claudius M. 1867. Das Gehörorgan von *Rhytina stelleri*. Mém. Acad. Imp. Sci. St. Pétersbourg, 7° Sér., 11 (5): 1-15.

Cockerell T., Miller I., Printz M. 1914. The auditory ossicles of American rodents. Bull. Amer. Mus. Nat. Hist., 33: 347-380.

Cooper E. R. A. 1948. The development of human auditory pathway from the cochlear ganglion to the medial geniculate body. Acta Anat., 5: 99.

Crowe S. J., Hughson W., Witting E. G. 1931. Function of the tensor tympani muscles. Arch. Otolaryngol., 14: 5.

Davies J.L. 1958. The Pinnipedia: an essay in zoogeography. Georg. Rev., 48 (4): 474-493.

Denker A. 1899. Vergleichend-anatomische Untersuchungen über das Gehörorgan der Säugetiere nach Corrosionspräparaten und Knochenschnitten. Leipzig, 1-115.

Denker A. 1901. Zur Anatomie des Gehörorgans der Monotremata. Denkschr. med.-naturwiss. Ges. Jena, 6: 635-662.

Denker A. 1902. Zur Anatomie des Gehörorgans der Cetacea. Anat. Hefte, I. Abt., Wiesbaden, 19 (63): 421-448.

Densert O. 1974. Adrenergic innervation in the rabbit cochlea. Acta Otolaryngol., 78: 345-356.

Densert M., Flock A. 1974. An electron-microscopic study of adrenergic innervation in the cochlea. Acta Otolaryngol., 77: 185-197.

Doran A. H. G. 1878. Morphology of the mammalian Ossicula auditus. Trans. Roy. Linn. Soc. London, 1: 371-497.

Drooglever-Fortuyn A.B. 1914. Cortical cell lamination of the hemispheres of some rodents. Arch. Neurol. Pathol., London, 6: 221-354.

Duckert L. 1974. The morphology of the cochlear aqueduct and periotic duct of the guinea pig – a light and electron microscopic study. Trans. Amer. Acad. Ophthalmol. Otolaryngol., 78: 21-48.

Dudok van Heel W.H. 1962. Sound and Cetacea. Netherland J. Sea Res., 1: 407-507.

Duffield Kulu D. 1972. Evolution and cytogenetics. Mammals of the sea. In: Ridgeway S.H. (Ed.), Springfield. Illinois: Publ. Ch. C. Thomas, 503-527.

Dunn R. A., Morest D. K. 1975. Receptor synapses without synaptic ribbons in the cochlea of the cat. Proc. Natn. Acad. Sci. USA, 72 (9): 3599-3603.

Dyban A.P., Puchkov V.F., Baranov V.S. etc. 1975. Laboratory mammals: The mouse, *Mus musculus*, the rat, *Rattus norvegicus*, the rabbit, *Oryctolagus cuniculus*, the hamster, *Cricetus griseous*. In: Subjects of Developmental Biology. Moscow: Nauka, 505-563 (In Russian).

Echandia E. L. R. 1967. An electron microscopic study on the cochlear innervation. I. The recepto-neural junctions at the outer hair cells. Z. Zellforsch., 78: 30-46.

Edinger T. 1955. Hearing and smell in cetacean history. Monatsh. Psychiatr. Neurol., 129: 37-58.

Engström H. 1951. Microscopic anatomy of the inner ear. Acta Otolaryngol., 40: 5.

Engström H. 1955. The structure of the basilar membrane. Acta Otolaryngol., 44: 531.

Engström H 1958. On the double innervation of the sensory epithelia of the inner ear. Acta Otolaryngol., 49: 109-118.

Engström H., Wersall J. 1958. Structure and innervation of the inner ear sensory epithelia. Int. Rev. Cytol., 7: 535.

Engström H., Ades H., Hawkins J. E. 1963. Cytoarchitecture of the organ of Corti. Acta Otolaryngol., 188: 92-99.

Eschricht D.F. 1849. Zoologische anatomisch-physiologische Untersuchungen über die nordischen Waltiere. Leipzig, 1-206.

Eschweiler R. 1899. Zur vergleichenden Anatomie der Muskeln und der Topographie des Mittelohres verschiedener Säugetiere. Arch. Mikrosk. Anat. u. Entwicklungsgesch., 53: 558-622.

Fay F.H., Rausch V.R., Felts E.T. 1967. Cytogenetical comparison of some pinnipeds (Mammalia: Eutheria). Can. J. Zool., 45 (5): 773-778.

Fernandez G. 1951. The innervation of the cochlea (Guinea pig). Laryngoscope, 61 (12): 1152.

Fick L. 1844. Über das Labyrinth des Elephantes. Arch. Anat. Phys. wiss. Med., Leipzig, Jg. 1844: 431-432.

Firbas W., Welleschik B. 1973. A quantitative study on the spiral ganglion of the Chiroptera. Period. Biol., 75 (1): 67-70.

Fleischer G. 1971. Über Schwingungsmessungen am Skelett des Mittelohres von *Halicore* (Sirenia). Z. Säugetierk., 36: 350-360.

Fleischer G. 1973a. On structure and function of the middle ear in the bottlenosed dolphin (*Tursiops truncatus)*. Proc. 9th Ann. Conf. Biol. Sonar and Diving Mammals. Standford Research Institute Press, 137-179.

Fleischer G. 1973b. Studien am Skelett des Gehörorgans der Säugetiere, einschließlich des Menschen. Säugetierk. Mitt., 21 (2-3): 131-239.

Fleischer G. 1975. Über das spezialisierte Gehörorgan von *Kogia breviceps* (Odontoceti). Z. Säugetierk., 40: 89-102.

Fleischer G. 1976. Über Beziehungen zwischen Hörvermögen und Schadelbau bei Walen. Säugetierk. Mitt., 24 (1): 48-59.
Flock A. 1965. Electron microscopic and electrophysiological studies on the lateral line canal organ. Acta Otolaryngol., Suppl., 99.
Fraser F.C. 1959. Some aquatic adaptations of whales and dolphins. Proc. Roy. Instn. Gr. Brit., 37 (3): 319-333.
Fraser F.C., Purves P.E. 1954. Hearing in cetaceans. Bull. Brit. Mus. Nat. Hist., Zool., 2 (5): 101-114.
Fraser F.C., Purves P.E. 1959. Hearing in whales. Endeavour, 18: 93-98.
Fraser F.C., Purves P.E. 1960. Hearing in cetaceans. Bull. Brit. Mus. Nat. Hist., Zool., 7 (1): 1-140.
Freund L. 1910. Zur Morphologie der äusseren Gehörorgans der Säugetiere. Beitr. Hals-, Nasen- u. Ohrenheilk., Berlin, 3: 1-34.
Frey F. 1911. Vergleichend-anatomische Studien über die Hammer-Amboss-Verbindung der Säuger. Anat. Hefte, I. Abt., Wiesbaden, 44 (132-134): 363-437.
Gacek R. 1961. The efferent cochlear bundle in man. Arch. Otolaryngol., 74: 690.
Gacek R., Rasmussen G.L. 1961. Fiber analysis of the stato-acoustic nerve in guinea pigs, cats and monkeys. Anat. Rec., 139: 455-463.
Gagloeva K.E. 1973. Development of the auditory and vestibular analyzers of man. In: Voprosy morfologii nervnoy sistemy (Problems of the Morphology of the Nervous System). Moscow: Medgiz, 54-60 (In Russian).
Gagloeva K.E. 1976. Topographo-anatomical relations of the peripheral part of the auditory analyzer in a comparative-anatomical aspect. In: Funktsionalnye structurnye osnovy sistemnoy deyatel'nosti i mekhanizmy plastichnosti mozga (Functional and Structural Foundations of Systemic Functioning and the Mechanisms of Brain Plasticity). Moscow, 5: 14-17 (In Russian).
Giraud-Sauveur D. 1969. Recherches biophysiques sur les osselets des cétacés. Mammalia, 33 (2): 285-340.
Gracham Sylvia F. 1967. Seal ears. Science, 155 (3761): 489.
Gray A. 1913. Notes on the comparative anatomy of the middle ear. J. Anat. Physiol., London, 47: 390-413
Gray O. 1953. The chorda tympani. J. Laryngol. a. Otol., 67: 128-138.
Gregory W.K. 1910. The orders of mammals. Bull. Amer. Mus. Nat. Hist., 27: 1-524.
Griffin D.R. 1958. Listening in the Dark. The Acoustic Orientation of Bats and Men. New Haven, Conn.: Yale Univ. Press, 413 p.
Griffin D.R. 1967. Discriminative echolocation by bats. In: Busnel R.G. (Ed.), Animal sonar systems. Biology and Bionics, INRA-NRS, 1: 273-300.
Guild S.R., Crowe S.J., Bunch C.C., Polvogt L.M. 1931. Correlations of differences in the density of innervation of the organ of Corti with differences in the acuity of hearing, including evidence as to the location in the human cochlea of the receptors of certain tones. Acta Otolaryngol., 15: 269-308.
Gutha C. 1961. La région temporale des édentés. Thèse, Le Puy, 191 p.
Haag E. 1933. Über den Bau Labyrinthkapsel einiger Affenarten. Arch. Ohren-, Nasen- u. Kehlkopfheilk., 136: 14-23.

Hagenbach E. 1835. Die Paukenhöhle der Säugetiere. Leipzig, 1-48.
Hall J.G., Lindeman H.H., Mathisen H. 1974. The cochlea and the cochlear nuclei in the seals *Cystophora cristata* and *Phocida groenlandica*. Neurobiology, 2: 105-115.
Hanke H. 1914. Ein Beitrag zur Kenntnis der Anatomie des äusseren und mittleren Ohres der Bartenwale. Jena. Z. Naturwiss., 51: 487-524.
Harrison J.M., Warr W.B., Irving R. 1962. Second order neurons in the acoustic nerve. Science, 138 (3543): 893-895.
Harrison J.M., Irving R. 1964. Nucleus of the trapezoid body: dual afferent innervation. Science, 143: 473-474.
Harrison R. J., King J. E. 1965. Marine mammals. N Y, Hilary House Publ.
Harrison R.J., Tomlinson D.W. 1963. Anatomical and physiological adaptations in diving mammals. In: Carthy J.D. & Duddington C.L. (Eds), Viewpoints in Biology, London: Butterworths, 2, 115-162.
Held H. 1897. Zur Kenntniss der peripheren Gehörleitung. Arch. Anat. Physiol, Anat., 1897: 331-360.
Held H. 1902. Untersuchungen über den feineren Bau des Ohrlabyrinthes der Wirbeltiere. 1. Zur Kenntnis des Cortischen Organs und der übrigen Sinnesapparate der Säugetiere. Abh. Sachs. Ges. Wiss., math.-phys. Kl., 28: 1.
Held H. 1909. Zur Entwicklungsgeschichte des Cortischen Organs und der Macula acustica bei Säugetieren und Vögeln. Abh. Sachs. Ges. Wiss., math.-phys. Kl., 31: 195.
Held H. 1926. Die Cochlea der Säuger und der Vögel, ihre Entwicklung und ihr Bau. In: Bethe's Handb. norm. u. pathol. Physiol., Berlin: Springer, 2 (Receptionsorgane), 467-534.
Henson M. N. 1973. Unusual nerve-fiber distribution in the cochlea of the bat *Pteronotus p. parnellii* (Gray). J. Acoust. Soc. Amer., 53 (6): 1739-1740.
Henson O. 1961. Some morphological and functional aspects of certain structures of the middle ear in bats and insectivores. Univ. Kansas Sci. Bull., 42 (3): 151-265.
Henson O. 1965. The activity and function of the middle ear muscles in echolocation of bats. J. Physiol., 180: 871-887.
Hentzen E. 1970. Histology studies of the normal mucosa in the middle ear, mastoid cavities and eustachian tube. Ann. Otol., Rhtnol. Laryngol., 79 (4): 825-833.
Heptner V.G., Chapsky K.K., Arsenyev V.A., Sokolov V.E. 1976. Mammals of the Soviet Union. Pinnipedia and Cetacea. Moscow: Vysshaya Shkola, 2 (3), 718 p. (In Russian).
Holz K. 1931. Vergleichende anatomische und topographische Studien über das Mittelohr der Säugetiere. Z. Anat. Entwicklungsgesch, Berlin, 94: 757-791.
Home E. 1812. An account of some peculiarities in the structure of the organ of hearing in *Balaena mysticetus* Linnaeus. Philos. Trans. Roy. Soc. London, 102: 83-88.
Hopson J. 1966. The origin of the mammalian middle ear. Amer. Zool., 6: 437-450.
Howell A.B. 1929. Contribution to the comparative anatomy of the eared and earless seals (genera *Zalophus* and *Phoca*). Proc. U.S. Natn. Mus., 73 (15): 1-142.

Howell A.B. 1930. Aquatic mammals. Baltimore, Springfield, 338 p.

Howell A.B. 1933. Aquatic mammals. Their adaptations to life in the water. Baltimore: Springfield.

Huber E. 1934. Anatomical notes on Pinnipedia und Cetacea. Publ. Carnegie Inst. Wash., 447: 105-136.

Hunter J. 1787. Observations on the whales. Philos. Trans. Roy. Soc. London, 57 (16): 306-351.

Hyden H. 1943. Protein metabolism in the nerve cell during growth and function. Acta Physiol. Scand., 6, Suppl. 17.

Hyrtl J. 1844. Vergleichend-anatomische Untersuchungen über das innere Gehörorgan des Menschen und der Säugetiere. Prage, 150 S.

Hyrtl J. 1845. Vergleichend-anatomische Untersuchungen über des innere Gehörorgan des Menschen und der Säugetiere. Prag, 1-139 S.

Hyrtl J. 1948. Zur vergleichenden Anatomie der Trommelhöhle. Abh. math.-naturw. Cl. Kaiserl. Akad. Wiss. Wien, 1: 1-17.

Ichichara T. 1964. Prenatal development of ear plug in ballen whale. Scient. Repts. Whales Res. Ins., 18: 29-48.

Ikeda Y., Yokote T. 1939. Über einige, teils bisher noch unbekannte Eigentümlichkeiten in der Schnecke einer Art von Fledermaus (*Rhinolophus nippon* Temminck). Nag. Igakkai Zai, 17: 1041-1060.

Ivashin M.V., Popov L.A., Stapko A.S. 1972. Marine mammals (guidebook). Moscow: Pishchevaya Promyshlennost', 303 p. (In Russian).

Iwata N. 1927. Über das Labyrinth der Fledermau mit besonderer Berücksichtigung des statischen Apparates. Aichi J. Exp. Med., Nagoya, 114: 41-173.

Jacobs S., Morgane P.J., Zilly, Campbell B. 1964. Analysis of the cranial nerves in the dolphin. Anat. Rec., 148: 2.

Jurato S. 1961. Submicroscopic structure of the membranous labyrinth. II. The epithelium of Corti's organ. Z. Zellforsch., 53: 259.

Jurato S. 1962. Efferent fibers to the sensory cells of Corti's organ. Exp. Cell Res., 27: 162-164.

Kampen P. 1905. Die Tympanalgegend des Säugetierschädels. Morphol. Jahrb., Leipzig, 34: 321-722.

Kappers A.C.U. 1921. Die vergleichende Anatomie des Nervensystems der Wirbeltiere und des Menschen. Haarlem, 1 & 2.

Kappers A. 1941. Kopfplacoden bei Wirbeltieren. Ergebn. Anat. u. Entwicklungsgesh., 33: 370.

Kasuya T. *1973*. Systematic consideration of recent toothed whales based on the morphology of tympano-periotic bone. Sci. Rep. Whales Res. Inst. Tokyo, 25: 1-103.

Keen J.A., Grobblaar C.S. 1941. The comparative anatomy of the tympanic bulla and auditory ossicles, with a note suggesting their function. Trans. Roy. Soc. S. Afr., 28: 307-329.

Keen J. 1942. The comparative anatomy of the promontory region and of the bony cochlea in mammals. Trans. Roy. Soc. S. Afr., 29: 49-58.

Kellerhals M., Engström H., Ades H.W. 1967. Die Morphologie des Ganglion spirale cocleae. Acta Otolaryngol., 67, Suppl. 220.

Kellogg R. 1922. Pinnipeds from Miocene and Pleistocene deposits of California. Univ. Calif. Bull., Dept. Geol. Sci., 13 (4): 23-132.

Kellogg R. 1928. The history of whales. Their adaptation to life in the water-quart. Rev. Biol., 3: 29-76.

Kellogg W.H. 1958. Echo ranging in porpoise. Science, 128 (3330): 982-988.

Kernan J. 1919. Bone conduction of sound in Cetacea and in relation to increased hone conduction in human beings. Laryngoscope, 29: 510-521.

Kiang N., Pfeiffer R., War W.B., Backus A. 1965. Stimulus coding in the cochlear nucleus. Ann. Oto-Rhino-Laryngol., 74 (2): 463-485.

King J. E. 1964. Seals of the World. Brit. Mus. (Nat. Hist.), London, 125-126.

Klaauw J. v. D. 1922. Über die Entwicklung des Entotympanicums. Diss. Thesis, Leiden, 40 S.

Klaauw J. 1930. On mammalian auditory bullae showing an indistincly complex structure in the adult. J. Mammal., 11: 55-60.

Kleinenberg S.E. 1958. To the problem of the origins of Cetacea. Dokl. Akad. Nauk SSSR, 122 (5): 950-952 (In Russian).

Kleinenberg S.E., Yablokov A.V., Belkovitch V.M., Tarasevitch M.N. 1964. The White Whale. Moscow: Nauka, 456 p. (In Russian).

Klumov S.K 1934. To the question of the distribution and harvest of the white whale in the White Sea. Rybn. Khoz. SSSR, 4 (In Russian).

Klumov S.K. 1936. White whale distribution (*Delphinapterus leucas* Pall.) in the European and Asian North. Polyarn. Komiss. AN SSSR, 27 (In Russian).

Kobrak G.G. 1963. The Middle Ear. Moscow: Medgiz (In Russian).

Köllicker A. 1850. Mikroskopische Anatomie. Leipzig, 2 (1): 1-183.

Köllicker A. 1861. Der embryonale Schnecken-Kanal und Beziehungen zu den Theilen der fertigen Cochlea. Würzburger naturwiss. Zeitschrift., 2: 1.

Kohlloffel L.U.E. 1974. Recordings from spiral ganglion neurons. In: Zwicker E. & Terhard E. (Eds), Psychophysical Models and Physiological Facts in Hearing. Berlin: Springer, 1 & 2.

Koichev K.A. 1975. The structure of the neurons of the spiral and vestibular ganglia and their change under the conditions of adequate effects. Arkhiv Anat., Gistol. i Embriol., 68 (2): 28-32 (In Russian).

Kolmer W. 1907. Beiträge zur Kenntnis des feineren Baues des Gehörorganes mit besonderer Berücksichtigung der Haussäugetiere. Arch. microsk. Anat., 70: 695-707.

Kolmer W. 1908. Über das häutige Labyrinth des Delphins. Anat. Anz., Jena, 32: 295-300.

Kolmer W. 1909. Histologische Studien am Labyrinth mit besonderer Berücksichtigung des Menschen, der Affen und Halbaffen. Arch. mikrosk. Anat. u. Entwicklungsgesch., Bonn, 74: 259-310.

Kolmer W. 1913. Studien am Labyrinth von Insectivoren. SB. Akad. Wiss., Wien, 122 (Abt. 3): 29-52.

Kolmer W. 1927. Gehörorgan. In: Mollendorf's Handbuch der microskopischen Anatomie des Menschen. Haut- und Sinnesorgane. Berlin: Springer, 3: 250-478.

Kolmer W., Eisinger O. 1923. Über das Gehörorgan von *Elephas indicus*. Monatsschr. Ohrenheilk. u. Laryngo-Rhinol., 57: 3, 1-3.

Konstantinov A.I., Movchan V.N. 1985. Sounds in Animal Life. Leningrad: LGU, 303 p. (In Russian).

Krause H. 1921. Mikroskopische Anatomie der Wirbeltiere in Einzeldarstellungen. I. Säugetiere. Berlin & Leipzig, 173 S.

Krylov V.I. 1965. Age estimation, the rate of growth and age structure analysis in carnages of the Pacific ocean walrus. In: Marine Mammals, Moscow: Nauka (In Russian).

Krylov V.I., Fedoseyev G.A., Shustov A.P. 1964. Pinnipeds of the Far East. Moscow: Pishchevaya promyshlennost (In Russian).

Kükenthal W. 1922. Zur Stammesgeschichte der Wale. Sitzungsher. Preuss. Akad. Wiss., Phys. Math., 1922: 72-96.

Künze W., Kietz H. 1949. Über Hörempfindungen in Ultraschallgebiet bei Knochenleitung. Arch. Ohren-, Nasen- u. Kehlkopfheilk., 135: 683-692

Kupin A.G. 1986. The ways of sound conduction in *Pusa caspica*. In: Zemsky V.A. (Ed.), Marine Mammals. Arkhangelsk, 357-358 (In Russian).

Lichtensteiger W., Spoendlin H. 1967. Adrenergic pathways to the cochlea of the cat. Life Sci., 6 (15): 1639-1645.

Lillie D.G. 1910. Cetacea, British Antarctic ("Terra nova") Expedition, Zoology, 1915, 1, 3: 85-124.

Lilly J. 1967. The Mind of the Dolphin. NY.

Lipatov N.V. 1972. Functional foundations of underwater hearing in mammals. Abstr. of the Vth All-Union Conf. on Marine Mammal Research. Makhachkala, 2: 137-140 (In Russian).

Lipatov N.V., Solntseva G.N. 1972. Some peculiarities of bio-mechanics of the middle ear of the dolphins. Abstr. of the Vth All-Union Conf. on Marine Mammal Research. Makhachkala, 2: 140-142 (In Russian).

Lorente de No R. 1927. Untersuchungen über die Anatomie und Physiologie des Ohrlabyrinthes und des Nervus octavus. Monatsschr. Ohrenheilk., 61: 1300.

Lorente de No R. 1937. The sensory endings in the cochlea. Laryngoscope, 47: 373-377.

Lorente de No R. 1976. Some unresolved problems concerning the cochlear nerve. Ann. Otol., Rhinol. a. Laryngol., 85 (6): 3.

Markov V.I., Tarchevskaya V.L. 1978. To an evaluation of the possibilities of the sound generation system in the bottlenose dolphin. Morskie mlekopitayuschiye. Resultaty i metody issledovanii (Marine Mammals. Results and Methods of Study). Moscow: Nauka, 142-156 (In Russian).

Mattew W.D. 1939. Climate and evolution. 2nd ed. Spec. Publ., NY Acad. Sci., 1: 1-233.

Maw A.R. 1973. Further morphological modifications of the concept of efferent cochlear innervations at an ultrastructural level. J. Laryngol. a. Otol., 77: 691-738.

McCormick J.G., Wever E. G., Palin J., Ridgway S.H. 1970. Sound conduction in the dolphin ear. J. Acoust. Soc. Amer, 48 (6, part II): 1418-1428.

McLaren J. A. 1960. Are the Pinnipedia biphyletic? Syst. Zool., 9 (1): 18-28.

Mednikov B.M., Shubina E.A. 1977. About the phylogeny of cetaceans. Priroda, 7: 47-48 (In Russian).

Meyer M. 1933. Die normale Anatomie der Labyrinthkapsel (Untersuchungen am Menschen und Laboratoriumstieren). Z. Hals-, Nasen- u. Ohrenheilk., Berlin, 34: 3-72.

Mohl B. 1967. Frequency discrimination in the common seal and discussion of the concept of upper hearing limit. In: Underwater Acoustics. NY: Plenum Press, 2: 43-54.

Mohl B. 1968. Auditory sensitivity of the common seal in air and water. J. Acoust. Res., 1: 27-38.

Montagna W., Noback C.R., Zak F.G. 1948. Pigment, lipids and other substances in the glands of the external auditory meatus of man. Amer. J. Anat., 83: 409-435.

Morrison D., Schindler R.A., Wersall J. 1975. A quantitative analyses of the organ of Corti in guinea pig. Acta Otolaryngol., 79: 1-2, 11-23.

Moskowitz N., Liu J.-Ch. 1969. Projections of the cochlear nerve fibers of the squirrel monkey (*Saimiri sciureus*). Anat. Rec., 163: 233-245.

Moskowitz N., Liu J.-Ch. 1972. Central projections of the spiral ganglion of the squirrel monkey. J. Exp. Neurol., 144 (3): 335-344.

Murariu D. 1976. Les glandes tégumantaires de certains insectivores de Roumanie. Anatomie, histotogie et histochimie. Trav. Hst. Nat. Gr. Antipa, 17: 387-413.

Murie J. 1873. On the organisation of the Caaing Whale (*Globicephalus melas*). Trans. Roy. Zool. Soc. London, 8 (4): 235-301.

Mygind S. 1945. Experimental histological studies on the labyrinth. Acta Otolaryngol., 33: 75-116.

Nakai Y., Igarashi M. 1974. Distribution of the crossed olivo-cochlear bundle terminals in the squirrel monkey cochlea. Acta Otolaryngol., 77 (6): 393-404.

Nielsen S.W. 1953. Glands of the canine skin — morphology and distribution. Amer. J. Vet. Res., 14 (52): 448-454.

Norris K.S. 1964. Some problems of echolocation in Cetacea. In: Marine bioacoustics/ Ed. by W. N. Tavolga. N. Y.: Pergamon Press., pp. 317-336.

Norris K.S. 1968. The evolution of acoustic mechanisms in Odontocete Cetaceans.Peabody Mus. Centenary Celebr. Vol., Yale Univ., New Haven.

Odend'hal S., Poulter T. C. 1966. Pressure regulation in the middle ear cavity of the sea lions: A possible mechanism. Science, 153: 768-769.

Oelschläger H.A. 1990. Evolutionary morphology and acoustics in the dolphin skull. In: Thomas J.A. & Kastelein R.A. (Eds), Sensory Abilities of Cetaceans. NY: Plenum Press, 137-162.

Ognev S.I. 1935. Wild animals of the USSR and adjacent countries. Carnivores and pinnipeds. Moscow, 3, 752 p. (In Russian).

Otis E.M., Brent R. 1954. Equivalent ages in mouse and human embryos. Anat. Rec., 120: 33-63

Paradiesgarten A., Spoendlin H. 1976. The unmyelinated nerve fibers of the cochlea. Acta Otolaryngol., 82 (3-4): 157-164.
Perkins R.E., Morest D.K. 1975. A study of cochlear innervation patterns in cats and rats with the Golgi method and Nomarski optics. J. Exp. Neurol., 163 (2): 129-158.
Perlov A.S. 1970. Distribution and numbers of Steller sea lions on the rookerys of the Kurile Islands. Izvestiya TINRO, 70 (In Russian).
Perlov A.S. 1972. The terms of puberty onset in the Steller sea lions. Trudy VNIRO, 82 (In Russian).
Pilleri G. *1987*. The Cetacea of the Italian Pliocene. Inst. Brain Anat., Univ. of Berne (Switzerland), 160 p.
Platzer W. 1960. Über eine dritte Spirallamelle in der Cochlea und zur Anatomie aller drei Spiralblätter unter besonderer Berücksichtigung der Schnecke von *Talpa europaea*. Acta Anat., 43: 62-95.
Platzer W. 1964. Zur vergleichenden Anatomie der Cochlea bei *Talpa europaea, Sorex araneus* und *Sorex alpinus*. Anat.Anz., Jena, 115: 117-118.
Popov L.A. 1959. Walrus distribution in the western sector of the Soviet Arctic in the summer-autumn period. Inform. Bull. VNIRO, 7 (In Russian).
Popov L.A. 1960. Material of walrus reproduction biology in the Laptev Sea. Byull. Mosk. Obshch. Ispyt. Prir., Biol., 65: 2 (In Russian).
Potelov V.A. 1966. Distribution of the beaded seal in the White, Barents and Kara seas. Abstr. 3rd All-Union Conf. on the study of marine mammals. Moscow & Leningrad: Nauka (In Russian).
Potelov V.A. 1967. About the terms of beaded seal molting. Materials of Fishery Studies of the Northern Basin, Murmansk, 9 (In Russian).
Poulter T.C., Thomas C. 1969. Sonar of penguins and fur seals. Proc. Calif. Acad. Sci., 36 (13): 363-380.
Provorov N.B. 1957. To the question of the biology and harvest of white whale in the Kandalaksha Bay of the White Sea. Trudy Arktich. Inst., 205 (In Russian).
Purves P.E. 1955. The wax plug in the external auditory meatus of the Mysticeti. Discovery Repts, 27: 293-302.
Purves P.E. 1958. The use of the ear plug in age determination in whales. Proc. 15th Int. Congr. Zool., Sect. 3: 35.
Purves P.E. 1966. Anatomy and physiology of the outer and middle ear in cetaceans. In: Norris K.S. (Ed.), Whales, Dolphins and Porpoises. Berkeley & Los Angeles: Univ. Calif. Press, 320-380.
Purves P.E. 1967. Anatomical and experimental observations on the cetacean sonar system. In: Busnel R.G. (Ed.), Animal sonar systems. Biology and Bionics. INRA-CNRS, 1: 197-270.
Purves P. E., Utrecht W. L. 1964. The anatomy and function of the ear of the bottlenosed dolphin, *Tursiops truncatus*. Beaufortia, 9: 241-255.
Pye A. 1966a. The Megachiroptera and Vespertilionoidea of the Microchiroptera. The structure of the cochlea in Chiroptera. J. Morphol., 118: 495-507.
Pye A. 1966b. The structure of the cochlea in Chiroptera. 1. Microchiroptera and Vespertilionoidea of the Microchiroptera. J. Morphol., 119: 101-109.

Pye A. 1967. The structure of the cochlea in Chiroptera. III. Microchiroptera. Phyllostomatoidea. J. Morphol., 121: 241-254.

Pye A. 1970. The structure of the cochlea in Chiroptera. A selection of Microchiroptera from Africa. J. Zool., London, 162: 335-343.

Pye A. 1973. The structure of the cochlea in Chiroptera from Africa. Period. Biol., 75: 83-87.

Pye J.D. 1968. Animal sonar in air. Ultrasonics, 6: 32-38.

Ragovitza E.G. 1903. Résultats du voyage S. Y. Belgica. Zoologie. Cétacés. Rapp. Sci., 9: 1-142.

Ramprashad F. 1975. Aquatic adaptations in the ear of the harp seal, *Pagophilus groenlandicus* (Erxleben, 1777). In: Ronald R. & Manswield A.W. (Eds), Biology of the Seal. Rapp. et Proc.-verb. Réun. Cons. Int. Explor. Mer, 169: 102-111.

Ramprashad F. 1976. Population and density of the bipolar ganglion cells in the cochlea of the harp seal (*Pagophilus groenladicus* Erxleben, 1777). Can. J. Zool., 54 (11): 1918-1926.

Ramprashad F., Coray S., Ronald K. 1971. The harp seal, *Pagophilus groenlandicus* (Erxleben, 1777). XIII. The gross microscopic structure of the auditory meatus. Can. J. Zool., 49 (2): 241-248.

Ramprashad F., Corey S., Ronald K. 1973. The harp seal, *Pagophilus groenlandicus* (Erxleben, 1777). XIV. The gross and microscopic structure of the middle ear. Can. J. Zool., 51 (6): 589-600.

Ramprashad F., Ronald K., Geraci J., Smith T.G. 1976. A comparative study of surface preparations of the organ of Corti of the harp seal (*Pagophilus groenlandicus* Erxleben, 1777) and the ringed seal (*Pusa hispida*). I. Sensory cell population and density. Can. J. Zool., 54 (1): 1-9.

Rasmussen G. L. 1942. An efferent cochlear bundle. Anat. Rec., 82: 44.

Rasmussen G.L. 1946. The olivary peduncle and other fiber projections of the superior olivary complex. J. Comp. Neurol., 84: 141-219.

Rasmussen G.L. 1953. Further observations of the efferent cochlear bundle. J. Comp. Neurol., 99: 61-74.

Rasmussen G.L. 1960. Efferent fibers of the cochlear nerve and cochlear nucleus. In: Neural Mechanisms of the Auditory and Vestibular Systems. In: Rasmussen G.L. & Windle W.F. (Eds), Springfield, III.: Ch. C. Thomas, 105-115.

Ray C.A. Geography of phocid evolution. Syst. Zool., 25 (4): 391-406.

Repenning C.A. 1975. Otarioid evolution. In: Ronald R. & Manswield A.W. (Eds), Biology of the Seal. Rapp. et Proc.-verb. Réun. Cons. Int. Explor. Mer, 169: 27-33.

Repenning C.A. 1976. Adaptive evolution of sea lions and walruses. Syst. Zool., 25 (4): 375-390.

Retzius G. 1881. Das Gehörorgan der Wirbeltiere. 1. Das Gehörorgan der Fische und Amphibien. Stockholm.

Retzius G. 1884. Das Gehörorgan der Wirbeltiere. II. Das Gehörorgan der Reptilien, der Vögel und der Säugetiere. Stockholm.

Reysenbach de Haan F.W. 1957a. De gehoorzin van Cetacea. Vakbl. biol., 37 (8): 117-127.

Reysenbach de Haan F.W. 1957b. Hearing in whales. Acta Otolaryngol., 134: 1-114.

Reysenbach de Haan F.W. 1966. Listening underwater: thoughts on sound and cetacean hearing: whales, dolphins and porpoises. Berkeley & Los Angeles: Univ. Calif. Press, 583-596.

Reznikov A.E. 1970. On ultrasonoscopy in dolphins. Abstr. 23rd Sci.-techn. Conf. LIAP. Leningrad (In Russian)

Rice D.W., Scheffer, V.B. 1968. A list of the marine mammals of the world. Univ. St Pish. Wild Serv. Spec. Sci. Rept. Fish., 579: 1-16.

Romer A. Sh. 1939. Paleontology of Vertebrates. Moscow & Leningrad: GONTI, 416 p. (In Russian).

Rosenbluth J. 1962. The fine structure of the acoustic ganglia in the rat. J. Cell Biol., 12: 329-359.

Rosenthal F. 1825. Über die Sinnesorgane der Seehunde. Nova Acta Leopold., 12: 673-694.

Ross M.D., Burkel W. 1973. Multipolar neurons in the spiral ganglion of the rat. Acta Otolaryngol., 76: 381-394.

Rossi G., Cortesina G. 1965. The efferent innervation of the inner ear. A historical-bibliographical survey. Laryngoscope, 75: 212-235.

Sando I. 1965. The anatomical interrelationships of the cochlear nerve fibers. Acta Otolaryngol., 59: 417-436.

Sarich V.M. 1969. Pinniped phylogeny. Syst. Zool., 18 (4): 416-422.

Schaffer J. 1940. Die Hautdrüsenorgane der Säugetiere mit besonderer Berücksichtigung ihres histologischen Aufbaues und Bemerkungen über die Proktodaldrüsen. Berlin: Urban Schwarzenberg.

Scheffer V.B. 1958. Seals, Sea Lions and Walruses. A Review of the Pinnipedia. Stanford Univ. Press, Calif., 179 p.

Schewill W.E., Lawrence B. 1953. Auditory response of the bottlenose porpoise *Tursiops truncatus* to frequencies above 100 kc. J. Exp. Zool., 124 (1): 147-165.

Schewill W.E., Watkins W.A., Ray C. 1963. Underwater sounds of pinnipeds. Science, 141 (3575): 50-53.

Schiefferdecker P. 1917. Die Hautdrüsen des Menschen und Säugetiere. Biol. Zentralbl., 37: 11-12.

Schmalhausen I.I. 1939. Pathways and regularities of the evolutionary process. Moscow & Leningrad: AN SSSR, 231 p. (In Russian).

Schmidt G.A. 1968. Types of embryogeny and their adaptive roles. Moscow: Nauka, 231 p. (In Russian).

Schuknecht H.F. 1953. Techniques for study of cochlear function and pathology in experimental animals. Arch. Otolaryngol., 58: 377-397.

Seal U.S., Erickson A.W., Siniff D.B., Hofman R.J. 1970. Biochemical, population genetic, phylogenetic and cytological studies of Antarctic seal species. In: Symposium on Antarctic Ice and Water Masses. Tokyo, 77-95.

Segall W. 1969. The auditory ossicles (malleus, incus) and their relationships to the tympanie in marsupials. Acta Anat., 73 (2): 176-191.

Severtsov A.N. 1939. Morphological regularities of evolution. Collected works. Moscow: AN SSSR (In Russian).

Shmelev L.A. 1973. Modelling the mechanisms of binaural hearing. PhD Thesis, Moscow (In Russian).

Simkin G.N. 1977. Acoustic orientation and contact of mammals. Abstr. of doctoral thesis. M., pp. 1-37 (In Russian).

Simpson G. 1945. The principles of classification and a classification of mammals. Bull. Amer. Mus. Nat. Hist., 85: 1-350.

Slijper E.J. 1936. Die Cetacea. Capita Zool., 7 (1-2): 1-591.

Smith C.A., Rasmussen G.L. 1968. Degeneration in the efferent nerve endings in the cochlea after axonal section. J. Cell. Biol., 26 (1): 63-77.

Smith C.A., Sjostrand F.S. 1961a. A sinaptic structure in the hair cells of the guinea pig cochlea. J. Ultrastruct. Res., 5: 184-192.

Smith C.A., Sjostrand F. A. 1961b. Structure of the nerve endings on the external hair cells of the guinea pig cochlea as studied by serial section. J. Ultrastruct. Res., 5: 523-556.

Sokolov V.E. 1973. Teguments of Mammals. Moscow: Nauka, 486 p. (In Russian).

Solntseva G.N. 1969. Comparative-anatomical and histological peculiarities of inner ear structure of some dolphins. Abstr. of 15th All-Union Conf. on Marine Mammal Research. Moscow, 54-57 (In Russian).

Solntseva G.N. 1971. Comparative-anatomical and histological peculiarities of outer and inner ear structures of some dolphins. In: Marine Mammal Research, Kaliningrad, 22: 237 (In Russian).

Solntseva G.N. 1972a. Features of afferent innervation of the external auditory meatus in the common dolphin. In: Application of Mathematical Methods and Computer Techniques to Medicine and Bionics. Conf. of Young Scientists. Leningrad, 63-64 (In Russian).

Solntseva G.N. 1972b. Comparative-anatomical peculiarities of middle ear structure in terrestrial, semi-aquatic and aquatic mammals. Abstr. 5th All-Union Conf. Marine Mammal Research. Makhachkala, 2: 216-220 (In Russian).

Solntseva G.N. 1972c. Morpho-functional peculiarities of the peripheral part of the auditory analyzer in the harp seal and nutria. Abstr. 5th All-Union Conf. Marine Mammal Research. Makhachkala, 2: 220-225 (In Russian).

Solntseva G.N. 1973a. Morpho-functional peculiarities of the outer ear of terrestrial, semi-aquatic and aquatic mammals. Abstr. 8th All-Union Acoustic Conf., Moscow, 127-128 (In Russian).

Solntseva G.N. 1973b. Biomechanical features of the middle ear in terrestrial, semi-aquatic and aquatic mammals. Abstr. 8th All-Union Acoustic Conf., Moscow, 128 (In Russian).

Solntseva G.N. 1973c. Morphological and biomechanic features of the middle ear of the Caspian seal (*Pusa caspica*). In: Dokl. sessii KaspNIRKH po rabotam za 1972. Astrakhan', 74-76 (In Russian).

Solntseva G.N. 1973d. Morpho-functional peculiarities of the outer ear of the Caspian seal (*Pusa caspica*). In: Dokl. sessii KaspNIRKH po rabotam za 1972. Astrakhan', 77-78 (In Russian).
Solntseva G.N. 1973e. Morpho-functional peculiarities of the outer ear for terrestrial, semi-aquatic, and aquatic mammals. Proc. 8th All-Union Acoustic Conf., Moscow, 25-28 (In Russian).
Solntseva G.N. 1973f. Biomechanical features of the middle ear for the terrestrial, semi-aquatic, and aquatic mammals Proc. 8th All-Union Acoustic Conf., Moscow, 29-32 (In Russian).
Solntseva G.N. 1973g .Morpho-functional peculiarities of the peripheral part of the auditory analyzer in terrestrial, semi-aquatic, and aquatic mammals. PhD Thesis, Moscow (In Russian).
Solntseva G.N. 1974. Auditory organ functioning under water. Priroda, 10: 21-25 (In Russian).
Solntseva G.N. 1975a. Morpho-functional peculiarities of the auditory organ in terrestrial, semi-aquatic and aquatic mammals. Zool. Zhurn., 44 (10): 1529-1539 (In Russian).
Solntseva G.N. 1975b. Tasks and perspectives of the research on sound propagation pathways in cetaceans. In: Marine Mammals, Kiev, 2: 93-95 (In Russian).
Solntseva G.N. 1977. Adaptive features of the cochlea of some mammals. In: Abstr. 1st All-Union Conf. on Research on the Mechanisms of Adaptation of Living Organisms as Influenced by Habitat Factors. Leningrad, 130-131 (In Russian).
Solntseva G.N. 1978. Features of auditory organ structure of some echolocating mammals. In: Abstr. 2nd All-Union Conf. Theriol. Soc., Moscow (In Russian).
Solntseva G.N., Chernova O.F. 1978. Morpho-functional analysis of auricular glands of some mammals in relation to the peculiarities of their ecology. In: Marine Mammals. Abstr. 7th All-Union Conf., Moscow, 305-307 (In Russian).
Solntseva G.N., Chernova O.F. 1980. Morpho-functional analysis of auricular glands in some mammals. Zool. Zhurn., 59 (8): 1230-1240 (In Russian).
Solntseva G.N. 1982. Development of the peripheral auditory analyzer of toothed cetaceans at different stages of embryogeny. In: Mammals of the USSR. Abstr. 3rd Conf. All-Union Theriol. Soc., Moscow, 2: 79 (In Russian).
Solntseva G.N. 1983. Early embryogenesis of the peripheral part of the auditory analyzer in a representative of toothed cetaceans, *Stenella attenuata*. Ontogenesis, 14 (3): 312-318 (In Russian).
Solntseva G.N. 1985a. Early embryogenesis of the peripheral part of the auditory analyzer in a representative of baleen cetaceans, *Balaenoptera acutorostrata*. Dokl. Akad. Nauk SSSR, 280 (6): 1428-1432 (In Russian).
Solntseva G.N. 1985b. Formation of the peripheral part of the auditory analyzer in a representative of true seals, the bearded seal *Erignathus barbatus*. Dokl. Akad. Nauk SSSR, 285 (6): 1504-1508 (In Russian).

Solntseva G.N. 1986a. Early embryogenesis of the peripheral part of the auditory analyzer of some pinnipeds. In: Marine Mammals. Abstr. 9th All-Union Conf. on the Study, Conservation and Rational Use of Marine Mammals. Arkhangelsk, 356-357 (In Russian).

Solntseva G.N. 1986b. Early embryogenesis of the peripheral part of the auditory analyzer of the walrus, *Odobenus rosmarus divergens*. Dokl. Akad. Nauk SSSR, 288 (4): 984-988 (In Russian).

Solntseva G.N. 1986c. Regularities in evolutionary change in the course of development of the peripheral auditory analyzer in some pinnipeds and cetaceans. Abstr. Conf. Trends in Living Organism Ontogeny, Moscow, 2: 129 (in Russian).

Solntseva G.N. 1987a. Directions in the evolutionary transformations of the peripheral part of the auditory analyzer in mammals with different ecologies. Zhurn. Obshch. Biol. (J. General Biol.), 158 (3): 403-410 (In Russian).

Solntseva G.N. 1987b. Structural-functional organization of the peripheral part of the auditory analyzer of the northern fur seal, *Callorchinus ursinus*. Dokl. Akad. Nauk SSSR, 293 (2): 450-458 (In Russian).

Solntseva G.N. 1988a. Morphological adaptations of the peripheral part of the auditory analyzer in echolocating cetaceans during pre- and postnatal development. Dokl. Akad. Nauk SSSR, 298 (1): 219-224 (In Russian).

Solntseva G.N. 1988b. Formation of the peripheral part of the auditory analyzer in a representative of true seals, the ringed seal *Pusa hispida*. Dokl. Akad. Nauk SSSR, 302 (6): 1489-1493 (In Russian).

Solntseva G. N. 1988c. Evolutionary transformations of the peripheral acoustic analyzer in mammals with different ecologies. Abstr. Int. Symp. Evolutionary Biology: Theory and Practice. Plzen, Czechoslovakia, 49.

Solntseva G.N. 1989. Formation of structural adaptions of the peripheral part of the auditory analyzer in aquatic and sound-locating mammals in ontogenesis. Abstr. 5th Int. Theriol. Congr., Roma, Italy, 394.

Solntseva G.N. 1990a. Regularities of evolutionary transformations and ontogenetic development of the peripheral part of the auditory analyzer in mammals with different ecologies. DrSc Thesis, Moscow (In Russian).

Solntseva G. N. 1990b. Evolutionary transformations of the peripheral acoustic analyzer in mammals with different ecologies. In: Evolutionary Biology: Theory and Practice. Proc. Int. Symp. Evolutionary Biology, Plzen, 22-26.08.1988. Prague: Czechoslov. Acad. Sci. Publ., 63-69.

Solntseva G.N. 1990c. Formation of an adaptive structure of the peripheral part of the auditory anatyzer in aquatic echolocating mammats during ontogenesis. In: Thomas J.A. & Kastelein R.A. (Eds), Sensory Abitities of Cetaceans. NY: Plenum Press, 363-383.

Solntseva G.N. 1992. Prenatal development of the peripheral part of the auditory system in mammals of different ecologies. In: Thomas J.A., Kastelein R.A. & Supin A. (Eds), Marine Mammal Sensory Systems. NY: Plenum Press, 179-195.

Solntseva G.N. 1993. A comparative-morphological analysis of development of the auditory organs in mammals. Ontogenesis, 24 (5): 62-79 (In Russian).

Solntseva G.N. 1995. The auditory organ of mammals in relation to the acoustic properties of the habitat and frequency tuning. In: Sensory Systems of Aquatic Mammals. Woerden, Netherlands: De Spil, 455-475.

Solntseva G.N. 1996. Development of the vestibular system in the aquatic mammal *Stenella attenuata*. Dokl. Biol. Nauk, 347: 146-149.

Solntseva G.N. 1997a. Early embryogenesis of the vestibular apparatus in the bearded seal (*Erignathus barbatus*) from the family Phocidae (true seals). Dokl. Biol. Nauk, 354: 260-263.

Solntseva G.N. 1997b. Prenatal development of the vestibular apparatus of the walrus (Pinnipedia: Odobenidae – *Odobenus rosmarus divergens*). Dokl. Biol. Nauk, 355: 414-417.

Solntseva G.N. 1997c. Structural-functional organization of the peripheral part of the auditory analyzer of the bottlenose dolphin (Odontoceti: *Tursiops truncatus*) in pre- and postnatal development. In: Sokolov V.E. & Romanenko E.V. (Eds), The Black Sea Bottlenose Dolphin. Moscow: Nauka, 420-441 (In Russian).

Solntseva G.N. 1998a Development of the auditory and vestibular systems of Steller 's sea lion, *Eumetopias jubatus,* a representative of eared seals. Dokl. Biol. Nauk, 358: 414-417.

Solntseva G.N. 1998b. Development of the vestibular apparatus in the little piked whale *Balaenoptera acutorostrata* (Cetacea: Mysticeti): Comparison with auditory apparatus development. Dokl. Biol. Nauk, 361: 337-340.

Solntseva G.N. 1998c. Structural-functional organization of the peripheral part of the auditory system of the northern fur seal in pre-and postnatal ontogenesis. In: Sokolov V.E. (Ed.), The Northern Fur Seal: Systematics, Morphology, Ecology, Behaviour. Moscow: Nauka, 303-319 (In Russian).

Solntseva G.N. 1999a. Comparison of the development of auditory and vestibular structures in a representative of toothed whales, the whale (*Delphinapterus leucas,* Cetacea: Odontoceti). Dokl. Biol. Nauk, 364: 78- 81.

Solntseva G.N. 1999b. Comparative analysis of the auditory and vestibular organ development in the ringed seal *Pusa hispida* (Pinnipedia: Phocidae). Dokl. Biol. Nauk, 365: 178-181.

Solntseva G.N. 1999c. Development of the auditory organ in terrestrial, semi-aquatic, and aquatic mammals. J. Aquatic Mammals, 25 (3): 135-148.

Solntseva G.N. 2000a. Prenatal development of the labyrinth in cetaceans. In: Marine Mammals of the Holarctic. Arkhangelsk: Pravda Severa, 359-363 (In Russian).

Solntseva G.N. 2000b. Vestibular apparatus development in the mammals with different ecologies. Dokl. Akad. Nauk SSSR, 371 (4): 569-573 (In Russian).

Solntseva G.N. 2001a. Structures, functions, and parameters of the bulla tympanica in the dolphin *Tursiops truncatus* in a three-dimensional presentation. Dokl. Biol. Nauk, 381: 551-554.

Solntseva G.N. 2001b. Comparative analysis of vestibular apparatus development in mammals with different ecologies. Ontogenesis, 32 (3): 212-224 (In Russian).

Solntseva G.N. 2001c. Structural-functional organization of the peripheral part of the auditory system and vestibular apparatus in pre- and postnatal development in walrus (Pinnipedia: *Odobenus rosmarus divergens*). In: Pavlov D.S. & Bychkov V.A. (Eds), The Walrus. Moscow: Nauka, 13-25 (In Russian).
Solntseva G.N. 2002a. Bulla tympanica of the dolphins *Tursiops truncatus* and *Delphinus delphis* in a three-dimensional presentation. In: Marine Mammals of the Holarctic. Irkutsk, 238-239 (In Russian).
Solntseva G.N. 2002b. Early embryogenesis of the vestibular apparatus in mammals with different ecologies. J. Aquatic Mammals, 28 (2): 159-169.
Solntseva G.N. 2006. The Auditory Organ of Marine Mammals. Moscow: KMK Sci. Press, 202 p. (In Russian).
Spoendlin H. 1957. Elektronenmikroskopische Untersuchungen am Cortischen Organ des Meerschweines. Prakt. Oto-Rhino-Laryngol., 10 (1-2): 192.
Spoendlin H. 1966. The organisation of the cochlear receptor. Basel, Karger.
Spoendlin H. 1969. Innervation patterns in the organ of Corti of the cat. Acta Otolaryngol., 67: 239.
Spoendlin H. 1972. Innervation densities of the cochlea. Acta Otolaryngol., 73: 235-247.
Spoendlin H. 1975. Retrograde degeneration of the cochlear nerve. Acta Otolaryngol., 79: 266-275.
Stosman I.M. 1978. Electric reactions of the auditory structures of sea-swine brainstem under mono- and binaural irritation by ultra-sound stimuli. In: Marine Mammals. Abstr. 7th All-Union Conf., Moscow, 308-309 (In Russian).
Sukhoruchenko M.N. 1971. The upper range of hearing in dolphin according to frequency. Trudy Inst. Akustiki, 17: 54-59 (In Russian).
Supin A.Ya., Sukhoruchenko M.N. 1970. Determination of the thresholds of audibility of dolphins by the method of conditioned galvanic skin reflex. Trudy Inst. Akustiki, 12: 194-199 (In Russian).
Supin A.Ya., Sukhoruchenko M.N. 1971. Study of the auditory sensitivity in dolphins. Abstr. 7th Acoustic Conf., Leningrad, 14 (In Russian).
Tandler J. 1899. Über ein Corpus cavernosum tympanicum beim Seehund. Monatsschr. Ohrenheilk., 33 (10): 437-440.
Teger I. 1906. Das Gehörgen. In: Ellenberger's Handb. vergl. mikroskop. Anat. der Haustiere, Berlin.
Terhune J.M. 1974. Directional hearing of a harbor seal in air and water. J. Acoust. Soc. Amer., 56 (6): 1862-1865.
Terhune J.M., Ronald K. 1975. Underwater hearing sensitivity of two ringed seals (*Pusa hispida*). Can. J. Zool., 53 (3): 227-266.
Thomas J.A., Pawloski J.L., Au W.W.L. 1990. Masked hearing abilities in the false killer whale (*Pseudorca crassidens*). In: Thomas J.A. & Kastelein R.A. (Eds), Sensory Abilities of Cetaceans. NY: Plenum Press, 395-404.
Titova L.K. 1968. Development of Receptor Structures in the Inner Ear of Vertebrates. Leningrad: Nauka, 217 p. (In Russian).
Tomilin A.G. 1957.. Wild Animals of the USSR and Neighbouring Countries. 9. Cetaceans. Moscow: AN SSSR, pp.14-756 (In Russian).

Tran-Ba-Hay P., Ohresser M. 1976. Anatomie macroscopique et ultra-microscopique de la cochlée. Sa vascularisation, son innervation. Rev. Med., 17: 1423-1438.

Trevisi M., Pagani F. A., Sirigu P. 1972. Comparative research on the cochlear ganglion in various species of mammals. Arch. Sci. Biol, 56 (2-4): 91-96.

Trevisi M. 1975. Sul ganglio spirale del Corti: rassegna delle esperienza compiute e prospetive di ulteriori ticerche. Arcispedale S. Ann. Ferrara, 28 (3-4): 195-205.

Turkewitsch B. 1934. Vergleichend anatomische Untersuchungen über das knochende innere Gehörorgan der Säugetiere (Cochlea). Z. wiss. Zool., 145: 39-65.

Turkewitsch B.G. 1935. Beiträge zur vergleichenden Anatomie der Cochlea der Säugetiere (Aquaeductus cochlea et Fenestra cochlea). Anat. Anz., 78: 155-166.

Van Valen L. 1968. Monophyly or diphyly in the origin of whales. Evolution, 22 (1): 36-41.

Vinnikov Ya.A., Titova L.K. 1959. A method of in vivo isolation of the membranaceous labyrinth (the cochlea and threshold), of preparation and description of plane preparations of the organ of Corti. Arkhiv Anat., Gistol. i Embriol., 36 (4): 82 (In Russian).

Vinnikov Ya.A., Titova L.K. 1961. The Organ of Corti. Moscow & Leningrad: Nauka (In Russian).

Warr W.B. 1975. Olivocochlear and vestibular efferent neurons of the feline brain stem: their location, morphology and number determinated by retrograde axonal transport and acetylcholinesterase histochemistry. J. Exp. Neurol., 161 (2): 159-182.

Wassif K. 1948. Studies on the structure of the auditory ossicles and tympanic bone in Egyptian Insectivora, Chiroptera and Rodentia. Bull. Fac. Sci., Fouad I Univ., Cairo, 27: 177-213.

Wassif K. 1950. The tensor tympani muscle of bats. Ann. Mag Nat. Hist., Ser. 12, 3: 811-812.

Weber M. 1886. Studien über Säugetiere. Ein Beitrag zur Frage nach dem Ursprung der Cetacean. Jena, 252 S.

Weber M. 1928. Die Säugetiere. Unterordung Carnivora Pinnipedia. Jena. Gustav Fischer, 2: 342-354.

Webster D. 1960. Auditory significance of the hypertrophied mastoid bullae in *Dipodomys*. Anat. Rec., 136: 299.

Webster D. 1961. The ear apparatus of the kangaroo rat, *Dipodomys*. Amer. J. Anat., 108: 123-147.

Webster D. 1966. Ear structure and function in modern mammals. Amer. J. Zool., 6: 451-466.

Welleschik B., Firbas W. 1973. Quantitative investigations on the organ of Corti of the Microchiroptera. Period. Biol., 75 (1): 97-100.

Werner Cl. F. 1960. Das Gehörorgan der Wirbeltiere und des Menschen. Leipzig: Georg Thieme, 410 S.

Wersall J., Frock A., Lundquist P.G. 1965. Structural basis for directional sensitivity in cochlear and vestibular sensory receptors. Cold Spring Harbor Symp. Quant. Biol., 30: 115-145.

Wever E.G. 1949. Theory of hearing. NY: Wiley & Sons, 291-293.

Wever E.G., McCormick J.G., Palin J., Ridgway S.H. 1971a. The cochlea of the dolphin, *Tursiops truncatus*: hair cells and ganglion cells. Proc. Nat. Acad. Sci. USA, 68: 2908-2912.

Wever E.G., McCormick J.G., Palin J., Ridgway S.H. 1971b. The cochlea of the dolphin, *Tursiops truncatus*: general morphology. Proc. Natn. Acad. Sci. USA, 68: 2381-2385.

Wever E.G., McCormick J.G., Palin J., Ridgway S H. 1972. Cochlear structure in the dolphin, *Lagenorhynchus obliquidens*. Proc. Natn. Acad. Sci. USA, 69: 657-661.

Wilson J.T. 1914. Observations upon young human embryos. J. Anat. Physiol., 48: 315.

Wüstenfeld E. 1957. Experimentelle Untersuchungen zum Problem der Schallanalyse im Innenohr. Z. mikr.-anat. Forsch., 63 (3): 327-387.

Yamada M. 1951. On the cetacean external auditory meatus, an independent sensory organ secondarily specialized (Speech in 1951, Abstr. Japan).

Yamada M. 1953. Contribution to the anatomy of the organ of hearing of whales. Sci. Repts Whales Res. Inst., 8: 1-79.

Yamada M., Yoshizaki F. 1959. Osseous labyrinth of Cetacea. Sci. Repts Whales Res. Inst., 14: 291-304.

Zaslavsky G.L., Babushina E.S. 1986. Sound conduction throughout the dolphin body. In: Marine Mammals. Abst. 9th All-Union Conf. on the Study, Conservation and Rational Use of Marine Mammals. Arkhangelsk, 141-142 (In Russian).

Zawitch-Ossenitz C. 1933. Das Talgdrüsenorgan des Gehörganges von Nagern und Insectivoren. Z. mikr.-anat. Forsch., 33: 625-675.

Zharskaya V.D., Solntseva G.N. 1986. Heterochrony in maturation of the structures of the auditory system, *Stenella attenuata* taken as an example. In: Zakonomernosti individual'nogo rasvitiya zhivykh organizmov (Patterns in the individual development of living organisms). Moscow, 1: 80 (In Russian).

Zharskaya V.D., Solntseva G.N. 1989.Heterochronic maturation of the central and peripheral auditory system of cetaceans. In: Problemy morfo-funkstionalnykh adaptatsii (Problems of morphological adaptation). Leningrad: LGU, 88-93 (In Russian).

Zwicker E., Scharf B. 1965. A model of loudness summation. Psychol. Rev., 72: 3-26.

Zwislocki J. 1969. Akustyka ucha. Arch. akust., 4 (4): 375-398.

www.ingramcontent.com/pod-product-compliance
Lightning Source LLC
LaVergne TN
LVHW020651300926
849161LV00004B/895